Lecture Notes in Mathematics 2150

Editors-in-Chief:
J.-M. Morel, Cachan
B. Teissier, Paris

Advisory Board:
Camillo De Lellis, Zurich
Mario di Bernardo, Bristol
Alessio Figalli, Austin
Davar Khoshnevisan, Salt Lake City
Ioannis Kontoyiannis, Athens
Gabor Lugosi, Barcelona
Mark Podolskij, Aarhus
Sylvia Serfaty, Paris and NY
Catharina Stroppel, Bonn
Anna Wienhard, Heidelberg

Vladislav Kharchenko

Quantum Lie Theory

A Multilinear Approach

Springer

Vladislav Kharchenko
Universidad Nacional Autónoma de México
Cuautitlán Izcalli
Estado de México, Mexico

ISSN 0075-8434 ISSN 1617-9692 (electronic)
Lecture Notes in Mathematics
ISBN 978-3-319-22703-0 ISBN 978-3-319-22704-7 (eBook)
DOI 10.1007/978-3-319-22704-7

Library of Congress Control Number: 2015958730

Mathematics Subject Classification: 17B37, 20G42, 16T20, 16T05, 17A50, 17B75, 17B81, 81R50

Springer Cham Heidelberg New York Dordrecht London

Printed on acid-free paper

Springer International Publishing AG Switzerland is part of Springer Science+Business Media (www.springer.com)

Je dédie ce livre à
Elená, Vadim, Ilya, Ludivine, et Andrei
mes petits monstres mais aussi ma source d'inspiration.

Моим маленьким монстрам:
Алёне, Вадиму, Илюше, Людмиле и Андрею.

Preface

The numerous attempts over the last 15–20 years to define a quantum Lie algebra as an elegant algebraic object with a binary "quantum" Lie bracket have not been evidently and widely accepted. Nevertheless, the q-deformations of the enveloping algebras introduced independently by Drinfeld and Jimbo have profoundly impacted the development of both the modern theory of quantum groups and the much older mathematical theory of Hopf algebras. Although the definition of the Drinfeld–Jimbo quantization is not simple, a clear common property unites all of these quantizations, as well as those that appeared later in different multiparameter versions articulated by Reshetikhin, Costantini, Varagnolo, Chin, Musson, and Benkart, with the universal enveloping algebras. Especially, these quantizations as Hopf algebras are generated by skew-primitive semi-invariants. This book is mainly concerned with Hopf algebras possessing this property. Because the action on a semi-invariant is defined by a character, we call such Hopf algebras *character Hopf algebras.*

We treat the character Hopf algebras as universal enveloping algebras of "quantum Lie algebras." The quantum Lie algebra must be an algebraic object located inside a character Hopf algebra. The Cartier–Kostant theorem asserts a category equivalence between Lie algebras (in characteristic zero) and connected co-commutative Hopf algebras. Given this equivalence, a Lie algebra corresponds to the space of primitive elements. This correspondence provides a clear idea to treat the space spanned by skew-primitive elements as a quantum Lie algebra.

To maintain the Cartier–Kostant category equivalence in characteristic $p > 0$, one must consider an additional unary operation $x \mapsto x^p$ on the Lie algebras. Thus, we must consider not only binary operations (brackets) but also operations involving one or various variables. In this manner, we develop the notion of *quantum Lie operation*, a polynomial in noncommutative skew-primitive variables with skew-primitive values. We thus consider the space spanned by the skew-primitive elements and equipped with the quantum Lie operations as a quantum analog of a Lie algebra.

There are many reasons motivating the extension of research to operations that replace the Lie bracket but that depend on greater numbers of variables, for example, operations of n-Lie algebras introduced by V.T. Filippov and then independently appearing under the name "Nambu–Lie algebras" in theoretical research on generalizations of Nambu mechanics.

Another group of problems requiring the generalization of Lie algebras corresponds to research on skew derivations of noncommutative algebras. A noncommutative version of the fundamental Dedekind algebraic independence lemma states that the algebraic structure of a Lie algebra and operators with "inner" action define all algebraic dependencies in ordinary derivations. This result was extended to the field of skew derivations by Chen-Lian Chuang. His fundamental theorem may be interpreted in the same manner, i.e., the algebraic structure and operators with "inner" action define all algebraic dependencies in skew derivations. Hence, the following question arises: Which algebraic structure corresponds to the skew derivation operators? This question requires the consideration of n-ary operations irreducible to bilinear operations.

A third group of problems concerning multivariable generalizations of the Lie bracket appeared in nonassociative algebra. P.O. Miheev and L.V. Sabinin demonstrated that a simply connected local analytic loop is determined by an algebraic system consisting of a series of multilinear operations. These systems are now called *Sabinin algebras*.

This book is intended as an introduction to the mathematics behind the phrase "quantum Lie algebra." Despite the complexity of the subject, we have attempted to make this exposition accessible to a wide audience. We assume a standard knowledge of linear algebra and some rudimentary knowledge of representation theory. Most of the text will be accessible to graduate students in mathematics who have completed an introductory course in linear algebra.

Chapter 1 is introductory in nature. It contains many basic definitions related to noncommutative algebra that are used in subsequent chapters. Starting with Gauss polynomials and Lyndon–Shirshov standard words, we discuss the foundations of Gröbner–Shirshov theory, which is the basic tool for investigating noncommutative algebras specified by generators and defining relations. In this "combinatorial paradigm," the Poincaré–Birkhoff–Witt theorem obeys an elegant proof, whereas the concepts of a skew group ring and crossed product can be perfectly analyzed. We then introduce the braid monoid and the permutation group and consider the set of shuffles as a transversal of a direct product of symmetric subgroups. Although representation theory is not used intensively in this book, we formulate the theorems of Maschke and Wedderburn as initial statements without proofs. The concept of a character Hopf algebra is central to this monograph. In the combinatorial paradigm, the free character Hopf algebra plays a crucial role. The notion of a combinatorial rank appears in the analysis of generators for Hopf ideals, which are the defining relations for Hopf algebras. We develop the bracket technique as an important tool for performing calculations that allows one to preserve and apply the intuition of the Lie algebra machinery. Coordinate differential calculi, filtered and associated graded spaces, and specific fundamental concepts from P.M. Cohn theory are developed as tools for further applications. We conclude the chapter with notes that provide the reader with an opportunity to learn more about the subjects we review in the introductory chapter. We have constructed this chapter to be as self-contained as possible. Some arguments are new, and the remaining chapters have not previously appeared in book form.

In the second chapter, we demonstrate that every character Hopf algebra has a PBW basis. Our proof intensively uses the coalgebraic structure, distinct from the known Lusztig's method, which uses the algebraic structure only. Because the coproduct may not differ between a polynomial with a zero value and a polynomial with a skew-primitive value, in establishing linear independence, we automatically obtain important information regarding the skew-primitive polynomials.

In the third chapter, we review possible quantum deformations of the universal enveloping algebras of Kac–Moody algebras. To this end, we associate a class $\mathfrak{A}$ with a given Kac–Moody algebra $\mathfrak{g}$. The class $\mathfrak{A}$ consists of all character Hopf algebras defined by the same number of relations and with the same degrees as $\mathfrak{g}$ has. $\mathfrak{A}$ contains all known quantizations of $\mathfrak{g}$. We demonstrate that Hopf algebras from $\mathfrak{A}$ have the so-called triangular decomposition as coalgebras. If the generalized Cartan matrix A of $\mathfrak{g}$ is indecomposable, then up to a finite number of exceptional cases, the algebraic structure is solely defined by one "continuous" parameter q related to the symmetrization of A and one "discrete" parameter $\mathfrak{m}$ related to the modular symmetrizations of A.

In the fourth chapter, consistent with the main concept of the book, we treat the skew-primitive polynomials as quantum Lie operations. We discuss linearization and specialization processes and criteria for a polynomial to be classified as a quantum Lie operation. We also classify multilinear quantum Lie operations in two, three, and four variables. Although generally a bilinear bracket there does not exists as an operation, a binary bracket exists that is an important and effective tool for the investigation. Specifically, all quantum Lie operations can be expressed in terms of that bracket. The bracket becomes a quantum operation only if characters that define the action of group-like elements satisfy a multiplicative skew-symmetry condition. In this case, the quantum Lie algebra transforms into a color Lie algebra.

The fifth chapter focuses on multilinear quantum Lie operations involving more than four variables. We establish a necessary and sufficient existence condition and the number of linearly independent operations that may exist and define the principle n-linear operation which by permutations of variables spans the space of all n-linear operations. The symmetric operations pose an opposite property, namely, in the context of permutations of variables, they do not change their values up to a scalar factor. We deduce that there are precisely $(n-2)!$ linearly independent symmetric generic quantum Lie operations and at least one principle generic n-linear operation. Although this chapter does not require specialized knowledge, it demands persistence from the reader.

The main goal of the sixth chapter is a detailed construction of free braided Hopf algebra and shuffle braided Hopf algebra on the tensor space of a given braided space. We define a Nichols algebra as a subalgebra of the shuffle braided Hopf algebra generated by the given braided space. All calculations are performed within the braid monoid but not in the braid group; therefore, the constructions remain valid for a noninvertible braiding. We then consider braided Hopf algebras that appear in the Radford decomposition of character Hopf algebras and discuss filtrations.

As previously mentioned, numerous definitions had been proposed for the binary quantum analog of a Lie algebra. It is likely that only the Gurevich–Manin generalization up to Lie τ-algebras represents a completely successful definition. In the seventh chapter we consider this generalization and its particular cases, specifically, Lie superalgebras and color Lie algebras. The PBW theorem for Lie τ-algebras transforms into a coalgebra isomorphism between universal enveloping algebras of Lie τ-algebras defined within the same braided space. We establish a τ-Friedrichs criterion and consider subalgebras of free Lie τ-algebra.

In the field of nonassociative algebras, there are known generalizations of Lie algebras with nonassociative envelopes. Many of these well-known generalizations involve only one or two operations. In the eighth chapter, we consider nonassociative primitive polynomials as operations for nonassociative Lie theory similar to how we considered skew-primitive polynomials as operations for quantum Lie theory. I.P. Shestakov and U.U. Umirbaev discovered infinitely many independent operations of that type. The proof constructed in this chapter demonstrates that Shestakov–Umirbaev primitive operations together with the commutator form a complete set of nonassociative Lie operations.

I am grateful to all who have offered suggestions or made corrections to the manuscript. I am pleased to express my thanks to Ivan Shestakov, Cristian Vay, Ualbai Umirbaev, Zbigniew Oziewicz, Robert Yamaleev, Mayra Lorena Díaz Sosa, David Tinoco Varela, José Luis Garza Rivera, Alma Virginia Lara Sagahón, Angélica Espinoza Godínez, Rodolfo Alvarado Cervantes, Alejandro Andrade Álvarez, and Ricardo Paramont Hernández García for valuable discussions and comments. Finally, I offer a special expression of thanks to my advisor, Leonid Bokut', who initiated me on the path toward understanding modern noncommutative algebra.

Acknowledgements The author was supported by PAPIIT, project IN 112913, and PIAPI of the FES-Cuautitlán, project VC06, UNAM, México.

Cuautitlán Izcalli, Mexico Vladislav Kharchenko
June 2014

Contents

Chapter 1
Elements of Noncommutative Algebra

Abstract The first chapter contains many basic definitions and proves related to noncommutative algebra that are used in subsequent chapters. Starting with Gauss polynomials and Lyndon-Shirshov standard words, we discuss the foundations of Gröbner–Shirshov theory, which is the basic tool for investigating noncommutative algebras specified by generators and defining relations. In this "combinatorial paradigm," the Poincaré-Birkhoff-Witt theorem obeys an elegant proof, whereas the concepts of a skew group ring and crossed product can be perfectly analyzed. We then introduce the braid monoid and the permutation group, and consider the set of shuffles as a transversal of a direct product of symmetric subgroups. The concept of a character Hopf algebra is central to this monograph. In the combinatorial paradigm, the free character Hopf algebra plays a crucial role. The notion of a combinatorial rank appears in the analysis of defining relations for Hopf algebras. We develop the bracket technique as an important tool for performing calculations that allows one to preserve and apply the intuition of the Lie algebra machinery. Coordinate differential calculi, filtered and associated graded spaces, and specific fundamental concepts from P.M. Cohn theory are developed as tools for further applications. Although representation theory is not used intensively in this book, we formulate the theorems of Maschke and Wedderburn as initial statements without proofs. We conclude the chapter with notes that provide the reader with an opportunity to learn more about the subjects we review in the introductory chapter.

This chapter contains the basic definitions and proves related to noncommutative algebra that are used in sequel. We discuss Gauss polynomials, Lyndon-Shirshov standard words, and the foundations of Gröbner–Shirshov theory, which is the basic tool for investigating noncommutative algebras specified by generators and defining relations. We then introduce the braid monoid and the permutation group. The concept of a character Hopf algebra is central to this monograph. In the combinatorial paradigm, the free character Hopf algebra plays a crucial role. The notion of a combinatorial rank appears in the analysis of defining relations for Hopf algebras. Coordinate differential calculi, filtered and associated graded spaces, and specific fundamental concepts from P.M. Cohn theory are developed as tools for further applications. Although representation theory is not used intensively in this book, we formulate the theorems of Maschke and Wedderburn. We conclude the

V. Kharchenko, *Quantum Lie Theory*, Lecture Notes in Mathematics 2150,
DOI 10.1007/978-3-319-22704-7_1

chapter with notes that provide the reader with an opportunity to learn more about the subjects we review in the introductory chapter.

1.1 Gauss Polynomials

Let x and y be variables subject to the relation $yx = qxy$, where q is a variable with values in the ground field $\mathbf{k}$. For future applications, we need to compute the powers of $x + y$. Expanding $(x + y)^n$, we see that the monomials in the expansion are all scalar multiples of monomials of the form $x^k y^{n-k}$. Therefore, for all $n > 0$ we have

$$(x+y)^n = \sum_{k=0}^{n} \begin{bmatrix} n \\ k \end{bmatrix}_q x^k y^{n-k}, \tag{1.1}$$

where $\begin{bmatrix} n \\ k \end{bmatrix}_q$, $0 \leq k \leq n$ are integer polynomials in q, called *Gauss polynomials*. We have $yx^k = q^k x^k y$, $k \geq 0$. Using these commutation rules, we may write

$$\begin{aligned}(x+y)\left(\sum_{k=0}^{n} \begin{bmatrix} n \\ k \end{bmatrix}_q x^k y^{n-k}\right) &= \sum_{k=0}^{n} \begin{bmatrix} n \\ k \end{bmatrix}_q x^{k+1} y^{n-k} + \sum_{k=0}^{n} \begin{bmatrix} n \\ k \end{bmatrix}_q yx^k y^{n-k} \\ &= x^{n+1} + \sum_{k=1}^{n} (\begin{bmatrix} n \\ k-1 \end{bmatrix}_q + q^k \begin{bmatrix} n \\ k \end{bmatrix}_q) x^k y^{(n+1)-k} + y^{n+1}.\end{aligned}$$

This equality and definition (1.1) with $n \leftarrow n + 1$ imply the following recurrence relation, called the *first q-Pascal identity*:

$$\begin{bmatrix} n+1 \\ k \end{bmatrix}_q = \begin{bmatrix} n \\ k-1 \end{bmatrix}_q + q^k \cdot \begin{bmatrix} n \\ k \end{bmatrix}_q, \quad \begin{bmatrix} n+1 \\ 0 \end{bmatrix}_q = \begin{bmatrix} n+1 \\ n+1 \end{bmatrix}_q = 1. \tag{1.2}$$

Similarly, starting with the decomposition $(x+y)^{n+1} = (x+y)^n \cdot (x+y)$, we obtain the *second q-Pascal identity*:

$$\begin{bmatrix} n+1 \\ k \end{bmatrix}_q = \begin{bmatrix} n \\ k-1 \end{bmatrix}_q \cdot q^{n-k+1} + \begin{bmatrix} n \\ k \end{bmatrix}_q. \tag{1.3}$$

The Gauss polynomials have the following rational representation

$$\begin{bmatrix} n \\ k \end{bmatrix}_q = \frac{q^{[n]} q^{[n-1]} \cdots q^{[n-k+1]}}{q^{[1]} q^{[2]} \cdots q^{[k]}}, \tag{1.4}$$

where by definition, $q^{[s]} = 1 + q + \cdots + q^{s-1}$, $q^{[0]} = 0$. To prove (1.4), it suffices to demonstrate that those rational functions satisfy recurrence relations (1.2).

We have $q^{[n+1]} = q^{[k]} + q^k \cdot q^{[n-k+1]}$. Therefore, $q^{[n+1]}/q^{[k]} = 1 + (q^k \cdot q^{[n-k+1]}/q^{[k]})$. This relationship implies the required decomposition:

$$\frac{q^{[n+1]}q^{[n]}\cdots q^{[n-k+2]}}{q^{[1]}q^{[2]}\cdots q^{[k]}} = \frac{q^{[n]}q^{[n-1]}\cdots q^{[n-k+2]}}{q^{[1]}q^{[2]}\cdots q^{[k-1]}}(1 + q^k \cdot \frac{q^{[n-k+1]}}{q^{[k]}})$$
$$= \frac{q^{[n]}q^{[n-1]}\cdots q^{[n-k+2]}}{q^{[1]}q^{[2]}\cdots q^{[k-1]}} + q^k \cdot \frac{q^{[n]}q^{[n-1]}\cdots q^{[n-k+1]}}{q^{[1]}q^{[2]}\cdots q^{[k]}}.$$

Future applications will require certain additional information about Gauss polynomials when $q^{[n]} = 0$. By multiplying the latter equality by $q-1$, we obtain $q^n = 1$. Hence, q is a primitive mth root of 1, and m is a divisor of n. The case $m = 1$ is also possible if the characteristic l of the ground field $\mathbf{k}$ is positive. Thus, we consider 1 to be a primitive 1st root of 1.

Lemma 1.1 *If q is a primitive mth root of 1 and m is a divisor of n, then*

$$\begin{bmatrix} n \\ k \end{bmatrix}_q = 0, \quad 1 \le k < m.$$

Proof In the rational representation (1.4), all factors $q^{[i]}$, $1 \le i \le k$ of the denominator have nonzero values because $q^i \ne 1$, $1 \le i < m$ by the definition of m. Each numerator has a factor $q^{[n]}$. Let $n = m \cdot s$. We have $q^{[n]} = q^{[ms]} = (q^m)^{[s]} \cdot q^{[m]}$. If $m \ne 1$, then $q^{[m]} = 0$. If $m = 1$, then there is nothing to prove. □

Lemma 1.2 *Let q be a primitive mth root of 1, and let $n = ml^k$, where $l = 1$ or $l = \operatorname{char} \mathbf{k} > 0$. If x, y are variables subject to the relation $yx = qxy$, then*

$$(x + y)^n = x^n + y^n.$$

Proof Due to the above Lemma and (1.1), we have $(x + y)^m = x^m + y^m$. In this case $y^m x^m = q^{m^2} x^m y^m = x^m y^m$. Hence we may apply the ordinary Newton binomial formula $(x^m + y^m)^{l^k} = x^{ml^k} + y^{ml^k} = x^n + y^n$. □

1.2 Lyndon — Shirshov Words

Let $X = \{x_i \mid i \in I\}$ be a set of variables. Assume that on X an order $\le$ is fixed such that X is a well-ordered set (every nonempty subset has a least element). Consider the set X to be an alphabet. On a set X^* of all words in this alphabet, define the *lexicographical order*: two words v and w are compared by moving from left to right until the first distinct letter is encountered. Otherwise, if one of the words is the beginning of another word, then the shorter word is assumed to be greater than the longer word. For example, all words of length at most two in two variables $x_1 > x_2$ respect the following order:

$$x_1 > x_1^2 > x_1x_2 > x_2 > x_2x_1 > x_2^2. \tag{1.5}$$

The lexicographical order is stable under left concatenations and unstable under right ones. Nevertheless, if $u > v$ and u is not a beginning of v, then the inequality is preserved under right concatenations, even by different words: $uw > vt$ for all w, t.

Even if the alphabet is finite, there exist infinite ascending and infinite descending chains of words (in particular, X^* is not a well-ordered set):

$$x_1 > x_1x_2 > x_1x_2^2 > \ldots > x_1x_2^m > \ldots ; \tag{1.6}$$

$$x_2 < x_1x_2 < x_1^2x_2 < \ldots < x_1^mx_2 < \ldots , \tag{1.7}$$

provided that $x_1 > x_2$. These chains make it impossible to perform induction (neither direct nor downward) on words using only the lexicographical order. Nevertheless, it is possible to perform induction based on two parameters, for example, the length of a word and its lexicographical position among words of the same length.

1.2.1 Standard Words

Definition 1.1 A word u is called *standard* (or a *Lyndon-Shirshov* word) if, for each decomposition $u = u_1u_2$, where u_1 and u_2 are nonempty words, the inequality $u > u_2u_1$ holds. For example, in (1.6), (1.7) all words are standard, whereas in (1.5), three words are standard: x_1, x_1x_2, and x_2.

If $u = x_{i_1}x_{i_2}\cdots x_{i_m}$ is a word, then the set of all possible words u_2u_1, where $u = u_1u_2$, is precisely the set of all cyclic permutations of u,

$$x_{i_1}x_{i_2}\cdots x_{i_m}, \quad x_{i_2}x_{i_3}\cdots x_{i_m}x_{i_1}, \quad \ldots \quad , x_{i_m}x_{i_1}x_{i_2}\cdots x_{i_{m-1}}. \tag{1.8}$$

Therefore, the word u is standard if and only if it is greater than each cyclic permutation of it. If the word u is not periodic, $u \neq v^h, h > 1$, then all words in (1.8) are different. Hence, there is precisely one standard word among the cyclic permutations of u. If the word u is periodic, $u = v^h$ with the maximal $h > 1$, then each cyclic permutation u' in (1.8) is periodic: $u' = (v')^h$, where v' is a cyclic permutation of v. In this case, the set of all cyclic permutations contains no standard words but has precisely h words of the form w^h with a standard w, whereas w is the standard cyclic permutation of v.

Lemma 1.3 *Let $u = sv$ be a standard word. If $s, v \neq \emptyset$, then v is not a beginning of u.*

Proof Suppose $u = vs'$. By definition, $sv = vs' > s'v$, i.e., $s > s'$. Similarly, $vs' = sv > vs$, hence $s' > s$ which is a contradiction. □

Lemma 1.4 *A word u is standard if and only if it is greater than each of its proper endings.*

Proof If the word u is standard and $u = vv_1$ then $vv_1 > v_1v$. According to Lemma 1.3, the word v_1 is not the beginning of vv_1; hence, $u = vv_1$ and v_1v differ already in their first $l(v_1)$ letters, where by definition $l(w)$ is the length of a word w. Therefore, $u > v_1$. Conversely, if $u = u_1u_2$ and $u > u_2$, then u is not the beginning of u_2, so the inequality $u > u_2$ holds when the right side is multiplied by u_1. □

Lemma 1.5 *Let u and v be standard words. If $u > v$, then $u^h > v$.*

Proof If u is not the beginning of v, then $u > v$ can be multiplied from the right by different words. Suppose that $v = u^k v'$ and that v' does not begin with u. If $k \geq h$, then $u^h > v$ as the beginning. If $k < h$ then v' is nonempty, and $v' < v < u$. It follows that $v = u^k \cdot v' < u^k \cdot u \cdot u^{h-k-1} = u^h$. □

Lemma 1.6 *If u, v are different standard words and u^n contains v^k as a sub-word, $u^n = cv^kd$, then u contains v^k as a sub-word, $u = bv^ke$.*

Proof Without loss of generality, we may suppose that $l(c) < l(u)$; otherwise, n can be diminished. In this case, $u = cv^s t$, $s \geq 0$. If $s < k$, then t is the beginning of v, $v = tt'$, and the ending of u. Then, according to Lemma 1.4, either $u > t > v$ or t is empty. In the latter case, $u > v$ because v is the ending of u. In turn, t' is the beginning of u^{n-1}; that is, $t' = u^r t''$, $r \geq 0$. Here, t'' is the ending of v and the beginning of u. Lemma 1.4 implies that either $v > t'' > u$ or t'' is empty. In the latter case $v > u$ because u becomes the ending of v. We note the contradiction that $u > v > u$. □

Lemma 1.7 *Let u and u_1 be standard words such that $u = u_3u_2$ and $u_2 > u_1$. Then*

$$uu_1 > u_3u_1, \quad uu_1 > u_2u_1. \tag{1.9}$$

Proof First we demonstrate that $u_2u_1 > u_1$. If u_1 does not begin with u_2, then the inequality follows immediately from $u_2 > u_1$. Assume that $u_1 = u_2^k \cdot u_1'$ and that u_2 is not the beginning of u_1'. Since u_1 is standard, it follows that $u_2^k u_1' > u_2^{k-1}u_1'$, i.e., $u_2u_1' > u_1'$. Hence, $u_2u_1 = u_2^k \cdot u_2u_1' > u_2^k \cdot u_1' = u_1$. Multiplying this inequality from the left by u_3 yields the first required inequality. Consider the second inequality. Because u is a standard word, $u_3u_2 > u_2$ according to Lemma 1.4. As u_3u_2 is not the beginning of u_2, we can multiply the latter inequality from the right by u_1. □

Lemma 1.8 *If u, v are standard words, and $u > v$, then uv is a standard word.*

Proof Using Lemma 1.4, it is sufficient to demonstrate that $uv = wt$, $w, t \neq \emptyset$ implies $uv > t$.

If $l(w) < l(u)$, then $u = wt'$, $t'v = t$. According to Lemma 1.4, we obtain $u > t'$. The word u is not the beginning of t' because $l(u) = l(w) + l(t')$. Therefore, $u > t'$ may be multiplied from the right by v. Thus, $uv > t'v = t$.

If $l(w) \geq l(u)$, then $w = ut'$ and $t't = v$. Applying Lemma 1.4, we obtain $v \geq t$, which implies that $u > v \geq t$. The inequality $u > t$ implies that $uv > t$ provided that u is not the beginning of t. Otherwise, $t = ut''$ implies that $v = t't = t'ut''$. Lemma 1.4 states that $v > t''$. Hence $uv > ut'' = t$. □

Theorem 1.1 *Each word u has a unique decomposition*

$$u = w_1^{n_1} \cdot w_2^{n_2} \cdot \ldots \cdot w_m^{n_m}, \tag{1.10}$$

where w_i, $1 \leq i \leq m$ are standard words and $w_1 < w_2 < \ldots < w_m$.

Proof The initial letter of u is a standard word of length one. Let v_1 be the longest beginning of u that is a standard word, $u = v_1 \cdot u_1$. Let v_2 be the longest beginning of u_1 that is a standard word, $u_1 = v_2 \cdot u_2$, and so on. In this manner, we find a decomposition $u = v_1 \cdot v_2 \cdot \ldots \cdot v_i \cdot \ldots$ with standard v_i, $i \geq 1$. In this decomposition, $v_1 \leq v_2 \leq \ldots \leq v_i \leq \ldots$ because, due to Lemma 1.8, the inequality $v_i > v_{i+1}$ implies that $v_i v_{i+1}$ is a standard beginning of u_i of length greater than $l(v_i)$.

If $v_1' \leq v_2' \leq \ldots \leq v_i' \leq \ldots$ and $u = v_1' \cdot v_2' \cdot \ldots \cdot v_i' \cdot \ldots$ is another decomposition with standard factors, $v_1' \neq v_1$, then v_1' is a proper beginning of v_1. In particular, $v_1 = v_1' \cdot \ldots \cdot v_i' t_i$, where $i \geq 1$ and t_i is a non-empty beginning of v_{i+1}'. The standard word v_1 is greater than its ending t_i and less than its beginning v_1'. In turn, v_{i+1}' is less than or equal to its beginning t_i. Thus, we have a contradiction:

$$t_i < v_1 < v_1' \leq \ldots \leq v_i' \leq v_{i+1}' \leq t_i.$$

□

Corollary 1.1 *Every standard word w of length greater than one has a decomposition $w = uv$, $u > v$ with standard u, v.*

Proof Let u be the longest proper standard beginning of w, $w = uv$. By Theorem 1.1, the word v has a decomposition $v = v_1 \cdot \ldots \cdot v_m$ with standard v_i, $1 \leq i \leq m$ and $v_1 \leq \ldots \leq v_m$. If $m > 1$, then uv_1 is a proper beginning of w. Therefore, uv_1 is not standard. By Lemma 1.8, this statement implies $u \leq v_1$. In this case, we have two different decompositions, i.e., $w = u \cdot v_1 \cdot \ldots \cdot v_m$ and $w = w$, that satisfy the conditions of Theorem 1.1. Thus, $m = 1$, and v is a standard word. □

1.2.2 Nonassociative Standard Words

The proven corollary and Lemma 1.8 make it possible to find all standard words step-by-step. First, all words of length one are standard. Next, if all standard words of length $< l$ are known, then every pair of different standard words u, v of length i and $l-i$, respectively, $1 \le i < l$, defines a standard word uv (if $u > v$) or vu (if $v > u$). In this way, all standard words appear.

In this process, certain standard words may appear several times. For example, if $x_1 > x_2 > x_3$, then $x_1x_2x_3 = x_1 \cdot x_2x_3$ and $x_1x_2x_3 = x_1x_2 \cdot x_3$. Recall that a *nonassociative word* is a word in which [,] are somehow arranged to show how the multiplication applies. We see that in the above process, a particular construction of a given standard word u is equivalent to an alignment of brackets.

Definition 1.2 If u is a standard word, then $[u]$ denotes a nonassociative word where the brackets are arranged by the following inductive algorithm. The factors v and w in the nonassociative decomposition $[u] = [[v][w]]$ are standard words such that $u = vw$, and v has the minimal possible length. The nonassociative word $[u]$ is called a *nonassociative standard word*.

For example, if $x_1 > x_2 > x_3$, then the words

$$x_1x_2x_3,\ \ x_1x_2^2,\ \ x_2^3x_3,\ \ x_1x_2x_3x_2,\ \ x_2x_3x_2x_3x_4,\ \ x_1x_2x_3^2x_2$$

are standard, and they define the following nonassociative standard words:

$$\begin{aligned}&[x_1x_2x_3] = [x_1[x_2x_3]],\ [x_1x_2^2] = [[x_1x_2]x_2],\ [x_2^3x_3] = [x_2[x_2[x_2x_3]]],\\ &[x_1x_2x_3x_2] = [[x_1[x_2x_3]]x_2],\ [x_2x_3x_2x_3x_4] = [[x_2x_3][x_2[x_3x_4]]],\\ &[x_1x_2x_3^2x_2] = [[x_1[[x_2x_3]x_3]]x_2].\end{aligned} \tag{1.11}$$

Proposition 1.1 *Let $u > u_1$ be standard words and $[u] = [[u_3][u_2]]$. Then, $[[u][u_1]]$ is a standard nonassociative word if and only if $u_2 \le u_1$.*

We prove this statement in two steps.

Lemma 1.9 *If $[[u][u_1]]$ is a standard nonassociative word, then $u_2 \le u_1$.*

Proof If $u_2 > u_1$, then u_2u_1 is a standard word, and we have a decomposition $uu_1 = u_3 \cdot u_2u_1$ where the length of the first factor is less than the length of $uu_1 = u_3u_2 \cdot u_1$. Hence, $[[u][u_1]]$ is not standard. □

Lemma 1.10 *If $[[u][u_1]]$ is not a standard nonassociative word then $u_2 > u_1$.*

Proof We perform induction on the length of uu_1. Let $X_1 = \{x_1, x_2, \ldots, x_n\}$ be the set of all letters that occur in the word uu_1, and assume that $x_1 > x_2 > \ldots > x_n$. Consider a set $Y = \{y_1, y_2, \ldots, y_{n-1}\}$ of new symbols. On the set $(X_1 \cup Y)^*$ of all words in the alphabet $X_1 \cup Y$, define the lexicographical order such that

$$x_1 > y_1 > x_2 > y_2 > \ldots > y_{n-1} > x_n. \tag{1.12}$$

For every word W in $X_1 \cup Y$, let $\xi(W)$ denote a word in X that results from W under the substitution $y_i \leftarrow x_i x_n,\ 1 \le i < n$. We note that the map $\xi : (X_1 \cup Y)^* \to X_1^*$ is an homomorphism of ordered monoids:

$$\xi(VW) = \xi(V)\xi(W), \quad V < W \Longleftrightarrow \xi(V) < \xi(W). \tag{1.13}$$

The former equality is evident. The latter condition follows from the fact that ξ preserves the order of letters: $x_1 > x_1x_n > x_2 > x_2x_n > \ldots > x_{n-1}x_n > x_n$.

A word $W \in (X \cup Y)^*$ is standard if and only if $\xi(W)$ is standard as a word in X because each cyclic permutation of $\xi(W)$ either starts with the smallest letter x_n or has the form $\xi(W')$ where W' is a cyclic permutation of W.

A decomposition $U = V \cdot W$ of a standard word U satisfies the conditions of the algorithm given in Definition 1.2 if and only if $\xi(U) = \xi(V) \cdot \xi(W)$ satisfies the same conditions as a decomposition of a word in X. Indeed, if $\xi(U) = v_1 \cdot w_1$, and $v_1,\ w_1$ are standard words such that v_1 has the minimal possible length, then either $v_1 = \xi(V_1),\ w_1 = \xi(W_1)$ for a suitable decomposition $U = V_1 \cdot U_1$, or w_1 starts with x_n. Because w_1 is a standard word, in the latter case, $w_1 = x_n$; however, in this case, the length of v_1 is greater than the length of $\xi(V)$.

If w is a standard word in X_1, then it does not start with the smallest letter x_n unless $w = x_n$. Let $\phi(w)$ be a word in $X_1 \cup Y$ that appears from w under replacements of all sub-words $x_i x_n$, with $y_i,\ 1 \le i < n$. Of course, we have

$$\xi(\phi(w)) = w.$$

If $w = w_1 w_2$ is a decomposition in the product of standard words, then

$$\phi(w) = \phi(w_1)\phi(w_2) \tag{1.14}$$

provided that $w_2 \neq x_n$. Equality (1.14) is still valid if $w_2 = x_n$ and w_1 ends with x_n.

Let us note that if $u_2 = x_n$, then u_3 ends with x_n. Indeed, $[u] = [[u_3]x_n]$ is a standard nonassociative word. If $[u_3] = [[u_4][u_5]]$, then by Lemma 1.9, we have $u_5 \le x_n$ which is possible only if the standard word u_5 equals x_n. This note and (1.14) imply

$$\phi(u) = \phi(u_3)\phi(u_2).$$

Similarly, we have

$$\phi(uu_1) = \phi(u)\phi(u_1)$$

because otherwise either the required condition, $u_2 > u_1$, holds, in which case we have nothing to prove, or $u_2 = u_1 = x_n$. In the latter case, (1.14) still applies.

Although the word $\phi(u_3)\phi(u_2)\phi(u_1)$ is a word in a new alphabet, its length is less the length of uu_1. Applying the induction hypothesis to the nonassociative word

$$[[[\phi(u_3)]\,[\phi(u_2)]]\,[\phi(u_1)]],$$

we obtain $\phi(u_2) > \phi(u_1)$, whereas condition (1.13) implies $u_2 > u_1$. □

Remark 1.1 The induction of the above lemma provides a dual algorithm of the alignment of brackets in a standard word w that results with the same standard nonassociative word $[w]$: first, we put the brackets on all sub words $[x_i x_n]$, $1 \leq i < n$, and we then consider these bracketed sub-words as new letters y_i with ordering (1.12). Next, we repeat the first step. This procedure is an alignment of brackets "from the bottom", while the algorithm given in Definition 1.2 is an alignment "from the top".

1.2.3 Deg-Lex Orders

As mentioned above, the lexicographic order does not satisfy either ACC nor DCC, see (1.6), (1.7), which makes it impossible to perform induction. To overcome this obstacle, one may introduce additional stratification of all words in groups so that each group has a finite number of words.

The simplest stratification is one given by the length. In this case, there appears the *Hall ordering* of words: $u <_h v$ if $l(u) < l(v)$, or $l(u) = l(v)$ and $u < v$. The Hall ordering is compatible with the concatenation product of words: If $u <_h v$, then $wut <_h wvt$ for all words w, t.

Another stratification is one given by the natural degrees. Let us assign natural degrees to the letters of the alphabet, $\deg x_i = d_i$, $i \in I$. As usual, the degree of a word is the sum of the degrees of its letters (normally, such a degree is called a *formal degree* with $\deg x_i = d_i$). In this way, there appears the so called *Deg-Lex ordering* of words: $u <_d v$ if $\deg u < \deg v$, or $\deg u = \deg v$ and $u < v$. The Deg-Lex ordering is also compatible with the concatenation product. Certainly, if all d_i equal 1, then the Dex-Lex order coincides with the Hall order.

If the alphabet is finite, then the set of words of fixed degree (or of fixed length) is finite. However, if the alphabet is infinite, than that set may be infinite. Therefore, it is useful to employ a more precise *stratification by constitution.*

Definition 1.3 A *constitution* of a word u in $X = \{x_i \mid i \in I\}$ is a family of nonnegative integers $\{m_i \mid i \in I\}$ such that u has m_i occurrences of x_i. The number m_i has a notation $m_i = \deg_i u$, which is called the *degree of u with respect to x_i*. In this terminology, the constitution of u is nothing more than the *multidegree of u*.

A constitution of a word has only a finite number of nonzero components. Therefore, the set of all words of a given constitution is finite. In fact, the order related to the stratification by constitution is precisely the Deg-Lex order if, instead of the natural degrees, we assign to the variables positive degrees from the free additive (commutative) monoid Γ generated by X.

This monoid consists of the formal finite linear combinations $\sum_{i\in I} n_i x_i$, where each n_i is a natural number or zero. Respectively, $\deg x_i = 1\cdot x_i$ and $\deg u = \sum m_i x_i$, where $\{m_i \mid i \in I\}$ is the constitution of u.

The monoid Γ is a well-ordered monoid with respect to the order

$$m_1 x_{i_1} + m_2 x_{i_2} + \ldots + m_k x_{i_k} > m_1' x_{i_1} + m_2' x_{i_2} + \ldots + m_k' x_{i_k} \tag{1.15}$$

provided that the first nonzero number from the left in

$$(m_1 - m_1', m_2 - m_2', \ldots, m_k - m_k')$$

is positive, while $x_{i_1} > x_{i_2} > \ldots > x_{i_k}$ in X.

The Deg-Lex order of words in X is compatible with the concatenation product: If $\deg u < \deg v$, then $\deg wut = \deg w + \deg u + \deg t < \deg w + \deg v + \deg t = \deg wvt$. If $\deg u = \deg v$, then v is not a beginning of u, and $u < v$ implies $wut < wvt$ for all words w, t.

1.3 Gröbner–Shirshov Systems of Defining Relations

In this section, we discuss the combinatorial representation of associative algebras by means of generators and relations. The crucial problem is that there does not exist a general algorithm to verify whether two polynomials are equal in the quotient algebra $\mathbf{k}\langle X\rangle/J$ (i.e., the equality problem for associative algebras is undecidable). Nevertheless, there is an algorithm that allows resolution of the equality problem and even the problem of the construction of a basis, provided that the system of defining relations satisfies an additional property (is closed with respect to the compositions). This algorithm is based on the Composition Lemma by A.I. Shirshov.

1.3.1 Composition Lemma

Let X be a set of variables, and let $\mathbf{k}\langle X\rangle$ be the free associative algebra freely generated by X. The free algebra $\mathbf{k}\langle X\rangle$ consists of noncommutative polynomials, the formal linear combinations $\sum_k \alpha_k w_k$ of words in the alphabet X, with the concatenation product. By definition, the algebra defined by the generators X and relations $F_i = 0$, $i \in I$, is the quotient algebra $\mathbf{k}\langle X\rangle/J$, where J is the ideal of $\mathbf{k}\langle X\rangle$ generated by F_i, $i \in I$.

We fix a well order $\prec$ on the set of words X^* such that $u \prec v$ implies $wut \prec wvt$ for all words $w, t \in X^*$. For example, $\prec$ may be the Hall order or one of the Deg-Lex orders described above.

Definition 1.4 A *leading word* of a polynomial $F = \sum_k \alpha_k w_k \in \mathbf{k}\langle X\rangle$, $\alpha_k \neq 0$ is the greatest word w of the finite set $\{w_k\}$.

Without loss of generality, we may assume that the coefficient at the leading word of each defining relation is equal to one. In this case, the relation $F_i = 0$ is equivalent to the relation $w_i = f_i$, where w_i is the leading word of F_i, so that f_i is a linear combination of lesser than w_i words, $F_i = w_i - f_i$. Consider a system of relations

$$w_i = f_i, \quad i \in I. \tag{1.16}$$

Lemma 1.11 *The set Σ of all words that have none of w_i, $i \in I$ as sub-words spans the algebra A defined by relations* (1.16).

Proof We must demonstrate that in A, each word is a linear combination of words from Σ. Let w be the minimal word that is not such a linear combination. In this case, $w \notin \Sigma$, and w has a sub-word w_i for a suitable $i \in I$; that is, $w = uw_iv$, where $u, v \in X^*$. In the algebra A, we have $w = uf_iv$. The leading term of the polynomial uf_iv is less than w because all words of f_i are less than w_i, which implies that all words of uf_iv are the required linear combinations. A contradiction. □

Definition 1.5 A system of relations (1.16) is said to be *closed with respect to the compositions* if the following conditions are met:

1. None of w_i contains w_s, $i \neq s \in I$ as a sub-word;
2. For each pair of words (not necessarily different) w_i, w_s such that some non-empty end of w_i coincides with a beginning of w_s; that is, $w_i = w_i'v$, $w_s = vw_s'$, the difference $f_iw_s' - w_i'f_s$, called a *composition*, has the following representation in the free algebra $\mathbf{k}\langle X\rangle$:

$$f_iw_s' - w_i'f_s = \sum_{k,t} \alpha_{kt}a_{kt}F_kb_{kt}, \ a_{kt}, \ b_{kt} \in X^*, \ \alpha_{kt} \in \mathbf{k}, \ \ a_{kt}w_kb_{kt} \prec w_i'vw_s'. \tag{1.17}$$

We stress that the first condition does not provide an essential restriction on the system of defining relations: if w_s is a sub-word of w_i, say, $w_i = aw_sb$, $a, b \in X^*$, then we may replace the relation F_i with $F_i - aF_sb$, diminishing the leading word.

Remark 1.2 Traditionally, a system of relations closed with respect to the compositions is called a *Gröbner–Shirshov basis* of the ideal generated by the F_i's. However, the word "basis" here is not perfect because some of the defining relations in a Gröbner-Shirshov "basis" may follow from the others. Additionally, this term sometimes leads to confusion between the set of relations (1.16) and the basis Σ

of the algebra defined by these relations, especially when instead of "the Gröbner–Shirshov basis of an ideal," they use "the Gröbner–Shirshov basis of the algebra" (defined by that ideal). For this reason, the term "*Gröbner-Shirshov system of relations*" seems to be more precise, at least in the context of this book.

Theorem 1.2 (Composition Lemma) *If system* (1.16) *is closed with respect to the compositions, then the set Σ of all words that have none of w_i, $i \in I$ as a sub-word is a basis of the algebra A defined by the relations* (1.16).

Proof Let us show that the leading word of each polynomial $F \in J$ has a sub-word w_i for a suitable $i \in I$. The polynomial $F \in J$ has a representation in the free algebra

$$F = \sum_{i,k} \alpha_{ik}\, a_{ik}\, F_i\, b_{ik}, \tag{1.18}$$

where $a_{ik}, b_{ik} \in X^*$, $0 \neq \alpha_{ik} \in \mathbf{k}$, but of course this representation is not unique. Among all representations, we consider only ones that share no similar terms: for each i, if $k \neq s$, then the pairs (a_{ik}, b_{ik}) and (a_{is}, b_{is}) are different. Let w be the maximal word among all words $a_{ik}\, w_i\, b_{ik}$. The word w is related to the given representation (1.18), and, of course, it may be different from the leading word of F. Among all representations (1.18) without similar terms, we choose one with the minimal possible word w. Our aim is to show that in this case, the word w (which contains the sub-word w_i) is the leading word of F.

The leading word of the term $a_{ik}\, F_i\, b_{ik}$ is the word $a_{ik}\, w_i\, b_{ik}$ because $\prec$ is stable under multiplications: $u \prec w_i$ implies $a_{ik}\, u\, b_{ik} \prec a_{ik}\, w_i\, b_{ik}$.

Let Π be the set of all pairs (i, k) such that $w = a_{ik}\, w_i\, b_{ik}$. We shall perform induction on the number of pairs in Π. If Π has only one pair, then all terms of (1.18) corresponding to other pairs are linear combinations of lesser than w words. Hence w is the leading word of F.

Assume that Π has more than one pair: $w = a_{ik}\, w_i\, b_{ik} = a_{st}\, w_s\, b_{st}$, $(s, t) \neq (i, k)$. Consider the following two cases.

(a) The sub-words w_i and w_s of the word w have no intersection, say, $w = aw_s dw_i b$, where $a, d, b \in X^*$. In this case, $a_{ik} = a_{st} w_s d$, $b_{st} = dw_i b_{ik}$. Let us modify the sum of two terms of (1.18) corresponding to the pairs (i, k) and (s, t):

$$\alpha_{ik} a_{ik}\, F_i\, b_{ik} + \alpha_{st} a_{st}\, F_s\, b_{st} = (\alpha_{ik} + \alpha_{st}) a_{ik} F_i b_{ik} - \alpha_{st} a_{st} F_s d(F_i - w_i) b_{ik}$$
$$+ \alpha_{st} a_{st} (F_s - w_s) d F_i b_{ik}.$$

All words that appear in the decomposition of the second and third summands are less than w, for $F_i - w_i = -f_i$, $F_s - w_s = -f_s$. Hence, the set Π for the modified representation (1.18) diminishes by one, if $\alpha_{ik} + \alpha_{st} \neq 0$, or by two, if $\alpha_{ik} + \alpha_{st} = 0$. The set Π for the modified representation is non-empty, because otherwise, the word w for the modified representation is less than for the initial one.

(b) The sub-words w_i and w_s have a non-empty intersection. Because neither of the words w_i, w_s is a sub-word of the other, some proper end of one of them equals a proper beginning of the other, say, $w_i = w_i'v$, $w_s = vw_s'$, whereas

$$w = a_{ik}w_i'vw_s'b_{st} \text{ and } a_{ik}w_i' = a_{st},\ w_s'b_{st} = b_{ik}.$$

In this case, we modify (1.18) as follows:

$$\alpha_{ik}a_{ik}F_i b_{ik} + \alpha_{st}a_{st}F_s b_{st} = (\alpha_{ik} + \alpha_{st})a_{ik}F_ib_{ik} + \alpha_{st}a_{ik}(w_i'F_s - F_iw_s')b_{st}$$
$$= (\alpha_{ik} + \alpha_{st})a_{ik}F_ib_{ik} - \alpha_{st}a_{ik}[f_iw_s' - w_i'f_s]b_{st}.$$

If we replace the resulting composition according to (1.17), then the set Π for the modified representation (1.18) diminishes by one, if $\alpha_{ik} + \alpha_{st} \neq 0$, or by two, if $\alpha_{ik} + \alpha_{st} = 0$. Again, the set Π for the modified representation is non-empty because otherwise, the word w for the modified representation is less than for the initial one.

Thus, we have demonstrated that the leading word of each polynomial $F \in J$ has a sub-word w_i for a suitable $i \in I$. The leading term of any linear combination F' of words from Σ belongs to Σ; hence, $F' \notin J$. □

Definition 1.5 does not provide an algorithm for how to check whether there exists a representation (1.17) for a given composition. The following statement demonstrates that the natural diminishing process always gives an answer.

Lemma 1.12 *A system of relations* (1.16) *is closed with respect to the compositions if and only if, first, none of w_i contains w_s, $i \neq s \in I$ as a sub-word and, second, each composition $f_iw_s' - w_i'f_s$ can be reduced to zero in the free algebra through a sequence of one-sided diminishing substitutions $w_t \leftarrow f_t$, $t \in I$.*

Proof Through the substitutions $w_t \leftarrow f_t$, $t \in I$ we may reduce any polynomial to a linear combination of words from Σ because each substitution diminishes the words. The value in A of a composition is zero, as $f_iw_s' - w_i'f_s = -(F_iw_s' - w_i'F_s)$ and the substitutions $w_t \leftarrow f_t$, $t \in I$ do not change the value of a polynomial in A. If (1.16) is closed with respect to the compositions, then the resulting linear combination of words from Σ must be empty because according to Theorem 1.2, the values of words from Σ form a basis of A.

Conversely, a substitution $w_t \leftarrow f_t$ transforms a word aw_tb to the linear combination of lesser words af_tb, and we have $af_tb = aw_tb - aF_tb$. If a polynomial P reduces to zero through one sided substitutions $w_t \leftarrow f_t$, then in the free algebra, we have the equality $0 = P - \sum_{ik} a_{ik}F_ib_{ik}$, where $a_{ik}w_ib_{ik}$ are the words to which the substitutions were applied. Because all words of the composition are less than $w_0 \stackrel{df}{=} w_i'vw_s'$, the new words appearing in the process are still less than w_0. Hence $P = \sum_{ik} a_{ik}F_ib_{ik}$ is the required representation. □

Remark 1.3 We stress that the succession of substitutions is not important: in any case, the diminishing process leads to the linear combination of words from the set Σ. Nevertheless, the shortest way to verify that a composition has representation (1.17) may include increasing steps, $aub \leftarrow a(u - \alpha^{-1}F_k)b$, when $aw_kb \prec w_i'vw_s'$ and the monomial $u \prec w_k$ occurs in F_k with the coefficient $\alpha \neq 0$.

Example 1.1 Denote by A_2^+ an algebra generated by variables x_1 and x_2 and defined by relations $x_1^2x_2 + \alpha x_1x_2x_1 + \beta x_2x_1^2 = 0$ and $x_1x_2^2 + \alpha x_2x_1x_2 + \beta x_2^2x_1 = 0$, where α, β are arbitrary elements from the ground field. If we fix the order $x_1 > x_2$, then the words of length three obey the lexicographical order:

$$x_1^3 > x_1^2x_2 > x_1x_2x_1 > x_1x_2^2 > x_2x_1^2 > x_2x_1x_2 > x_2^2x_1 > x_2^3.$$

Consequently, the defining relations in the form (1.16) are:

$$x_1^2x_2 = -\alpha \cdot x_1x_2x_1 - \beta \cdot x_2x_1^2; \tag{1.19}$$

$$x_1x_2^2 = -\alpha \cdot x_2x_1x_2 - \beta \cdot x_2^2x_1. \tag{1.20}$$

In other words, we have $w_1 = x_1^2x_2$, $w_2 = x_1x_2^2$, and $f_1 = -\alpha \cdot x_1x_2x_1 - \beta \cdot x_2x_1^2$, $f_2 = -\alpha \cdot x_2x_1x_2 - \beta \cdot x_2^2x_1$.

To apply Theorem 1.2, we shall analyze all possible compositions. The word w_2 has the endings x_2, x_2^2. Certainly, no one of them is a beginning of w_1 or w_2. The word w_1 has the endings x_2, x_1x_2. One of them, $v = x_1x_2$, is also a beginning of w_2. Hence, we have only one composition, which corresponds to the relation of the leading words,

$$x_1^2x_2 \cdot w_2' = w_1' \cdot x_1x_2^2,$$

with $w_1' = x_1$, $w_2' = x_2$:

$$f_1w_2' - w_1'f_2 = -\alpha \cdot x_1x_2x_1x_2 - \beta \cdot x_2\underline{x_1^2x_2} + \alpha \cdot x_1x_2x_1x_2 + \beta \cdot \underline{x_1x_2^2}x_1.$$

The first and the third terms cancel each other, whereas the words of the second and fourth ones contain (underlined) subwords w_1 and w_2. Applying the diminishing substitutions $w_1 \leftarrow f_1$, $w_2 \leftarrow f_2$, we obtain

$$\equiv \beta \cdot x_2(\alpha \cdot x_1x_2x_1 + \beta \cdot x_2x_1^2) - \beta \cdot (\alpha \cdot x_2x_1x_2 + \beta \cdot x_2^2x_1)x_1 = 0.$$

By Lemma 1.12, the system of relations (1.19), (1.20) is closed with respect to the compositions. Theorem 1.2 implies that the set Σ of all words containing not one of the subwords $x_1^2x_2$, $x_1x_2^2$ forms a basis of A_2^+, whereas the natural diminishing process provides a decomposition of words (and polynomials) in that basis. It is easy to see that

$$\Sigma = \{x_2^m (x_1x_2)^n x_1^k \mid m, n, k \geq 0\}.$$

Remark 1.4 If the diminishing process of the verification of a composition ends with a nontrivial linear combination $\tilde{F}_{is}$ of words from Σ, then we may add the relation $\tilde{F}_{is} = 0$ to system (1.16). Then, the very composition we start with has a required representation with respect to the extended system of relations, but many compositions with the new relation may appear. If we are lucky, some number of such extensions will result with a system of relations closed with respect to the compositions, which allows us to find a basis of the algebra A.

Example 1.2 Let us consider a more complicated example of a two-parameter family of algebras. Denote by B_2^+ an algebra generated by variables x_1 and x_2 and defined by relations

$$x_1^2x_2 + \alpha x_1x_2x_1 + \beta x_2x_1^2 = 0; \tag{1.21}$$

$$x_1x_2^3 + \gamma x_2x_1x_2^2 + \delta x_2^2x_1x_2 + \varepsilon x_2^3x_1 = 0, \tag{1.22}$$

where

$$\beta = \mu^2,\ \gamma = \alpha + \mu,\ \delta = \gamma\mu,\ \varepsilon = \mu^3, \tag{1.23}$$

and α, μ are arbitrary nonzero elements from the ground field. If we fix the order $x_1 > x_2$, then the above defining relations in the form (1.16) are:

$$x_1^2x_2 = -\alpha \cdot x_1x_2x_1 - \beta \cdot x_2x_1^2; \tag{1.24}$$

$$x_1x_2^3 = -\gamma \cdot x_2x_1x_2^2 - \delta \cdot x_2^2x_1x_2 - \varepsilon \cdot x_2^3x_1; \tag{1.25}$$

that is, we have $w_1 = x_1^2x_2$, $w_2 = x_1x_2^3$, and $f_1 = -\alpha \cdot x_1x_2x_1 - \beta \cdot x_2x_1^2$, $f_2 = -\gamma \cdot x_2x_1x_2^2 - \delta \cdot x_2^2x_1x_2 - \varepsilon \cdot x_2^3x_1$.

Let us analyze all possible compositions. The word w_2 has the endings x_2, x_2^2, x_2^3. Certainly no one of them is a beginning of w_1 or w_2. The word w_1 has the endings x_2, x_1x_2, and $v = x_1x_2$, is a beginning of w_2. Hence, we have only one composition, which corresponds to the relation of the leading words,

$$x_1^2x_2 \cdot w_2' = w_1' \cdot x_1x_2^3,$$

with $w_1' = x_1$, $w_2' = x_2^2$:

$$f_1w_2' - w_1'f_2 = -\alpha \cdot x_1x_2x_1x_2^2 - \beta \cdot x_2\underline{x_1^2x_2}x_2 + \gamma \cdot x_1x_2x_1x_2^2 + \delta \cdot x_1x_2^2x_1x_2 + \varepsilon \cdot \underline{x_1x_2^3}x_1.$$

The words of the second and fifth terms contain (underlined) subwords w_1 and w_2. Applying the diminishing substitutions $w_1 \leftarrow f_1,\ w_2 \leftarrow f_2$, we obtain

$$\begin{aligned}
&\equiv (\gamma-\alpha)x_1x_2x_1x_2^2 + \beta x_2(\alpha\cdot x_1x_2x_1 + \beta\cdot x_2x_1^2)x_2 + \delta x_1x_2^2x_1x_2 \\
&\quad - \varepsilon(\gamma\cdot x_2x_1x_2^2 + \delta\cdot x_2^2x_1x_2 + \varepsilon\cdot x_2^3x_1)x_1 \\
&= (\gamma-\alpha)x_1x_2x_1x_2^2 + \beta\alpha\, x_2x_1x_2x_1x_2 + \beta^2x_2^2\underline{x_1^2x_2} + \delta\, x_1x_2^2x_1x_2 \\
&\quad - \varepsilon\gamma\, x_2x_1x_2^2x_1 - \varepsilon\delta\, x_2^2x_1x_2x_1 - \varepsilon^2x_2^3x_1^2.
\end{aligned}$$

The word of the third term contains (underlined) subword w_1. Applying the diminishing substitution $w_1 \leftarrow f_1$, we may continue

$$\begin{aligned}
&\equiv (\gamma-\alpha)x_1x_2x_1x_2^2 + \beta\alpha\, x_2x_1x_2x_1x_2 + \delta\, x_1x_2^2x_1x_2 \\
&\quad -\varepsilon\gamma x_2x_1x_2^2x_1 - (\varepsilon\delta+\alpha\beta^2)\, x_2^2x_1x_2x_1 - (\varepsilon^2+\beta^3)x_2^3x_1^2.
\end{aligned} \tag{1.26}$$

None of the remaining words contains w_1 or w_2 as a subword. Consequently, due to Lemma 1.12, the system of relations (1.24), (1.25) is not closed with respect to the compositions.

The polynomial that appears in (1.26) is a relation of the algebra B_2^+. If we add that relation,

$$\begin{aligned}
x_1x_2x_1x_2^2 &= -\mu^{-1}\beta\alpha\, x_2x_1x_2x_1x_2 - \mu^{-1}\delta\, x_1x_2^2x_1x_2 + \mu^{-1}\varepsilon\gamma x_2x_1x_2^2x_1 \\
&\quad + \mu^{-1}(\varepsilon\delta+\alpha\beta^2)\, x_2^2x_1x_2x_1 + \mu^{-1}(\varepsilon^2+\beta^3)x_2^3x_1^2 \stackrel{df}{=} f_3,
\end{aligned} \tag{1.27}$$

$\mu = \gamma-\alpha$, to the two initial defining relations, then obviously the composition (1.26) reduces to zero after the diminishing substitution $x_1x_2x_1x_2^2 \leftarrow f_3$. We must analyze if the extended system (1.24), (1.25), (1.27) is closed.

In this case, there appeared a new leading word, $w_3 = x_1x_2x_1x_2^2$. The endings of this word are $x_2,\ x_2^2,\ x_1x_2^2,\ x_2x_1x_2^2$. Among them, $x_1x_2^2$ is a beginning of w_2. The beginnings of w_3 are $x_1,\ x_1x_2,\ x_1x_2x_1x_2$. Among them, x_1x_2 is an ending of w_1. That is, we have to analyze two more compositions.

1. $\underline{x_1x_2x_1x_2^2}\cdot x_2 = x_1x_2\cdot \underline{x_1x_2^3},\ w_2' = x_2,\ w_3' = x_1x_2$. We have

$$\begin{aligned}
f_3w_2' - w_3'f_2 &= -\mu^{-1}\beta\alpha\, \underline{x_2x_1x_2x_1x_2^2} - \mu^{-1}\delta\, \underline{\underline{x_1}}\,\underline{\underline{x_2}}^2\underline{\underline{x_1}}\,\underline{\underline{x_2}}^2 + \mu^{-1}\varepsilon\gamma\, \underbrace{x_2x_1x_2^2x_1x_2} \\
&\quad + \mu^{-1}(\varepsilon\delta+\alpha\beta^2)\boxed{x_2^2x_1x_2x_1x_2} + \mu^{-1}(\varepsilon^2+\beta^3)x_2^3\underline{x_1^2x_2}, \\
&\quad + \gamma\cdot \underline{\underline{x_1}}\,\underline{\underline{x_2}}^2\underline{\underline{x_1}}\,\underline{\underline{x_2}}^2 + \delta\cdot \underline{x_1x_2^3}x_1x_2 + \varepsilon\cdot \underline{x_1x_2^3}x_2x_1.
\end{aligned}$$

The second and the sixth terms are similar, whereas the first, fifth, seventh, and eighth terms have leading words w_3, w_1, w_2 as subwords. Applying the diminishing substitutions, we obtain

$$
\begin{aligned}
-\mu^{-1}\beta\alpha\, x_2\underline{x_1x_2x_1x_2^2} &\equiv (\mu^{-1}\beta\alpha)^2\, \boxed{x_2^2x_1x_2x_1x_2} + \beta\alpha\mu^{-2}\delta\, \underbrace{x_2x_1x_2^2x_1x_2} \\
&\quad -\mu^{-2}\beta\alpha\varepsilon\gamma\, \underline{\underline{x_2^2x_1x_2^2x_1}} - \beta\alpha\mu^{-2}(\varepsilon\delta+\alpha\beta^2)\, x_2^3x_1x_2x_1 \\
&\quad -\mu^{-2}\beta\alpha(\varepsilon^2+\beta^3)\mathbf{x}_2^4\mathbf{x}_1^2; \\
\mu^{-1}(\varepsilon^2+\beta^3)\cdot x_2^3\underline{x_1^2x_2} &\equiv -\mu^{-1}(\varepsilon^2+\beta^3)\alpha\cdot x_2^3x_1x_2x_1 - \mu^{-1}(\varepsilon^2+\beta^3)\beta\cdot \mathbf{x}_2^4\mathbf{x}_1^2 \\
\delta\cdot \underline{x_1x_2^3}x_1x_2 &\equiv -\delta\gamma\cdot \underbrace{x_2x_1x_2^2x_1x_2} - \delta^2\cdot \boxed{x_2^2x_1x_2x_1x_2} - \delta\varepsilon\cdot x_2^3\underline{x_1^2x_2}; \\
\varepsilon\cdot \underline{x_1x_2^3}x_2x_1 &\equiv -\varepsilon\gamma\cdot x_2\underline{x_1x_2^3}x_1 - \varepsilon\delta\cdot \underline{\underline{x_2^2x_1x_2^2x_1}} - \varepsilon^2\cdot x_2^3x_1x_2x_1.
\end{aligned}
$$

Further,

$$
-\delta\varepsilon\cdot x_2^3\underline{x_1^2x_2} \equiv \delta\varepsilon\alpha\cdot x_2^3x_1x_2x_1 + \delta\varepsilon\beta\cdot \mathbf{x}_2^4\mathbf{x}_1^2,
$$

and

$$
-\varepsilon\gamma\cdot x_2\underline{x_1x_2^3}x_1 \equiv \varepsilon\gamma^2\cdot \underline{\underline{x_2^2x_1x_2^2x_1}} + \gamma\varepsilon\delta\cdot x_2^3x_1x_2x_1 + \gamma\varepsilon^2\cdot \mathbf{x}_2^4\mathbf{x}_1^2.
$$

Using values $\beta = \mu^2$, $\gamma = \alpha + \mu$, $\delta = \gamma\mu$, $\varepsilon = \mu^3$ of the coefficients given in (1.23), we may find coefficients at the remaining six words.

$$
\begin{aligned}
\underline{\underline{x_1x_2^2x_1x_2^2}} &: -\mu^{-1}\delta + \gamma = -\mu^{-1}\gamma\mu + \gamma = 0; \\
\underbrace{x_2x_1x_2^2x_1x_2} &: \mu^{-1}\varepsilon\gamma + \beta\alpha\mu^{-2}\delta - \delta\gamma = \mu^{-1}\mu^3\gamma + \alpha\gamma\mu - \gamma\mu\gamma \\
&\quad = \gamma\mu(\mu + \alpha - \gamma) = 0; \\
\boxed{x_2^2x_1x_2x_1x_2} &: \mu^{-1}(\varepsilon\delta + \alpha\beta^2) + (\mu^{-1}\beta\alpha)^2 - \delta^2 = \mu^{-1}(\mu^3\gamma\mu + \alpha\mu^4) + \mu^{-2}\mu^4\alpha^2 \\
&\quad - \gamma^2\mu^2 = \mu^2(\gamma\mu + \alpha\mu + \alpha^2 - \gamma^2) = \mu^2(\alpha+\gamma)(\mu + \alpha - \gamma) = 0; \\
\underline{\underline{x_2^2x_1x_2^2x_1}} &: -\mu^{-2}\beta\alpha\varepsilon\gamma - \varepsilon\delta + \varepsilon\gamma^2 = \varepsilon\gamma(-\alpha - \mu + \gamma) = 0; \\
\mathbf{x}_2^4\mathbf{x}_1^2 &: -\mu^{-2}\beta\alpha(\varepsilon^2 + \beta^3) - \mu^{-1}(\varepsilon^2 + \beta^3)\beta + \delta\varepsilon\beta + \gamma\varepsilon^2 \\
&\quad = \mu^6(-2\alpha - 2\mu + 2\gamma) = 0; \\
x_2^3x_1x_2x_1 &: -\beta\alpha\mu^{-2}(\varepsilon\delta + \alpha\beta^2) - \mu^{-1}(\varepsilon^2 + \beta^3)\alpha - \varepsilon^2 + \delta\varepsilon\alpha + \gamma\varepsilon\delta \\
&\quad = \mu^4(-\alpha(\gamma + \alpha) - 2\mu\alpha - \mu^2 + \gamma\alpha + \gamma^2) \\
&\quad = \mu^4(-\alpha^2 - 2\mu\alpha - \mu^2 + \gamma^2) = 0.
\end{aligned}
$$

Thus, the first additional composition reduces to zero in the free algebra by diminishing substitutions. Similarly we consider the second additional composition.

2. $x_1 \cdot \underline{x_1x_2x_1x_2^2} = \underline{x_1^2x_2} \cdot x_1x_2^2,\ w_1' = x_1,\ w_3' = x_1x_2^2.$

$$\begin{aligned} f_1w_3' - w_1'f_3 = &-\alpha \cdot x_1x_2\underline{x_1x_1x_2}x_2 - \beta \cdot x_2x_1\underline{x_1^2x_2}x_2 + \mu^{-1}\beta\alpha\, x_1\, x_2\, x_1\, x_2\, x_1\, x_2 \\ &+ \mu^{-1}\delta\, \underline{x_1^2x_2}x_2x_1x_2 - \mu^{-1}\varepsilon\gamma \underline{x_1x_2x_1x_2^2}x_1 \\ &- \mu^{-1}(\varepsilon\delta + \alpha\beta^2)\, x_1x_2^2x_1x_2x_1 - \mu^{-1}(\varepsilon^2 + \beta^3)\underline{x_1x_2^3}x_1^2. \end{aligned}$$

The first, second, fourth and sixth terms have the leading words w_1, w_3, and w_2 as subwords. Applying the diminishing substitutions, we obtain

$$\begin{aligned} -\alpha \cdot x_1x_2\underline{x_1^2x_2}x_2 &\equiv \alpha^2 \cdot x_1\, x_2\, x_1\, x_2\, x_1\, x_2 + \alpha\beta \cdot x_1x_2^2\underline{x_1^2x_2} \\ -\beta \cdot x_2x_1\underline{x_1^2x_2}x_2 &\equiv \beta\alpha \cdot x_2\underline{x_1^2x_2}x_1x_2 + \beta^2 \cdot x_2x_1x_2\underline{x_1^2x_2} \\ \mu^{-1}\delta\, \underline{x_1^2x_2}x_2x_1x_2 &\equiv -\mu^{-1}\delta\alpha \cdot x_1\, x_2\, x_1\, x_2\, x_1\, x_2 - \mu^{-1}\delta\beta \cdot x_2\underline{x_1^2x_2}x_1x_2; \\ -\mu^{-1}\varepsilon\gamma \underline{x_1x_2x_1x_2^2}x_1 &\equiv \mu^{-2}\varepsilon\gamma\beta\alpha\, \underbrace{x_2x_1x_2x_1x_2x_1} + \mu^{-2}\varepsilon\gamma\delta\, x_1x_2^2x_1x_2x_1 \\ &\quad -\mu^{-2}\varepsilon^2\gamma^2 \boxed{x_2x_1x_2^2x_1^2} - \mu^{-2}\varepsilon\gamma(\varepsilon\delta + \alpha\beta^2)\, x_2^2x_1x_2x_1^2 \\ &\quad -\mu^{-2}\varepsilon\gamma(\varepsilon^2 + \beta^3)\mathbf{x}_2^3\mathbf{x}_1^3; \\ -\mu^{-1}(\varepsilon^2 + \beta^3) \cdot \underline{x_1x_2^3}x_1^2 &\equiv \mu^{-1}(\varepsilon^2 + \beta^3)\gamma \cdot \boxed{x_2x_1x_2^2x_1^2} + \mu^{-1}(\varepsilon^2 + \beta^3)\delta \cdot x_2^2x_1x_2x_1^2 \\ &\quad +\mu^{-1}(\varepsilon^2 + \beta^3)\varepsilon \cdot \mathbf{x}_2^3\mathbf{x}_1^3. \end{aligned}$$

Four of the new terms contain the leading words as subwords:

$$\begin{aligned} \alpha\beta \cdot x_1x_2x_2\underline{x_1^2x_2} &\equiv -\alpha^2\beta \cdot x_1x_2^2x_1x_2x_1 - \alpha\beta^2 \cdot \underline{x_1x_2^3}x_1^2 \\ \beta\alpha \cdot x_2\underline{x_1^2x_2}x_1x_2 &\equiv -\beta\alpha^2 \cdot x_2x_1x_2\underline{x_1^2x_2} - \beta^2\alpha \cdot x_2^2x_1\underline{x_1^2x_2}. \\ \beta^2 \cdot x_2x_1x_2\underline{x_1^2x_2} &\equiv -\alpha\beta^2 \cdot \underbrace{x_2x_1x_2x_1x_2x_1} - \beta^3 \cdot \boxed{x_2x_1x_2^2x_1^2} \\ -\mu^{-1}\delta\beta \cdot x_2\underline{x_1^2x_2}x_1x_2 &\equiv \mu^{-1}\delta\beta\alpha \cdot x_2x_1x_2\underline{x_1^2x_2} + \mu^{-1}\delta\beta^2 \cdot x_2^2x_1\underline{x_1^2x_2}, \end{aligned}$$

and five more appearing terms admit the diminishing substitutions:

$$\begin{aligned} -\alpha\beta^2 \cdot \underline{x_1x_2^3}x_1^2 &\equiv \alpha\beta^2\gamma \cdot \boxed{x_2x_1x_2^2x_1^2} + \alpha\beta^2\delta \cdot x_2^2x_1x_2x_1^2 + \alpha\beta^2\varepsilon \cdot \mathbf{x}_2^3\mathbf{x}_1^3; \\ -\beta\alpha^2 \cdot x_2x_1x_2\underline{x_1^2x_2} &\equiv \beta\alpha^3 \cdot \underbrace{x_2x_1x_2x_1x_2x_1} + \beta^2\alpha^2 \cdot \boxed{x_2x_1x_2^2x_1^2}; \end{aligned}$$

$$
\begin{aligned}
-\beta^2\alpha \cdot x_2^2x_1\underline{x_1^2x_2} &\equiv \beta^2\alpha^2 \cdot x_2^2\underline{x_1^2x_2}x_1 + \beta^3\alpha \cdot x_2^2x_1x_2x_1^2;\\
\mu^{-1}\delta\beta\alpha \cdot x_2x_1x_2\underline{x_1^2x_2} &\equiv -\mu^{-1}\delta\beta\alpha^2 \cdot \underbrace{x_2x_1x_2x_1x_2x_1} - \mu^{-1}\delta\beta^2\alpha \cdot \boxed{x_2x_1x_2^2x_1^2};\\
\mu^{-1}\delta\beta^2 \cdot x_2^2x_1\underline{x_1^2x_2} &\equiv -\mu^{-1}\delta\beta^2\alpha \cdot x_2^2\underline{x_1^2x_2}x_1 - \mu^{-1}\delta\beta^3 \cdot x_2^2x_1x_2x_1^2.
\end{aligned}
$$

Finally,

$$
\begin{aligned}
\beta^2\alpha^2 \cdot x_2^2\underline{x_1^2x_2}x_1 &\equiv -\beta^2\alpha^3 \cdot x_2^2x_1x_2x_1^2 - \beta^3\alpha^2 \cdot \mathbf{x}_2^3\mathbf{x}_1^3\\
-\mu^{-1}\delta\beta^2\alpha \cdot x_2^2\underline{x_1^2x_2}x_1 &\equiv \mu^{-1}\delta\beta^2\alpha^2 \cdot x_2^2x_1x_2x_1^2 + \mu^{-1}\delta\beta^3\alpha \cdot \mathbf{x}_2^3\mathbf{x}_1^3.
\end{aligned}
$$

Now, we are ready to calculate coefficients at the remaining six words using values $\beta = \mu^2$, $\gamma = \alpha + \mu$, $\delta = \gamma\mu$, $\varepsilon = \mu^3$ fixed in (1.23).

$$
\begin{aligned}
\underline{\underline{\underline{x_1\,x_2\,x_1\,x_2\,x_1\,x_2}}} &: \mu^{-1}\beta\alpha + \alpha^2 - \mu^{-1}\delta\alpha = \mu\alpha + \alpha^2 - \gamma\alpha = (\mu + \alpha - \gamma)\alpha = 0;\\
\underline{\underline{x_1x_2^2x_1x_2x_1}} &: -\mu^{-1}(\varepsilon\delta + \alpha\beta^2) + \mu^{-2}\varepsilon\gamma\delta - \alpha^2\beta = \mu^2(-\mu\gamma - \alpha\mu + \gamma^2 - \alpha^2)\\
&= \mu^2(\gamma + \alpha)(-\mu + \gamma - \alpha) = 0;\\
\underbrace{x_2x_1x_2x_1x_2x_1} &: \mu^2\varepsilon\gamma\beta\alpha - \alpha\beta^2 + \beta\alpha^3 - \mu^{-1}\delta\beta\alpha^2\\
&= \mu^2\alpha(\mu - \alpha)(\mu\gamma - \mu^2 + \alpha^2 - \gamma\alpha)\\
&= \mu^2\alpha(\mu - \alpha)(\gamma - \mu - \alpha) = 0;\\
\boxed{x_2x_1x_2^2x_1^2} &: -\mu^{-2}\varepsilon^2\gamma^2 + \mu^{-1}(\varepsilon^2 + \beta^3)\gamma + \alpha\beta^2\gamma - \beta^3 + \beta^2\alpha^2 - \mu^{-1}\delta\beta^2\alpha\\
&= \mu^4(-\gamma^2 + 2\mu\gamma + \alpha\gamma - \mu^2 + \alpha^2 - \gamma\alpha)\\
&= \mu^4(-(\gamma - \mu)^2 + \alpha^2) = 0;\\
x_2^2x_1x_2x_1^2 &: -\mu^{-2}\varepsilon\gamma(\varepsilon\delta + \alpha\beta^2) + \mu^{-1}(\varepsilon^2 + \beta^3)\delta + \alpha\beta^2\delta + \beta^3\alpha\\
&\quad - \mu^{-1}\delta\beta^3 - \beta^2\alpha^3 + \mu^{-1}\delta\beta^2\alpha^2\\
&= \mu^4(-\mu\gamma^2 - \mu\gamma\alpha + 2\mu^2\gamma + \mu\alpha\gamma + \mu^2\alpha - \mu^2\gamma - \alpha^3 + \gamma\alpha^2)\\
&= \mu^4(-\mu\gamma^2 + \mu^2\gamma + \mu^2\alpha + \alpha^2(-\alpha + \gamma))\\
&= \mu^5(-\gamma^2 + \mu\gamma + \mu\alpha + \alpha^2)\\
&= \mu^5(\gamma + \alpha)(-\gamma + \mu + \alpha) = 0\\
\mathbf{x}_2^3\mathbf{x}_1^3 &: -\mu^{-2}\varepsilon\gamma(\varepsilon^2 + \beta^3) + \alpha\beta^2\varepsilon - \beta^3\alpha^2 + \mu^{-1}\delta\beta^3\alpha + \mu^{-1}(\varepsilon^2 + \beta^3)\varepsilon\\
&= \mu^6(-2\mu\gamma + \alpha\mu - \alpha^2 + \gamma\alpha + 2\mu^2)\\
&= \mu^6(-2\mu\gamma + \alpha\mu + (-\alpha + \gamma)\alpha + 2\mu^2)\\
&= 2\mu^7(-\gamma + \alpha + \mu) = 0.
\end{aligned}
$$

Because both additional compositions reduce to zero, Lemma 1.12 implies that the system of relations (1.24), (1.25), (1.27) is closed with respect to the compositions. Theorem 1.2 implies that the set Σ of all words containing none of the subwords $x_1^2x_2$, $x_1x_2^3$, $x_1x_2x_1x_2^2$ forms a basis of B_2^+, and the natural diminishing process provides a decomposition of polynomials on that basis. It is easy to see that

$$\Sigma = \{x_2^m(x_1x_2x_2)^n(x_1x_2)^k x_1^s \mid m, n, k, s \geq 0\}.$$

Remark 1.5 If instead of the free algebra $\mathbf{k}\langle X\rangle$, we consider the polynomial algebra $\mathbf{k}[X]$, the free commutative algebra, then in this way, each system of relations may be closed. This process is precisely the Buchberger algorithm that resolves the equality problem for commutative algebras.

The Gröbner-Shirshov system of relations is not uniquely defined by a given set of defining relations. First, it depends essentially on the chosen order of words $\prec$. Moreover, even if the order is fixed, there may exist various Gröbner-Shirshov systems that define the same ideal of relations. The simplest example with $x_1 > x_2 > x_3$ is as follows:

$$\mathbf{k}\langle x_1, x_2, x_3 \mid x_1 = 0,\ x_2 = 0\rangle = \mathbf{k}\langle x_1, x_2, x_3 \mid x_2 = 0,\ x_1 = -x_2\rangle.$$

Nevertheless, the set of the leading words of the Gröbner-Shirshov system is uniquely defined by the ideal (or, equivalently, by the initial defining relations), provided that $\prec$ is a fixed Deg-Lex order.

Proposition 1.2 *If S_1 and S_2 are different Gröbner-Shirshov systems of a given ideal I with respect to the same Deg-Lex order, then the sets of the leading words of relations from S_1 and S_2 are the same. In particular, the number of elements in a Gröbner-Shirshov system is an invariant of the ideal.*

Proof We note that due to the first property of the Definition 1.5, none of the leading words may appear twice in the same Gröbner-Shirshov system. Denote by $W(S)$ the set of all leading words of a Gröbner-Shirshov system S. We have to demonstrate that $W(S_1) = W(S_2)$. Let us chose the minimal word w in the set $(W(S_1) \setminus W(S_2)) \cup (W(S_2) \setminus W(S_1))$. Let $w \in W(S_1) \setminus W(S_2)$ and $w = f$ is the relation from S_1 with the leading word w. Because $w - f \in I$ and S_2 is a Gröbner-Shirshov system for I, it follows that the word w contains some $v \in W(S_2)$ as a proper subword. In this case, v does not belong to $W(S_1)$ due to the first property given in Definition 1.5. So $v \in W(S_2) \setminus W(S_1)$ and $v \prec w$, which contradicts the choice of w. □

The following proposition shows that to some extent the Gröbner-Shirshov systems are indifferent to the non-leading terms of the relations.

Proposition 1.3 *Let $\{w_t = f_t \mid t \in T\}$ be a Gröbner-Shirshov system of an ideal I. If f_t', $t \in T$ are arbitrary polynomials such that $w_t - f_t' \in I$ and all monomials of f_t' are less than w_t, then $\{w_t = f_t' \mid t \in T\}$ is a Gröbner-Shirshov system for I.*

Proof By Lemma 1.12 it suffices to check that each composition can be reduced to zero in free algebra by means of the one-sided diminishing substitutions $w_t \leftarrow f_t'$. Certainly, every composition belongs to I. By induction on the leading word, we shall show that each polynomial $F \in I$ satisfies the required property. Because F belongs to I, its leading word has one of the words w_t, $t \in T$ as a subword. After the substitution $w_t \leftarrow f_t'$, the obtained polynomial F_1 still belongs to I because $w_t - f_t' \in I$. At the same time the leading word of F_1 is less than that of F. The induction applies. □

Thus, the set of all leading words of a Gröbner-Shirshov system and the set Σ related to it are the basic Gröbner-Shirshov invariants of an ideal I provided that a Deg-Lex order $\prec$ is fixed.

1.3.2 Noncommutative G-Polynomials

Let G be a group. We would like to discuss the construction of the algebra of *noncommutative G-polynomials* $G\langle X\rangle$, which admit coefficients from G and satisfy certain commutation rules,

$$x_i g = \lambda_g^i g x_i, \quad \lambda_g^i \in \mathbf{k}, \quad g \in G.$$

In the above context, we may introduce $G\langle X\rangle$ as an algebra defined by the generators $x_i \in X$, $g \in G$ and the relations

$$x_i g = \lambda_g^i g x_i, \quad gh = f, \; g, h, f \in G. \tag{1.28}$$

The latter group of relations is precisely the table of multiplication of G; that is, for each pair of elements $g, h \in G$ we have one relation $gh = f$ where f is the product of g and h in G.

On the set of all words in $X \cup G$ we consider the Hall order with respect to an arbitrary ordering of the variables with the only restriction that $g < x$ for all $g \in G$, $x \in X$. In this case (1.28) are relations of the form (1.16) with the leading words $x_i g$, gh. The set Σ related to these words is the set of all words that contains no one of $x_i g$, gh as sub-words; that is, Σ consists of the words gu, where $g \in G$, $u \in X^*$.

To be sure that these words are linearly independent in $G\langle X\rangle$, we must check all possible compositions for (1.28). There are only two types of compositions, which correspond to the following pairs of relations: $x_i g = \lambda_g^i g x_i$, $gh = f$; and $gh = f$, $hs = t$, where $s, t \in G$. We have

$$\lambda_g^i g x_i \cdot h - x_i \cdot f \equiv \lambda_g^i \lambda_h^i g h x_i - \lambda_f^i f x_i \equiv (\lambda_g^i \lambda_h^i - \lambda_f^i) f x_i.$$

In the first step, we apply the diminishing substitutions $x_i h \leftarrow \lambda_h^i h x_i$ and $x_i f \leftarrow \lambda_f^i f x_i$, whereas in the second step, $gh \leftarrow f$. We see that this composition reduces to zero only if $\lambda^i_{gh} = \lambda^i_g\lambda^i_h$; that is, the map $\chi^i : g \mapsto \lambda^i_g$ must be a character of the group G. Otherwise, in $G\langle X\rangle$, we have a relation $fx_i = 0$, which implies $x_i = 0$.

Similarly,

$$f \cdot s - g \cdot t \equiv s_1 - t_1,$$

where we apply the diminishing substitutions $fs \leftarrow s_1$ and $gt \leftarrow t_1$ defined by the relations $fs = s_1$ and $gt = t_1$. As the group G is associative, the equalities $s_1 = fs = (gh)s = g(hs) = gt = t_1$ are valid in the group G. Hence, $s_1 - t_1 = 0$ in $\mathbf{k}\langle G \cup X\rangle$.

1.3.3 Skew Group Rings

One may generalize the above construction, assuming that the variables are not free. In this way, a construction of a *skew group ring* appears. Let G be a group acting on an algebra R by linear transformations $g : a \mapsto a^g,\ a, a^g \in R$ (not necessarily faithfully). The skew group ring $R * G$ is defined as a space of formal sums $\sum_i g_i a_i,\ g_i \in G,\ a_i \in R$, with a multiplication induced by commutation rules

$$ag = ga^g, \quad g \in G, \ a \in R.$$

This multiplication is associative only if G acts by automorphisms, $(ab)^g = a^g b^g$:

$$g(ab)^g = (ab)g = a(bg) = a(gb^g) = (ag)b^g = (ga^g)b^g = g(a^g b^g).$$

Of course, it is more or less evident that monomials ga_i, where $g \in G$ and a_i runs through a basis $\{a_i \mid i \in I\}$ of R, are linearly independent in $R * G$. Nevertheless, to be certain, we may apply the Composition Lemma.

In this context, we may introduce the skew group ring $R*G$ as an algebra defined by the generators $a_i,\ i \in I,\ g \in G$ and the relations

$$a_i g = g a_i^g, \quad gh = f, \quad a_i a_s = \sum_{k\in I} \alpha^k_{i,s} a_k, \quad g, h, f \in G, \ i, s \in I. \tag{1.29}$$

Here, in the first group of relations, ga_i^g means a linear combination $\sum_s \alpha_i^s g a_s$, where $\sum_s \alpha_i^s a_s$ is a decomposition of a_i^g in the basis $\{a_i\}$; the second group of relations is the table of multiplication of G; and the third group of relations is the table of multiplication of R with coefficients $\alpha^k_{i,s}$, called the *structural constants*, from the ground field $\mathbf{k}$.

On the set of all words in $\{a_i\} \cup G$ we consider the Hall order with respect to an arbitrary ordering of the variables such that $g < a_i,\ g \in G,\ i \in I$. In this case (1.29) are relations of the form (1.16) with the leading words $a_i g,\ gh,\ a_i a_s$. The set Σ related to this system of relations consists of the words ga_i, where $g \in G,\ i \in I$.

There are four types of compositions, and they correspond to the following values of the word $w_i'vw_s' = w_i \cdot w_s' = w_i' \cdot w_s$ appeared in (1.17):

$$a_i g \cdot h = a_i \cdot gh, \quad a_i a_s \cdot g = a_i \cdot a_s g, \quad gh \cdot f = g \cdot hf, \quad a_i a_s \cdot a_t = a_i \cdot a_s a_t.$$

Consider the related compositions one by one. We have

$$ga_i^g \cdot h - a_i f \equiv gh(a_i^g)^h - fa_i^f \equiv f[(a_i^g)^h - a_i^f] = 0.$$

In the first step we apply the diminishing substitutions $a_i^g h \leftarrow h(a_i^g)^h$ and $a_i f \leftarrow fa_i^f$, whereas in the second step $gh \leftarrow f$. This composition reduces to zero because by definition of the action, the element $f = gh$ acts on R as a superposition of g and h; that is, a_i^f equals $(a_i^g)^h$ as a linear combination of the a_i's.

Considering that G acts by homomorphisms, the equality $a_i^g a_s^g = (a_i a_s)^g$, which is valid in R, implies that the expression $a_i^g a_s^g$ by application of the table of multiplication in the basis $\{a_i\}$ reduces to the linear combination $(\sum_k \alpha_{i,s}^k a_k)^g$; that is, $a_i^g a_s^g \equiv \sum_k \alpha_{i,s}^k a_k^g$. Hence, we have

$$\sum_k \alpha_{i,s}^k a_k \cdot g - a_i \cdot g a_s^g \equiv \sum_k \alpha_{i,s}^k g a_k^g - g a_i^g a_s^g \equiv g \sum_k \alpha_{i,s}^k a_k^g - g \sum_k \alpha_{i,s}^k a_k^g = 0.$$

The third type of compositions is already considered in the above subsection. Let us examine the fourth:

$$\sum_k \alpha_{i,s}^k a_k \cdot a_t - a_i \cdot \sum_k \alpha_{s,t}^k a_k \equiv \sum_k \alpha_{i,s}^k \sum_r \alpha_{k,t}^r a_r - \sum_k \alpha_{s,t}^k \sum_r \alpha_{i,k}^r a_r.$$

The value in R of the latter linear combination equals $a_i a_s \cdot a_t - a_i \cdot a_s a_t = 0$. Because the a_r's are linearly independent in R, this combination remains zero provided that the a_r's are considered to be the free variables.

1.3.4 Poincaré-Birkhoff-Witt Theorem

We conclude this section with an elegant proof of the Poincaré-Birkhoff-Witt theorem due to L.A. Bokut' based on the Composition Lemma. Recall that a *Lie algebra* is a linear space L endowed with a bilinear operation $[\,,\,] : L^{\otimes 2} \to L$ that satisfies the antisymmetry and Jacoby identities:

$$[u, u] = 0; \quad [[u, v], w] + [[v, w], u] + [[w, u], v] = 0.$$

If L is a Lie algebra and $u, v \in L$, then $0 = [u+v, u+v] = [u, u] + [u, v] + [v, u] + [v, v] = [u, v] + [v, u]$. Therefore, the antisymmetry identity implies

$$[u, v] = -[v, u].$$

A fundamental example of a Lie algebra appears from an associative algebra R when in place of the bilinear operation, one considers the commutator $[u, v] = uv - vu$. This Lie algebra is denoted by $R^{(-)}$. Every Lie algebra is isomorphic to a Lie subalgebra of a Lie algebra $R^{(-)}$ if, in place of R, we take the *universal enveloping algebra* $U(L)$ of L. The algebra $U(L)$ has the following construction in terms of generators and defining relations.

Let us fix a well-ordered basis $B = \{u_i \mid i \in I\}$ of L. Consider this basis to be a set of free variables and define an associative algebra $U(L)$ by the relations

$$u_i u_s = u_s u_i + [u_i, u_s], \quad u_i > u_s, \tag{1.30}$$

where $[u_i, u_s]$ is a linear combination $\sum_t \alpha_{i,s}^t u_t$ that equals $[u_i, u_s]$ in L.

Theorem 1.3 (Poincaré-Birkhoff-Witt) *The set of all monomials*

$$u_1^{n_1} u_2^{n_2} \cdots u_k^{n_k}, \quad u_1 < u_2 < \ldots < u_k, \quad u_i \in B, \; 1 \leq i \leq k$$

form a basis of $U(L)$.

Proof On the set of all words in $\{u_i\}$, we consider the Hall order defined by the fixed above ordering of $\{u_i\}$. In this case, (1.30) are relations of the form (1.16) with the leading words $u_i u_s, \; u_i > u_s$. The set Σ related to (1.30) is precisely the set of monomials mentioned in the theorem. Therefore, it remains to analyze the compositions of (1.30). There exists only one type of compositions when in (1.17), we have

$$w_i' v w_s' \leftarrow u_i u_s \cdot u_t = u_i \cdot u_s u_t, \qquad u_i > u_s > u_t.$$

We have

$$\begin{aligned}
&(u_s u_i + [u_i, u_s]) u_t - u_i (u_t u_s + [u_s, u_t]) \\
&\quad = u_s \underline{u_i u_t} - \underline{u_i u_t} u_s + [u_i, u_s] u_t - u_i [u_s, u_t] \\
&\quad \equiv u_s (u_t u_i + [u_i, u_t]) - (u_t u_i + [u_i, u_t]) u_s + [u_i, u_s] u_t - u_i [u_s, u_t] \\
&\quad = \underline{u_s u_t} u_i - u_t \underline{u_i u_s} + u_s [u_i, u_t] - [u_i, u_t] u_s + [u_i, u_s] u_t - u_i [u_s, u_t] \\
&\quad \equiv (u_t u_s + [u_s, u_t]) u_i - u_t (u_s u_i + [u_i, u_s]) + u_s [u_i, u_t] \\
&\qquad - [u_i, u_t] u_s + [u_i, u_s] u_t - u_i [u_s, u_t] \\
&\quad = [u_s, u_t] u_i - u_t [u_i, u_s] + u_s [u_i, u_t] - [u_i, u_t] u_s + [u_i, u_s] u_t - u_i [u_s, u_t].
\end{aligned} \tag{1.31}$$

Let $[u_s, u_t] = \sum_k \alpha_{s,t}^k u_k$. In this case, $[u_s, u_t]u_i - u_i[u_s, u_t] = \sum_k \alpha_{s,t}^k (u_k u_i - u_i u_k)$. If, for a given k, we have $u_k > u_i$, then $\underline{u_k u_i} - u_i u_k \equiv [u_k, u_i]$, whereas if $u_k < u_i$, then $u_k u_i - \underline{u_i u_k} \equiv -[u_i, u_k] = [u_k, u_i]$ due to the antisymmetry identity of L. If $u_k = u_i$, then it is still true that $u_k u_i - u_i u_k = 0 = [u_k, u_i]$. Therefore, in all cases, we have

$$[u_s, u_t]u_i - u_i[u_s, u_t] \equiv \sum_k \alpha_{s,t}^k [u_k, u_i] = [\sum_k \alpha_{s,t}^k u_k, u_i] = [[u_s, u_t], u_i],$$

where the last two equalities are equalities of linear combinations of the u_r's in L. In the perfect analogy, we obtain

$$-u_t[u_i, u_s] + [u_i, u_s]u_t \equiv [[u_i, u_s], u_t],$$

and

$$u_s[u_i, u_t] - [u_i, u_t]u_s \equiv [u_s, [u_i, u_t]] = -[[u_i, u_t], u_s] = [[u_t, u_i], u_s].$$

Applying these equivalences to (1.31), we see that by the diminishing process the composition reduces to a linear combination

$$[[u_i, u_s], u_t] + [[u_s, u_t], u_i] + [[u_t, u_i], u_s].$$

This linear combination equals zero in L due to the Jacobi identity. Since the u_i's are linearly independent in L, it follows that this combinations is still zero in the free algebra $\mathbf{k}\langle u_i \mid i \in I\rangle$. □

1.4 Braid Monoid and Permutation Group

By definition the *braid monoid* B_n is generated by braids $s_1, s_2, \ldots, s_n$ subject to the relations

$$s_k s_{k+1} s_k = s_{k+1} s_k s_{k+1}, \quad s_i s_k = s_k s_i, \quad |i - k| > 1. \tag{1.32}$$

As usual, we assume that B_n has the unit element 1 (the empty product of braids), and $B_0 = \{1\}$. Considering that the above defining relations are invariant with respect to the substitutions $s_i \leftarrow s_{n-i}$, $1 \leq i \leq n$, there exists an automorphism $\imath : B_n \to B_n$ such that $\imath(s_i) = s_{n-i}$, $1 \leq i \leq n$. Similarly, an involution $* : B_n \to B_n$ is well-defined on B_n which acts as follows

$$(s_{i_1} s_{i_2} \cdots s_{i_{k-1}} s_{i_k})^* = s_{i_k} s_{i_{k-1}} \cdots s_{i_2} s_{i_1}.$$

Lemma 1.13 *In the braid monoid, the following commutation rules hold*

$$(s_k s_{k+1} \cdots s_{t-1} s_t) \cdot (s_r s_{r-1} \cdots s_m) = (s_{r+1} s_r \cdots s_{m+1}) \cdot (s_k s_{k+1} \cdots s_{t-1} s_t), \qquad (1.33)$$

provided that $k \leq m \leq r < t$, and

$$(s_t s_{t-1} \cdots s_{k+1} s_k) \cdot (s_r s_{r-1} \cdots s_m) = (s_{r-1} s_{r-2} \cdots s_{m-1}) \cdot (s_t s_{t-1} \cdots s_{k+1} s_k) \qquad (1.34)$$

provided that $k \leq m-1 \leq r \leq t$. In the latter formula, if $r = m-1$, then we postulate $s_r s_{r-1} \cdots s_m = s_{r-1} s_{r-2} \cdots s_{m-1} = 1$ as the empty products.

Proof If $k \leq i < t$, then s_i commutes with $s_{i+2}, s_{i+3}, \ldots, s_t$. Therefore we have

$$\begin{aligned}(s_k s_{k+1} \cdots s_t) \cdot s_i &= s_k s_{k+1} \cdots s_{i-1} \underline{s_i s_{i+1} s_i} \cdot s_{i+2} \cdots s_t \\ &= s_k s_{k+1} \cdots s_{i-1} \underline{s_{i+1} s_i s_{i+1}} \cdot s_{i+2} \cdots s_t = s_{i+1} \cdot (s_k s_{k+1} \cdots s_t)\end{aligned}$$

because s_{i+1} commutes with $s_{i-1}, s_{i-2}, \ldots, s_k$. Applying the resulting equality to $i = r, r-1, \ldots, m$, we obtain (1.33). The proof of (1.34) is quite similar. □

Let S_n denote the permutation group of indices $\{1, 2, \ldots, n\}$. It is well-known that S_n is generated by the elementary transpositions $t_i = (i, i+1)$, $1 \leq i < n$. It is easy to check that the elementary transpositions satisfy the braid relations (1.32) with t_i in place of s_i. Therefore, the map $s_i \mapsto t_i$ can be extended to a homomorphism $\omega : B_n \to S_n$ of monoids.

Theorem 1.4 *The group S_n as a monoid is defined by the generators t_i, $1 \leq i < n$ and the relations*

$$t_i t_{i+1} t_i = t_{i+1} t_i t_{i+1}, \quad t_i^2 = 1, \quad t_i t_k = t_k t_i, \quad k \neq i \pm 1. \qquad (1.35)$$

Proof Let $\tilde{S}_n$ be the monoid defined by relations (1.35) with $\tilde{t}_i$ in place of t_i. Because the elementary transpositions satisfy all relations (1.35), there is a natural homomorphism $\tilde{\omega} : \tilde{t}_i \mapsto t_i$. We shall demonstrate that $\tilde{\omega}$ is an isomorphism.

For each permutation $\pi \in S_n$, we fix an element $\pi^b \in \tilde{S}_n$ as follows. If $\pi \in S_0 = \{1\}$, then $\pi^b = 1$. Assume that π^b is already defined for $\pi \in S_{n-1}$. If $\pi \in S_n \setminus S_{n-1}$, then the permutation $\pi t_{\pi(n)} t_{\pi(n)+1} \cdots t_{n-2} t_{n-1}$ belongs to S_{n-1}, and we put

$$\pi^b = (\pi t_{\pi(n)} t_{\pi(n)+1} \cdots t_{n-2} t_{n-1})^b \, \tilde{t}_{n-1} \tilde{t}_{n-2} \cdots \tilde{t}_{\pi(n)+1} \tilde{t}_{\pi(n)}. \qquad (1.36)$$

Using induction on n, it is easy to verify that $\tilde{\omega}(\pi^b) = \pi$, $\pi \in S_n$:

$$\tilde{\omega}(\pi^b) = \tilde{\omega}[(\pi t_{\pi(n)} \cdots t_{n-1})^b] t_{n-1} \cdots t_{\pi(n)} = \pi t_{\pi(n)} \cdots t_{n-1} \cdot t_{n-1} \cdots t_{\pi(n)} = \pi. \qquad (1.37)$$

Let us demonstrate that each element $\tilde{\pi} \in \tilde{S}_n$ has a representation $\tilde{\pi} = \pi^b$ for a suitable $\pi \in S_n$. By definition, we have $\tilde{t}_i = t_i^b$. In particular, S_n^b contains all generators $\tilde{t}_i$, $1 \le i < n$. Hence, it suffices to prove that S_n^b is closed with respect to the right multiplications by the $\tilde{t}_i$'s. By induction on n, we shall prove the following identity

$$\pi^b \tilde{t}_i = (\pi t_i)^b. \tag{1.38}$$

If $n = 1$, then we have $1 \cdot \tilde{t}_1 = t_1^b$, and $\tilde{t}_1 \cdot \tilde{t}_1 = 1 = (t_1^2)^b$. If $\pi \in S_n \setminus S_{n-1}$, then the induction supposition implies

$$(\pi t_{\pi(n)} \cdots t_{n-1})^b \tilde{t}_i = (\pi t_{\pi(n)} \cdots t_{n-1} t_i)^b, \quad 1 \le i < n-1. \tag{1.39}$$

Consider the following four cases.

1. $\pi(n) < i < n$. In this case, $(\pi t_i)(n) = \pi(n)$. As both the t_i's and $\tilde{t}_i$'s satisfy braid relations (1.32), the commutation rules from Lemma 1.13 are valid in $\tilde{S}_n$ and in S_n. In particular, (1.33) with $t \leftarrow n-1$, $k \leftarrow \pi(n)$, $r \leftarrow i-1$, $m \leftarrow i-1$ implies

$$t_i \cdot t_{\pi(n)} \cdots t_{n-1} = t_{\pi(n)} \cdots t_{n-1} \cdot t_{i-1},$$

respectively, (1.34) with $t \leftarrow n-1$, $k \leftarrow \pi(n)$, $r \leftarrow i$, $m \leftarrow i$ yields

$$\tilde{t}_{i-1} \cdot \tilde{t}_{n-1} \cdots \tilde{t}_{\pi(n)} = \tilde{t}_{n-1} \cdots \tilde{t}_{\pi(n)} \cdot \tilde{t}_i.$$

Considering these relations, we have

$$\begin{aligned}(\pi t_i)^b &= (\pi t_i t_{\pi(n)} \cdots t_{n-1})^b \tilde{t}_{n-1} \cdots \tilde{t}_{\pi(n)} = (\pi t_{\pi(n)} \cdots t_{n-1} t_{i-1})^b \tilde{t}_{n-1} \cdots \tilde{t}_{\pi(n)} \\ &\overset{(1.39)}{=} (\pi t_{\pi(n)} \cdots t_{n-1})^b \tilde{t}_{i-1} \tilde{t}_{n-1} \cdots \tilde{t}_{\pi(n)} = (\pi t_{\pi(n)} \cdots t_{n-1})^b \tilde{t}_{n-1} \cdots \tilde{t}_{\pi(n)} \tilde{t}_i \\ &= \pi^b \tilde{t}_i.\end{aligned}$$

2. $i = \pi(n)$. In this case $(\pi t_i)(n) = \pi(n) + 1$, and we have

$$(\pi t_i)^b = (\pi t_i t_{\pi(n)+1} \cdots t_{n-1})^b \tilde{t}_{n-1} \cdots \tilde{t}_{\pi(n)+1} \cdot \tilde{t}_{\pi(n)} \cdot \tilde{t}_{\pi(n)} = \pi^b \tilde{t}_i.$$

3. $i = \pi(n) - 1$. In this case $(\pi t_i)(n) = \pi(n) - 1$, and $t_i t_{\pi(n)-1} = 1$, which implies

$$(\pi t_i)^b = (\pi t_i t_{\pi(n)-1} t_{\pi(n)} \cdots t_{n-1})^b \tilde{t}_{n-1} \cdots \tilde{t}_{\pi(n)} \tilde{t}_{\pi(n)-1} = \pi^b \tilde{t}_i.$$

4. $1 \le i < \pi(n) - 1$. In this case, again, $(\pi t_i)(n) = \pi(n)$, and $\tilde{t}_i$ commutes with all $\tilde{t}_k$, $\pi(n) \le k < n$, whereas t_i commutes with all t_k, $\pi(n) \le k < n$. Additionally, the inequality $i < \pi(n) - 1$ shows that (1.39) is valid. Using these arguments,

we have

$$(\pi t_i)^b = (\pi t_i t_{\pi(n)} \cdots t_{n-1})^b \tilde{t}_{n-1} \cdots \tilde{t}_{\pi(n)} = (\pi t_{\pi(n)} \cdots t_{n-1} t_i)^b \tilde{t}_{n-1} \cdots \tilde{t}_{\pi(n)}$$
$$= (\pi t_{\pi(n)} \cdots t_{n-1})^b \tilde{t}_i \tilde{t}_{n-1} \cdots \tilde{t}_{\pi(n)} = (\pi t_{\pi(n)} \cdots t_{n-1})^b \tilde{t}_{n-1} \cdots \tilde{t}_{\pi(n)} \tilde{t}_i = \pi^b \tilde{t}_i.$$

Finally, if $\tilde{\omega}(\tilde{\pi}) = \tilde{\omega}(\tilde{\mu})$, $\tilde{\pi}, \tilde{\mu} \in \tilde{S}_n$, then $\tilde{\pi} = \pi^b$, $\tilde{\mu} = \mu^b$, for suitable $\pi, \mu \in S_n$. In view of (1.37), the equality $\tilde{\omega}(\tilde{\pi}) = \tilde{\omega}(\tilde{\mu})$ reduces to $\pi = \mu$. Therefore we have $\tilde{\pi} = \pi^b = \mu^b = \tilde{\mu}$. Thus, $\tilde{\omega}$ is an isomorphism.

□

Definition 1.6 For two indices m, k, let $[m; k]$ designates a monotonous cycle starting with m up to k, for example $[2; 4] = (2, 3, 4)$, whereas $[4; 2] = (4, 3, 2)$. For $m < k$ we have decompositions in the symmetric group S_n:

$$[m; k] = t_{k-1} t_{k-2} \cdots t_{m+1} t_m, \quad [k; m] = t_m t_{m+1} \cdots t_{k-2} t_{k-1}.$$

In B_n, we maintain similar notation: $[k; k] = 1$;

$$[m; k] = s_{k-1} s_{k-2} \cdots s_{m+1} s_m, \quad [k; m] = s_m s_{m+1} \cdots s_{k-2} s_{k-1}, \quad m < k. \tag{1.40}$$

In these designations, the commutation rules from Lemma 1.13 after the substitution $t \leftarrow t-1$, $r \leftarrow r-1$ take the following form:

$$[t; k] \cdot [m; r] = [m+1; r+1] \cdot [t; k], \quad k \le m \le r < t, \tag{1.41}$$
$$[k; t] \cdot [m; r] = [m-1; r-1] \cdot [k; t], \quad k \le m \le r \le t. \tag{1.42}$$

Lemma 1.14 *Consider an arbitrary partition of the set* $\{1, 2, \ldots, n\}$ *into two subsets* $K = \{k_1 < k_2 < \ldots < k_r\}$ *and* $I = \{i_1 < i_2 < \ldots < i_{n-r}\}$. *In the braid monoid* B_n *the following relation holds*

$$[1; k_1][2; k_2] \cdots [r; k_r] = [n; i_{n-r}][n-1; i_{n-r-1}] \cdots [r+2; i_2][r+1; i_1]. \tag{1.43}$$

Proof First, we note that this relation is valid if one of the sets is empty because in this case, one side is the empty product, and the other side is a product of elements $[m; m] = 1$.

We perform induction on n. Let K_n^r and $I_n^{(n-r)}$ denote the left- and right-hand sides of (1.43), respectively. Assume, first, that $n \in K$. In this case, we have $k_r = n$, and $K_n^r = (K \setminus \{n\})_{n-1}^{r-1}[r; n]$, whereas $I_n^{(n-r)}$ takes the form

$$(s_{i_{n-r}} s_{i_{n-r}+1} \cdots \underline{s_{n-1}})(s_{i_{n-r-1}} s_{i_{n-r-1}+1} \cdots \underline{s_{n-2}}) \cdots (s_{i_2} s_{i_2+1} \cdots \underline{s_{r+1}})(s_{i_1} s_{i_1+1} \cdots \underline{s_r}).$$

Using the relations $s_a s_b = s_b s_a$, $|a-b| > 1$, we may move all underlined braids to the right margin position, yielding $I_n^{(n-r)} = I_{n-1}^{(n-r)}[r\,;n]$. The induction supposition states that $(K \setminus \{n\})_{n-1}^{r-1} = I_{n-1}^{(n-r)}$, and hence $K_n^r = I_n^{(n-r)}$.

If $n \in I$, then $i_{n-r} = n$, and $[n; i_{n-r}] = 1$. Hence, $I_n^{(n-r)} = (I \setminus \{n\})_{n-1}^{(n-1-r)}$, whereas $K_n^r = K_{n-1}^r$, and induction applies. □

1.4.1 Co-sets and Shuffles

Given a group G and a subgroup H, the *right co-sets* of H in G are classes of the equivalence relation $\pi \sim \nu \leftrightarrow \pi\nu^{-1} \in H$, so that we obtain a partition of the group G in the form

$$G = \bigcup_{\nu \in A} H\nu$$

called the *right co-set decomposition*. The subset A of G containing a single element from each right co-set is called a *right transversal* of H in G.

Let $S_n^{(r)}$ be the subgroup of permutations leaving fixed all indices $1, 2, \ldots, r$. Of course, $S_n^{(r)}$ is isomorphic to the symmetric group S_{n-r}. Consider a subgroup $H = S_r \times S_n^{(r)}$ of S_n generated by S_r and $S_n^{(r)}$.

Definition 1.7 A permutation $\pi \in S_n$ is called an *r-shuffle* if

$$\pi(1) < \pi(2) < \ldots < \pi(r); \quad \pi(r+1) < \pi(r+2) < \ldots < \pi(n). \tag{1.44}$$

Below, Sh_n^r denotes the set of all r-shuffles.

Clearly, if $r = n$, then the set Sh_n^r contains only identical permutation. An r-shuffle π is uniquely defined by the set

$$Y = \{\pi(1), \pi(2), \ldots, \pi(r)\}$$

because $\pi(r+1)$ is the smallest number of the interval $1, 2, \ldots, n$ that does not belong to Y; $\pi(r+2)$ is the next element with the same property, and so on. Thus, the total number of r-shuffles equals $n!/r!(n-r)!$, the number of all r-element subsets of $\{1, 2, \ldots, n\}$.

Theorem 1.5 *The set Sh_n^r of all r-shuffles is a right transversal of $S_r \times S_n^{(r)}$ in S_n.*

Proof The index of the subgroup $S_r \times S_n^{(r)}$ in the group S_n equals the total number of r-shuffles: $|S_n|/(|S_r||S_n^{(r)}|) = n!/r!(n-r)!$. Therefore, it suffices to check that two r-shuffles π, ν are equivalent only if they are equal to each other. Let $\pi\nu^{-1} = h \in H$.

Considering that $h \in S_r \times S_n^{(r)}$, we have

$$\{1, 2, \ldots, r\} = \{h(1), h(2), \ldots, h(r)\}$$

and

$$\{r+1, r+2, \ldots, n\} = \{h(r+1), h(r+2), \ldots, h(n)\}.$$

Due to $\pi = h\nu$, these equalities imply

$$\{\nu(1), \nu(2), \ldots, \nu(r)\} = \{\pi(1), \pi(2), \ldots, \pi(r)\} \tag{1.45}$$

and

$$\{\nu(r+1), \nu(r+2), \ldots, \nu(n)\} = \{\pi(r+1), \pi(r+2), \ldots, \pi(n)\}. \tag{1.46}$$

Because ν and π are r-shuffles, in both sets of (1.45), the elements increase; hence, $\nu(1) = \pi(1), \ldots, \nu(r) = \pi(r)$. In the perfect analogy, the equality (1.46) implies $\nu(r+1) = \pi(r+1), \ldots, \nu(n) = \pi(n)$. □

Lemma 1.15 *The following recurrence relation is valid*

$$Sh_n^r = Sh_{n-1}^r \cup [n; r] \cdot Sh_{n-1}^{r-1}, \quad 1 < r < n. \tag{1.47}$$

Proof If π is an r-shuffle, then the inequalities (1.44) imply that either $\pi(n) = n$ or $\pi(r) = n$. In the former case, clearly $\pi \in Sh_{n-1}^r$. Let $\pi(r) = n$. Consider the element $\pi' = [r; n]\pi$. Obviously, $\pi'(n) = n$, whereas $\pi'(i) = \pi(i)$ when $1 \leq i < r$, and $\pi'(i) = \pi(i+1)$ when $r \leq i < n$. In particular, $\pi' \in S_{n-1}$ and inequalities (1.44) with $\pi \leftarrow \pi'$, $r \leftarrow r-1$ remain valid; that is, $\pi' \in Sh_{n-1}^{r-1}$ and $\pi = [n; r]\pi' \in [n; r]Sh_{n-1}^{r-1}$. Thus, the left-hand side of (1.47) is a subset of the right-hand side.

If $r = n$, then both sides contain only the identical permutation. If $r \neq n$, then the union of the right-hand side of (1.47) is disjunctive. Due to the Pascal equality,

$$\frac{n!}{r!(n-r)!} = \frac{(n-1)!}{(r-1)!(n-r-1)!}\left(\frac{1}{r} + \frac{1}{n-r}\right)$$
$$= \frac{(n-1)!}{r!(n-r-1)!} + \frac{(n-1)!}{(r-1)!(n-r)!},$$

both sides have the same number of elements. □

Lemma 1.16 *Each r-shuffle π has a decomposition*

$$\pi = [\pi(r); r][\pi(r-1); r-1] \cdots [\pi(2); 2][\pi(1); 1]. \tag{1.48}$$

Conversely, if $1 \le k_1 < k_2 < \ldots < k_r \le n$, *then*

$$\mu = [k_r; r][k_{r-1}; r-1] \cdots [k_2; 2][k_1; 1]$$

is an r-shuffle and $\mu(i) = k_i$, $1 \le i \le r$.

Proof To prove (1.48), one may use the recurrence relation (1.47) for obvious induction on n because $\pi \in [n; r] \cdot Sh_{n-1}^{r-1}$ implies $\pi(r) = n$.

Conversely, if we define $\pi(i) = k_i$, $1 \le i \le r$, and $\pi(r+1)$ is the smallest number of the interval $1, 2, \ldots, n$ that does not belong to $\{k_1, k_2, \ldots, k_r\}$; $\pi(r+2)$ is the next element with the same property, and so on, then clearly $\pi \in Sh_n^r$. By (1.48), we have $\mu = \pi$. □

Corollary 1.2 *The group* S_n *has the following right and left co-set decompositions:*

$$S_n = \bigcup_{1 \le k_1 < k_2 < \cdots < k_r \le n} S_r \times S_n^{(r)} \cdot [k_r; r] \cdots [k_2; 2][k_1; 1],$$

$$S_n = \bigcup_{1 \le k_1 < k_2 < \cdots < k_r \le n} [1; k_1][2; k_2] \cdots [r; k_r] \cdot S_r \times S_n^{(r)}.$$

Proof The former decomposition follows from Theorem 1.5 and Lemma 1.16. The latter decomposition follows from the former decomposition by application the involution $\pi \mapsto \pi^{-1}$ because $[i; k_i]^{-1} = [k_i; i]$, $1 \le i \le r$. □

1.5 Hopf Algebras

We are reminded that a *tensor product* of two linear spaces A and B over the ground field $\mathbf{k}$ can be defined as a linear space with a basis of formal tensors $a_i \otimes b_s$, where a_i and b_s run through fixed bases of A and B, respectively. In this case, the symbol $\otimes$ is extended to a bilinear map

$$\otimes : A \times B \to A \otimes B, \quad a \otimes b = \sum_{i,s} \alpha_i \beta_s \, a_i \otimes b_s,$$

where $a = \sum_i \alpha_i a_i$ and $b = \sum_s \beta_s b_s$ are decompositions of the elements a, b in the bases $\{a_i\}$ and $\{b_s\}$, respectively.

We shall frequently use a functorial property of the tensor product: If $\varphi : A \to A'$ and $\psi : B \to B'$ are linear maps, then the map

$$\varphi \otimes \psi : A \otimes B \to A' \otimes B', \quad (\varphi \otimes \psi)(a \otimes b) = \varphi(a) \otimes \psi(b)$$

is a well-defined linear map, in which case

$$\ker(\varphi \otimes \psi) = A \otimes \ker \psi + \ker \varphi \otimes B. \tag{1.49}$$

This property implies the following statement: If

$$a_1 \otimes b_1 + a_2 \otimes b_2 + \cdots + a_n \otimes b_n = 0, \quad a_t \in A,\ b_t \in B,\ 1 \le t \le n,$$

and $b_1, b_2, \ldots, b_n$ are linearly independent, then $a_t = 0,\ 1 \le t \le n$. Symmetrically, if $a_1, a_2, \ldots, a_n$ are linearly independent, then $b_t = 0,\ 1 \le t \le n$. Indeed, if $b_1, b_2, \ldots, b_n$ are linearly independent in B, then there exists a linear map $\varphi : B \to \mathbf{k}$ such that $\varphi(b_1) = 1,\ \varphi(b_2) = 0,\ \varphi(b_3) = 0,\ \ldots,\ \varphi(b_n) = 0$. We have

$$0 = (\mathrm{id} \otimes \varphi)(a_1 \otimes b_1 + a_2 \otimes b_2 + \cdots + a_n \otimes b_n) = a_1 \otimes 1,$$

and hence, $a_1 = 0$ because $A \otimes \mathbf{k} \cong A$. Similarly $a_t = 0,\ t = 2, 3, \ldots, n$.

By a *Hopf algebra*, we mean an associative algebra H over a ground field $\mathbf{k}$ equipped with a homomorphism

$$\Delta : H \to H \otimes H, \tag{1.50}$$

called a *coproduct*, which is coassociative and has a counit (a homomorphism $\varepsilon : H \to \mathbf{k}$) and an antipode (an antihomomorphism $\sigma : H \to H$). We use the Sweedler notations for the coproduct,

$$\Delta(a) = \sum_{(a)} a_{(1)} \otimes a_{(2)}. \tag{1.51}$$

In these notations, the coassociativity takes the form

$$\Delta(a) = \sum_{(a)} \Delta(a_{(1)}) \otimes a_{(2)} = \sum_{(a)} a_{(1)} \otimes \Delta(a_{(2)}), \tag{1.52}$$

whereas the counit and the antipode by definition satisfy

$$\sum_{(a)} \varepsilon(a_{(1)}) a_{(2)} = \textstyle\sum_{(a)} a_{(1)} \varepsilon(a_{(2)}) = a; \tag{1.53}$$

$$\sum_{(a)} \sigma(a_{(1)}) a_{(2)} = \textstyle\sum_{(a)} a_{(1)} \sigma(a_{(2)}) = \varepsilon(a) \cdot 1. \tag{1.54}$$

For example, each group algebra $\mathbf{k}[G]$ becomes a Hopf algebra if we define

$$\Delta(g) = g \otimes g,\ \ \varepsilon(g) = 1,\ \ \sigma(g) = g^{-1},\ \ g \in G.$$

1.5.1 Group-Like and Primitive Elements

Definition 1.8 A nonzero element g of a Hopf algebra H is said to be a *group-like* element if $\Delta(g) = g \otimes g$.

Proposition 1.4 *The set G of all group-like elements is a group. Different group-like elements are always linearly independent; that is, the group-like elements span a Hopf subalgebra that is isomorphic to the group algebra* $\mathbf{k}[G]$.

Proof If $\Delta(g) = g \otimes g$, then by definition of the counit, we have $\varepsilon(g)g = g$; hence, $\varepsilon(g) = 1$. The definition of the antipode yields $g\sigma(g) = \sigma(g)g = \varepsilon(g) \cdot 1 = 1$; that is, g is invertible. If $h \in G$, then $\Delta(gh) = \Delta(g)\Delta(h) = (g \otimes g)(h \otimes h) = gh \otimes gh$, hence $gh \in G$. Finally, $1 \otimes 1 = \Delta(1) = \Delta(gg^{-1}) = \Delta(g)\Delta(g^{-1})$, and therefore $\Delta(g^{-1}) = g^{-1} \otimes g^{-1}$, which implies $g^{-1} \in G$.

Let $g = \alpha_1 g_1 + \ldots + \alpha_n g_n$, where $\alpha_i \in \mathbf{k}$, $g, g_i \in G$ and elements $g_1, \ldots, g_n$ are linearly independent. In this case, we have

$$g \otimes g = \Delta(g) = \alpha_1 \Delta(g_1) + \ldots + \alpha_n \Delta(g_n) = \alpha_1 g_1 \otimes g_1 + \ldots + \alpha_n g_n \otimes g_n,$$

or

$$\sum_{i,s} \alpha_i \alpha_s g_i \otimes g_s = \sum_i \alpha_i g_i \otimes g_i.$$

Because the elements $g_i \otimes g_s$, $1 \le i, s \le n$ are linearly independent in $H \otimes H$, we have $\alpha_i \alpha_s = 0$ at $i \neq s$ and $\alpha_i^2 = \alpha_i$. This statement is possible only if one of the coefficients α_i is equal to 1 and the others are zero. Thus, the initial dependence assumes the form $g = g$, as required. □

Definition 1.9 A nonzero element u of a Hopf algebra H is said to be a *primitive element* if $\Delta(u) = u \otimes 1 + 1 \otimes u$.

The universal enveloping algebra $U(L)$ of any Lie algebra L has a Hopf algebra structure so that all elements from L are primitive. More precisely, according to the Poincaré-Birkhoff-Witt theorem, the basis of $U(L)$ consists of the monomials

$$W = u_1^{n_1} u_2^{n_2} \cdots u_k^{n_k}, \tag{1.55}$$

where $u_1 < u_2 < \ldots < u_k$ run through a fixed well-ordered basis of L. The coproduct is defined as follows:

$$\Delta(u_1^{n_1} u_2^{n_2} \cdots u_k^{n_k}) = \sum_{r_i+s_i=n_i,\, 1\le i\le k} \binom{n_1}{r_1}\binom{n_2}{r_2}\cdots\binom{n_k}{r_k} u_1^{r_1} u_2^{r_2} \cdots u_k^{r_k} \otimes u_1^{s_1} u_2^{s_2} \cdots u_k^{s_k},$$

whereas $\varepsilon(W) = 0$ for all non-empty basis monomials W, and $\varepsilon(\emptyset) = 1$, where as usual, the empty monomial is the unit of $U(L)$. The antipode is given by

$$\sigma(u_1^{n_1} u_2^{n_2} \cdots u_k^{n_k}) = (-1)^{n_1+n_2+\cdots+n_k} u_k^{n_k} \cdots u_2^{n_2} u_1^{n_1}.$$

Of course, the coproduct and the counit, as homomorphisms, are uniquely defined by $\Delta(u) = u\otimes 1 + 1\otimes u$, $\varepsilon(u) = 0$, whereas the antipode, as an antihomomorphism, is uniquely defined by $\sigma(u) = -u$, $u \in L$. In this particular case, the group G is trivial, $G = \mathrm{id}$.

If the characteristic of the ground field is $p > 0$, then the restricted universal enveloping algebra $U_p(L)$ of each *restricted Lie algebra* L has the structure of Hopf algebra with the same Δ, ε, σ also. Recall that according to the restricted version of the Poincaré-Birkhoff-Witt theorem the basis of $U_p(L)$ consists of increasing monomials (1.55) with the additional condition $n_i < p$, $1 \le i \le k$.

Consider an arbitrary Hopf algebra H. The set L of all primitive elements of H is closed with respect to the Lie brackets $[a, b] = ab - ba$:

$$\begin{aligned}\Delta([a,b]) &= \Delta(ab-ba) = \Delta(a)\Delta(b) - \Delta(b)\Delta(a)\\ &= (1\otimes a + a\otimes 1)(1\otimes b + b\otimes 1) - (1\otimes b + b\otimes 1)(1\otimes a + a\otimes 1)\\ &= 1\otimes ab + b\otimes a + a\otimes b + ab\otimes 1 - 1\otimes ba + a\otimes b\\ &\quad + b\otimes a + ba\otimes 1\\ &= (ab-ba)\otimes 1 + 1\otimes(ab-ba) = [a,b]\otimes 1 + 1\otimes [a,b].\end{aligned}$$

In other words, L is a Lie algebra. If $\mathbf{k}$ is of characteristic $p > 0$, then L is closed with respect to an additional unary operation $a \mapsto a^p$:

$$\Delta(a^p) = (1\otimes a + a\otimes 1)^p = \sum_{k=0}^{p} \binom{p}{k} a^k \otimes a^{p-k} = 1\otimes a^p + a^p\otimes 1,$$

and L is a restricted Lie algebra with respect to the operations $[a, b]$, a^p.

Proposition 1.5 *A subalgebra P generated by all primitive elements of a Hopf algebra H is isomorphic to the universal enveloping algebra $U(L)$ if the characteristic of* $\mathbf{k}$ *is zero; otherwise, it is isomorphic to the restricted universal enveloping algebra.*

Proof If $\Delta(u) = u\otimes 1 + 1\otimes u$, then $\varepsilon(u)\cdot 1 + \varepsilon(1)\cdot u = u$, which implies $\varepsilon(u) = 0$. Similarly $\sigma(u)\cdot 1 + \sigma(1)\cdot u = \varepsilon(u)$, implies $\sigma(u) = -u$.

Let us fix a well-ordered basis $\{u_i \mid i \in I\}$ of L. As observed above, $[u_i, u_s] \in L$ for all $i, s \in I$. Therefore $[u_i, u_s] = \sum_{k\in I} \alpha_k^{i,s} u_s$, and the following relations hold:

$$u_i u_s = u_s u_i + \sum_{k\in I} \alpha_{i,s}^k u_k, \quad u_i > u_s.$$

If char $\mathbf{k} = p > 0$, then $u_i^p \in L$ and we have the additional relations

$$u_i^p = \sum_{s \in I} \alpha_i^s u_s, \quad i \in I.$$

These relations define a diminishing procedure with respect to the Hall order that allows one to decompose each word in $\{u_i \mid i \in I\}$ into a linear combination of words with no $u_i u_s$, $u_i > u_s$ (and no u_i^p if char $\mathbf{k} = p > 0$) as sub-words; see Lemma 1.11. A word has no $u_i u_s$, $u_i > u_s$ as sub-words if and only if it has the form (1.55). Additionally, it has no sub-words u_i^p if and only if in (1.55) we have $n_i < p$, $1 \leq i \leq k$. Hence increasing monomials (1.55) (with $n_i < p$ if char $\mathbf{k} = p > 0$) span P.

It remains to demonstrate that different increasing monomials (1.55) (with $n_i < p$ if char $\mathbf{k} = p > 0$) are linearly independent. We perform induction on the length. Assume that all increasing (restricted if $p > 0$) words of length $< M$ are linearly independent. Let $W = \sum_i W_i$, where the length of W equals M and $W_i \prec W$ with respect to the Hall ordering of words. Applying the coproduct to both sides of the latter equality, we obtain

$$\sum_{W' \circ W'' = W} W' \otimes W'' = \sum_i \alpha_i \left(\sum_{W_i' \circ W_i'' = W_i} W_i' \otimes W_i'' \right).$$

In this formula, $\circ$ is the commutative product of increasing monomials:

$$u_1^{n_1} u_2^{n_2} \cdots u_k^{n_k} \circ u_1^{m_1} u_2^{m_2} \cdots u_k^{m_k} = u_1^{n_1+m_1} u_2^{n_2+m_2} \cdots u_k^{n_k+m_k}.$$

Cancelling the sums $\sum_i \alpha_i W_i \otimes 1 + \sum_i 1 \otimes \alpha_i W_i$ from both sides, we obtain

$$\sum_{W' \circ W'' = W,\, W', W'' \neq \emptyset} W' \otimes W'' = \sum_i \sum_{W_i' \circ W_i'' = W_i,\, W_i', W_i'' \neq \emptyset} \alpha_i W_i' \otimes W_i''.$$

According to the induction supposition, all different words W', W'', W_i', W_i'' are linearly independent. Therefore, all tensors in the above equality must be cancelled.

Let $W = u_1^{n_1} \tilde{W}$, where $\tilde{W} = u_2^{n_2} \cdots u_k^{n_k}$, $u_1 < u_2 < \ldots < u_k$. In the left-hand side, the tensor $u_1^{n_1-1} \tilde{W} \otimes u_1$ is encountered exactly n_1 times. In the right-hand side, tensors of the type $u_1^{n_1-1} \tilde{W} \otimes u_s$ appear only under decomposition of the coproduct of words $u_s u_1^{n_1-1} \tilde{W}$, where $u_s < u_1$. The requirement of cancelling all terms results in

$$u_1^{n_1-1} \tilde{W} \otimes (n_1 u_1 - \sum_{s \in S} \alpha_s u_s) = 0,$$

which is impossible, as n_1 is an invertible element of the ground field $\mathbf{k}$, and the elements $u_1, u_s,\ s \in S$ are linearly independent in H. □

The group G acts by conjugations on L:

$$\Delta(g^{-1}ug) = (g^{-1} \otimes g^{-1})(u \otimes 1 + 1 \otimes u)(g \otimes g) = g^{-1}ug \otimes 1 + 1 \otimes g^{-1}ug.$$

Hence $g^{-1}ug = \sum_i \alpha_i u_i$, where as above $\{u_i \mid i \in I\}$ is a fixed basis of L. These relations determine the following commutation rules:

$$u_i g = \sum_i \alpha_i g u_i, \quad g \in G,\ i \in I.$$

By Lemma 1.11, the products gW, where $g \in G$ and W is a (restricted) increasing monomial, span the subalgebra generated by all primitive and group-like elements. Moreover, arguments similar to the proof of the above Lemma demonstrate that these products are linearly independent in H. In other words, the subalgebra generated by primitive and group-like elements is isomorphic to a *smash product* of Hopf algebras $\mathbf{k}[G]\#U(L)$ or, in the case of a finite characteristic, $\mathbf{k}[G]\#U_p(L)$. We are reminded that the well-known Kostant-Sweedler theorem states that each cocommutative Hopf algebra over an algebraically closed field of zero characteristic is generated by the primitive and group-like elements and is therefore precisely the smash product $\mathbf{k}[G]\#U(L)$.

1.5.2 Character Hopf Algebras

Let H be an arbitrary Hopf algebra over a field $\mathbf{k}$ with comultiplication Δ, counit ε, and antipode σ. Denote by G the group of all group-like elements.

Definition 1.10 Given $h, f \in G$, an element $a \in H$ is called *(h,f)-primitive* if

$$\Delta(a) = a \otimes h + f \otimes a. \tag{1.56}$$

If h, f are not specified, the element a is called *skew-primitive*.

If $g \in G$ and a is skew-primitive , then both ga and ag are also skew-primitive:

$$\Delta(ga) = \Delta(g)\Delta(a) = ga \otimes gh + gf \otimes ga.$$

An element a is called *semi-invariant* if ga and ag are proportional for all $g \in G$.

Definition 1.11 A Hopf algebra H is referred to as a *character Hopf algebra* if the group G of all group-like elements is commutative and H is generated over $\mathbf{k}[G]$ by skew primitive semi-invariants.

Remark 1.6 Since the coproduct is a homomorphism of algebras, there exists a clear procedure to find a coproduct of a word (product) of skew-primitive elements $w = a_1 a_2 \cdots a_n$. This coproduct is a sum of 2^n tensors $w^A \otimes w^{\bar{A}}$, where A runs through all subsets of $\{1, 2, \ldots, n\}$, whereas w^A is a word that appears from w upon replacement of all a_i, $i \in A$ with f_i, respectively, and $w^{\bar{A}}$ appears from w upon replacement of all a_i, $i \notin A$ with h_i.

Given an Abelian group G, the *character group* $\hat{G}$ is the set of all homomorphisms $\chi : G \to \mathbf{k}^*$ from G to the multiplicative group $\mathbf{k}^*$ of the ground field $\mathbf{k}$ with the multiplication $(\chi \cdot \chi^1)(g) = \chi(g) \cdot \chi^1(g)$.

Lemma 1.17 *Every character Hopf algebra is graded by the character group* $\hat{G}$*:*

$$H = \bigoplus_{\chi \in \hat{G}} H^\chi, \quad H^\chi = \{a \in H \,|\, g^{-1} a g = \chi(g) a \text{ for all } g \in G\}.$$

Proof By definition, every semi-invariant a satisfies the commutation rules $ag = \alpha_g^a ga$, $\alpha_g^a \in \mathbf{k}$ for all $g \in G$. In this case, the map $\chi^a : g \mapsto \alpha_g^a$ is a character of the group G, whereas $a \in H^{\chi^a}$. Moreover, if $u = a_1 a_2 \cdots a_n$ then $gu \in H^\chi$, where $\chi = \chi^{a_1} \chi^{a_2} \cdots \chi^{a_n}$, $g \in G$. Hence, the subspaces H^χ, $\chi \in \hat{G}$ span H.

It remains to verify that nonzero elements u_i from different subspaces H^{χ^i} are linearly independent. Suppose instead that $\sum_{l=1}^n \alpha_i u_i = 0$. Conjugation by any $g \in G$ yields $\sum_i \chi^i(g) \alpha_i u_i = 0$. By the Dedekind theorem on the linear independence of characters, there exist $g_1, g_2, \ldots, g_n \in G$ such that the determinant of the $n \times n$ matrix $||\chi^i(g_s)||$, $1 \le i, s \le n$ is not zero. The system of equalities $\sum_i \alpha_i u_i \chi^i(g_s) = 0$, $1 \le s \le n$ may be written in matrix form as follows:

$$(\alpha_1 u_1, \alpha_2 u_2, \ldots, \alpha_n u_n) ||\chi^i(g_s)|| = 0,$$

which implies $(\alpha_1 u_1, \alpha_2 u_2, \ldots, \alpha_n u_n) = 0$, for $||\chi^i(g_s)||$ is invertible. □

Lemma 1.18 *If* $\Delta(a) = a \otimes h + f \otimes a$, $h, f \in G$, *then* $\varepsilon(a) = 0$, $\sigma(a) = -f^{-1} a h^{-1}$.

Proof By the definition of the counit, we have $\varepsilon(a)h + \varepsilon(f)a = a$, which implies $\varepsilon(a) = 0$. The antipode axiom yields $\sigma(a)h + \sigma(f)a = \varepsilon(a) \cdot 1 = 0$. Therefore, $\sigma(a) = -\sigma(f) a h^{-1} = -f^{-1} a h^{-1}$. □

Lemma 1.19 *If* $a \in \mathbf{k}[G]$ *is skew-primitive, then* $a = \alpha(h - f)$, $\alpha \in \mathbf{k}$, $h, f \in G$.

Proof Let $a = \sum_i \alpha_i g_i$ be an (h,f)-primitive element. By definition, we have

$$\sum_i \alpha_i g_i \otimes g_i = \sum_i \alpha_i g_i \otimes h + f \otimes \sum_i \alpha_i g_i. \tag{1.57}$$

Because different group-like elements are linearly independent, there exists a linear map $\pi_i : \mathbf{k}[G] \to \mathbf{k}[G]$ such that $\pi_i(g_s) = \delta^i_s g_s$. Applying $\pi_i \otimes \pi_i$ to both sides of (1.57), we obtain $\alpha_i g_i \otimes g_i = \alpha_i g_i \otimes \delta_h^{g_i} + \delta_f^{g_i} \otimes \alpha_i g_i$. This equality implies that for each i either $g_i = h$ or $g_i = f$; that is, $a = \alpha h + \beta f$ and

$$\alpha h \otimes h + \beta f \otimes f = \alpha h \otimes h + \beta f \otimes h + \alpha f \otimes h + \beta f \otimes f,$$

which yields $0 = (\alpha + \beta) f \otimes h$; that is, $\beta = -\alpha$, and $a = \alpha(h - f)$. □

Below, $L_{h,f}$ denotes the space of all (h,f)-primitive elements:

$$L_{h,f} = \{a \mid \Delta(a) = a \otimes h + f \otimes a\}, \quad h, f \in G.$$

Note that the spaces $L_{h,f}$ related to different pairs (h,f) are not independent. First, $L_{h,f} \cap L_{f,h}$ contains $h - f$:

$$\Delta(h-f) = (h-f) \otimes h + f \otimes (h-f) = (h-f) \otimes f + h \otimes (h-f).$$

More generally, if $u_1 = h - f_1,\ u_2 = f_1 - f_2,\ \ldots,\ u_n = f_{n-1} - f$, then

$$\sum_{i=1}^{n} u_i = (h - f_1) + (f_1 - f_2) + \cdots + (f_{n-1} - f) = h - f.$$

The following lemma implies, in particular, that all linear dependences between skew-primitive elements related to different pairs have the above form.

Lemma 1.20 *Let u_i, $1 \le i \le n$ be (h_i,f_i)-primitive elements of a Hopf algebra H, and let $(h_i,f_i) \neq (h_s,f_s)$ for $i \neq s$. If $\sum_i u_i \in \mathbf{k}[G]$, then $u_i = \alpha_i(h_i - f_i)$, $1 \le i \le n$, where $\alpha_i \in \mathbf{k}$.*

Proof Due to Lemma 1.19, it suffices to demonstrate that $u_i \in \mathbf{k}[G]$, $1 \le i \le n$. Assume to the contrary that, say, $u_1 \notin \mathbf{k}[G]$. We have

$$u_1 = -\sum_{i=2}^{n} u_i + \sum_s \alpha_s g_s, \quad g_s \in G,\ \alpha_s \in \mathbf{k}.$$

We may suppose that the elements u_i, $1 < i \le n$ are linearly independent modulo $\mathbf{k}[G]$, because otherwise, the number n can be diminished. Applying the coproduct, we obtain

$$\sum_{i=2}^{n} (-u_i) \otimes (h_i - h_1) + (f_i - f_1) \otimes (-u_i) + \sum_s \alpha_s g_s \otimes g_s = 0. \tag{1.58}$$

As u_i, $1 < i \le n$ are linearly independent modulo $\mathbf{k}[G]$, there exists a linear map $\pi : H \to \mathbf{k}$ such that $\pi(u_2) = 1$, $\pi(\mathbf{k}[G]) = 0$, $\pi(u_i) = 0$, $2 < i \le n$. Applying id $\otimes\, \pi$ to (1.58), we obtain $f_2 = f_1$. Similarly, the application of $\pi \otimes$ id implies

$h_2 = h_1$, which contradicts the condition that all pairs (h_i, f_i), $1 \le i \le n$ are different. □

The evident equality $gL_{h,f} = L_{gh,gf}$ allows one to normalize the skew primitive elements. Given $g \in G$, we define $L_g = \{a \in H \mid \Delta(a) = a \otimes 1 + g \otimes a\} = L_{1,g}$. This set forms a linear space, and we call its elements *g-primitive*, or *normalized skew-primitive* if g is not specified. Linear spaces L_g are independent; that is, their linear span Prin (H) is a direct sum

$$\text{Prin}\,(H) \stackrel{\text{df}}{=} \sum_{g \in G} L_g = \bigoplus_{g \in G} L_g, \quad L_g = \{a \in H \mid \Delta(a) = a \otimes 1 + g \otimes a\}. \tag{1.59}$$

Indeed, if $\sum_i u_i = 0$, $u_i \in L_{g_i}$, then by Lemma 1.20, we have $g_i = \alpha_i(1-g_i)$, which implies $\sum_i \alpha_i \cdot 1 - \sum_i \alpha_i g_i = 0$, and $\alpha_i = 0$, $1 \le i \le n$ because distinct group-like elements are linearly independent.

The space Prin (H) is invariant under conjugations by the group-like elements:

$$L_h^g \stackrel{\text{df}}{=} g^{-1} L_h g = L_{g^{-1}hg}, \quad g,\ h \in G.$$

In other words, Prin (H) is an *Yetter–Drinfeld module* over the Hopf algebra $\mathbf{k}[G]$.

1.5.3 Free Character Hopf Algebra

Let $Y = \{y_i \mid i \in I\}$ be a set of free variables, and let G be an Abelian group. We wish to consider Y as a set of free skew-primitive generators of a character Hopf algebra. To this end, we must introduce the group-like elements $h_i, f_i \in G$ and define a coproduct

$$\Delta(y_i) = y_i \otimes h_i + f_i \otimes y_i, \quad \Delta(h_i) = h_i \otimes h_i, \quad \Delta(f_i) = f_i \otimes f_i. \tag{1.60}$$

According to the definition of a character Hopf algebra, the group G must act on the space V spanned by the y_i's via diagonal transformations: $g^{-1} y_i g = \lambda_g^i y_i$, $\lambda_g^i \in \mathbf{k}$ or, equivalently, we must postulate commutation rules

$$y_i g = \lambda_g^i g y_i, \quad \lambda_g^i \in \mathbf{k}, \quad i \in I, \quad g \in G. \tag{1.61}$$

These commutation rules define the algebra of noncommutative G-polynomials only if the maps $\chi^i : g \mapsto \lambda_g^i$ are characters of the group G, see Sect. 1.3.2. We consider the algebra of noncommutative G-polynomials $G\langle Y\rangle$ with commutation rules (1.61) as the *free character Hopf algebra* with counit $\varepsilon(y_i) = 0$, $\varepsilon(g) = 1$, and antipode $S(y_i) = -f_i^{-1} y_i h_i^{-1}$, $S(g) = g^{-1}$, $g \in G$, $i \in I$.

Thus, to define the free character Hopf algebra, we must associate with each variable y_i a character χ^i, and two group elements h_i, f_i. In this case, the character

Hopf subalgebra of $G\langle Y\rangle$ generated by Y is $G_0\langle Y\rangle$, where G_0 is a subgroup of G generated by $h_i, f_i,\ i \in I$.

The Hopf algebra $G_0\langle Y\rangle$ is defined by the parameters q_{ik}, q'_{ik}, that appear in the commutation rules for h_i, f_i:

$$y_i h_k = q_{ik} h_k y_i, \quad y_i f_k = q'_{ik} f_k y_i, \quad i, k \in I. \tag{1.62}$$

Moreover, $G_0\langle Y\rangle$ as a Hopf algebra is completely defined by the parameters

$$p_{ik} = q_{ik}^{-1} q'_{ik}. \tag{1.63}$$

Indeed, consider the set of normalized generators $x_i = h_i^{-1} y_i,\ i \in I$. We have

$$\Delta(x_i) = x_i \otimes 1 + g_i \otimes x_i, \quad g_k^{-1} x_i g_k = p_{ik} x_i, \quad i, k \in I, \tag{1.64}$$

where $g_i = h_i^{-1} f_i$. In this case, we have the equality of Hopf algebras

$$G_0\langle Y\rangle = G_0\langle X\rangle, \text{ and } G\langle Y\rangle = G\langle X\rangle. \tag{1.65}$$

1.5.4 Brackets

Let $\chi^i : G \to \mathbf{k}^*$ be the character associated with the variable $y_i,\ i \in I$. For every word w in $Y \cup G$, let h_w denote an element of G that appears from w by replacing each y_i with h_i. Similarly, f_w denotes a group-like element that appears from w by replacing each y_i with f_i, whereas χ^w denotes a character that appears from w by replacing each $g \in G$ with 1 and y_i with χ^i. Because both the group G and the group of characters $\hat{G}$ are commutative, the values h_w, f_w, χ^w are independent of the order of variables y_i, g in the word w. For this reason, we may extend χ^w on the set of all homogeneous elements $w \in G\langle Y\rangle$ in each $y_i,\ i \in I$. Similarly, h_w, f_w have extensions on the set of all homogeneous polynomials w in each $y_i \in Y$ and each $g \in G$.

In terms of the characters, the commutation rules take the following form:

$$wg = \chi^w(g) g w, \tag{1.66}$$

where w is an arbitrary element that is homogeneous in each $y_i \in Y$.

Let u, v be homogeneous polynomials in each $y_i \in Y$ and each $g \in G$. We define brackets by the following formula:

$$[u, v] = \chi^v(h_u) uv - \chi^u(f_v) vu, \quad [u, v]^* = \chi^u(h_v^{-1}) uv - \chi^v(f_u^{-1}) vu. \tag{1.67}$$

For example, $[y_i, y_k] = q_{ki}y_iy_k - q'_{ik}y_ky_i$, whereas $[x_i, x_k] = x_ix_k - p_{ik}x_kx_i$ because

$$[h_i^{-1}y_i, h_k^{-1}y_k] = \chi^k(h_i^{-1}h_i)x_ix_k - \chi^i(h_k^{-1}f_k)x_kx_i = x_ix_k - p_{ik}x_kx_i.$$

Similarly, $[y_i, y_k]^* = q_{ik}^{-1}y_iy_k - (q'_{ki})^{-1}y_ky_i$, and $[x_i, x_k]^* = x_ix_k - p_{ki}^{-1}x_kx_i$, for

$$[h_i^{-1}y_i, h_k^{-1}y_k]^* = \chi^i(h_kh_k^{-1})x_ix_k - \chi^k(h_if_i^{-1})x_kx_i = x_ix_k - p_{ki}^{-1}x_kx_i.$$

We did not define the brackets if $u = y_i + x_i$, because $y_i + h_i^{-1}y_i$ is not homogeneous with respect to $h_i^{-1} \in G$ unless $h_i = 1$.

Lemma 1.21 *For every homogeneous u, we have*

$$h_i[u, x_i] \sim [u, y_i], \quad h_i[x_i, u] \sim [y_i, u], \quad i \in I. \tag{1.68}$$

Here $a \sim b$ is the projective equality: $a = \alpha b$, $\alpha \in \mathbf{k}$, $\alpha \neq 0$.

Proof By the definition (1.67) and the commutation rules (1.62), we have

$$\begin{aligned} h_i[u, x_i] &= \chi^i(h_u)h_iux_i - \chi^u(h_i^{-1}f_i)h_ix_iu = \chi^i(h_u)h_iuh_i^{-1}y_i - \chi^u(h_i^{-1}f_i)y_iu \\ &= \chi^i(h_u)\chi^u(h_i)^{-1}uy_i - \chi^u(h_i)^{-1}\chi^u(f_i)y_iu = \chi^u(h_i)^{-1}[u, y_i]. \end{aligned}$$

Similarly,

$$h_i[x_i, u] = h_ix_iu - \chi^i(f_u)\chi^u(h_i)^{-1}uh_ix_i \sim [y_i, u].$$

□

1.5.5 Defining Relations

If $R_1, R_2, \ldots, R_m$ are elements from the free character Hopf algebra $G\langle Y\rangle$, then

$$G\langle y_1, y_2, \ldots, y_n \,||\, R_1, R_2, \ldots, R_m\rangle \stackrel{\text{df}}{=} G\langle Y\rangle/\mathrm{Id}(R_1, R_2, \ldots, R_m),$$

the G-algebra defined by generators $y_1, \ldots, y_n$ and relations $R_i = 0$, $1 \leq i \leq m$ retains the Hopf algebra structure only if the ideal J generated by the R_i's is a Hopf ideal.

Definition 1.12 A subspace J of a Hopf algebra (or more generally of a coalgebra) H is said to be *coideal* if $\Delta(J) \subseteq J \otimes H + H \otimes J$ and $\varepsilon(J) = 0$. It is *antipode stable* if $\sigma(J) \subseteq J$. An ideal that is an antipode stable coideal is called a *Hopf ideal*.

Lemma 1.22 *Each coideal J of $\mathbf{k}[G]$ is spanned by its skew-primitive elements.*

Proof Assume, to the contrary, that $u = \sum_{i=1}^{m} \alpha_i g_i \in J$ and the element u is not a linear combination of skew-primitive elements from J. Among all those elements, we choose the one with the minimal m. If $m = 1$, then $0 = \varepsilon(u) = \alpha_1$ and $u = 0$.

Suppose that $m > 1$. We have

$$g_1 \equiv \sum_{i=2}^{m} \beta_i g_i, \ (\mathrm{mod}\, J), \quad g_i \in G, \ \ \beta_i = -\alpha_1^{-1}\alpha_i, \ \ 2 \le i \le m.$$

The elements $g_i, \ 1 < i \le m$ are linearly independent by modulo J because otherwise, one can diminish the number m. Applying the coproduct, we obtain

$$g_1 \otimes g_1 \equiv \sum_{i=2}^{m} \beta_i g_i \otimes g_i \ (\mathrm{mod}\, J \otimes \mathbf{k}[G] + \mathbf{k}[G] \otimes J). \tag{1.69}$$

Because $g_i, \ 1 < i \le m$ are linearly independent modulo J, there exists a linear map $\pi : \mathbf{k}[G] \to \mathbf{k}$ such that $\pi(g_2) = 1, \ \pi(J) = 0, \ \pi(g_i) = 0, \ 2 < i \le m$. Applying $\mathrm{id} \otimes \pi$ to (1.69), we obtain $g_1 \pi(g_1) \equiv \beta_2 g_2 \ (\mathrm{mod}\, J)$, which implies

$$\pi(g_1) = \varepsilon(g_1 \pi(g_1)) = \varepsilon(\beta_2 g_2) = \beta_2$$

because $\varepsilon(J) = 0$. Thus, $g_1 - g_2 \in J$. Now, the element $u' = u - \alpha_1(g_1 - g_2) \in J$ has a representation $u' = (\alpha_2 + \alpha_1)g_2 + \sum_{i=3}^{m} \alpha_i g_i$ with $m - 1$ summands. Hence u' is a linear combination of skew-primitive elements from J. In this case, so is $u = u' + \alpha_1(g_1 - g_2)$. □

At first glance, to introduce a character Hopf algebra defined by the relations $R_i = 0, \ 1 \le i \le m$, it looks reasonable to consider the smallest Hopf ideal that contains the R_i's. However, the smallest Hopf ideal does not always exist because the intersection of two Hopf ideals is not necessarily a Hopf ideal. For example, if $h_i = f_i = 1, \ i = 1, 2$ then the intersection of two Hopf ideals $\mathrm{Id}(y_1) \cap \mathrm{Id}(y_2)$ is not a Hopf ideal. We shall discuss when an ideal generated by a given system of relations is automatically a Hopf ideal.

Proposition 1.6 *An ideal generated by an antipode stable coideal is a Hopf ideal.*

Proof Let S be an antipode stable coideal. An arbitrary element T of the ideal J generated by S is a sum $T = \sum_i a_i s_i b_i, \ a_i, b_i \in H, \ s_i \in S$. We have $\sigma(T) = \sum_i \sigma(b_i)\sigma(s_i)\sigma(a_i) \in J$ as σ is an antihomomorphism. Further,

$$\Delta(T) = \sum_i \Delta(a_i)\Delta(s_i)\Delta(b_i) \in H(S \otimes H + H \otimes S)H \in J \otimes H + H \otimes J,$$

and $\varepsilon(T) = \sum_i \varepsilon(a_i)\varepsilon(s_i)\varepsilon(b_i) = 0$. □

Proposition 1.7 *Each ideal generated by skew-primitive elements is a Hopf ideal.*

Proof If $R_1, R_2, \ldots, R_m$ are skew-primitive, then by Lemma 1.18 the elements $fR_ih, f, h \in G, 1 \leq i \leq m$ span an antipode stable coideal, and Proposition 1.6 applies. □

1.5.6 Combinatorial Rank

The majority of the known examples of character Hopf algebras are defined by skew-primitive relations. Nevertheless, a Hopf ideal is not always generated by the skew-primitive elements as ideal. The simplest example is given below.

Example 1.3 Consider the free character Hopf algebra in two variables defined by parameters $p_{11} = p_{22} = p_{21} = -1$, $p_{12} = 1 \neq -1$. Let I be an ideal of $G\langle x_1, x_2\rangle$ generated by x_1^2, x_2^2, $x_1x_2x_1x_2 + x_2x_1x_2x_1$. Then, I is a Hopf ideal, which is not generated by its skew-primitive elements.

Nevertheless, the skew-primitive relations play a permanent role in the construction of character Hopf algebras due to the following important statement.

Theorem 1.6 *Every nonzero Hopf ideal of a character Hopf algebra has a nonzero skew-primitive element.*

Recall that in Definition 1.3, we have defined a constitution (multidegree) of a word u in $X = \{x_i \mid i \in I\}$ as a family of nonnegative integers $\{m_i \mid i \in I\}$ such that u has m_i occurrences of x_i. We consider the set D of all families $\{m_i \mid i \in I\}$ where m_i are natural numbers or zero as a partially ordered additive monoid

$$\{m_i \mid i \in I\} + \{n_i \mid i \in I\} = \{m_i + n_i \mid i \in I\},$$
$$\{m_i \mid i \in I\} \preceq \{n_i \mid i \in I\} \text{ if and only if } m_i \leq n_i,\ i \in I,$$

whereas the constitution $D(u)$ is considered to be a degree of the word u. We extend this degree on all G-monomials via $D(gu) = D(u)$; that is, we set $D(G) = 0$. Clearly, the formula $D(uv) = D(u) + D(v)$ is valid for G-monomials u, v. The coproduct of a monomial has a decomposition $\Delta(u) = \sum u_{(1)} \otimes u_{(2)}$ where $u_{(1)}, u_{(2)}$ are monomials such that $D(u) = D(u_{(1)}) + D(u_{(2)})$.

Let H be an arbitrary character Hopf algebra generated over $\mathbf{k}[G]$ by skew-primitive elements $a_i, i \in I$. There exists a Hopf algebra homomorphism

$$\xi : G\langle X\rangle \to H,\ \xi(x_i) = a_i,\quad i \in I,\ \xi(g) = g,\ g \in G. \tag{1.70}$$

For each $\gamma \in D$, let H_γ denotes a space generated by the values in H of all monomials that have D-degree less than or equal to γ, and let H_γ^- denotes a space generated by the values of all monomials that have D-degree less than γ. In particular, $H_0 = \mathbf{k}[G]$, whereas $H_0^- = 0$.

Definition 1.13 We say that a nonzero element S is *D-minimal* with respect to a subset $J \subsetneq H$ if $S \in H_\gamma \cap J$ and $H_\gamma^- \cap J = 0$ for some positive $\gamma \in D$.

Because D as a partially ordered set satisfies the Ascending Chain Condition, Theorem 1.6 follows from the more general statement below.

Lemma 1.23 *Each D-minimal element in a nonzero coideal J of H is a linear combination of skew-primitive elements from J.*

Proof For all $\gamma > 0$, we have $\mathbf{k}[G] \cap J \subseteq H_\gamma^- \cap J$. Therefore, if $\mathbf{k}[G] \cap J \neq 0$, then all D-minimal elements of J belong to $\mathbf{k}[G]$, and we may apply Lemma 1.22 to $\mathbf{k}[G] \cap J$. Thus, we may assume $\mathbf{k}[G] \cap J = 0$; that is, all group like elements are linearly independent modulo J.

Let d be a D-minimal element in J. The element d has a decomposition

$$d = \alpha w + \sum_{i=1}^{m} \alpha_i w_i + b, \tag{1.71}$$

where $\alpha_i, \alpha \neq 0$, $b \in H_\gamma^-$ and w, w_i are G-monomials of D-degree γ. Among all of the D-minimal elements in J let us choose one that does not have the representation required in the theorem and has the decomposition (1.71) with the smallest m. Let us show that in the decomposition (1.71) of this element all w_i, $1 \leq i \leq m$ are linearly independent modulo $J \oplus H_\gamma^-$, provided that $m > 0$.

Indeed, the dependence

$$\sum_{i=1}^{m} \beta_i w_i = d_1 + b_1, \quad \beta_i \neq 0, \ d_1 \in J, \ b_1 \in H_\gamma^-$$

implies

$$\beta_1^{-1} d_1 = w_1 + \sum_{i=2}^{m} \beta_1^{-1} \beta_i w_i - \beta_1^{-1} b_1.$$

Therefore d_1 has representation (1.71) with smaller m; that is, there exists the representation for d_1 required by the theorem. Moreover,

$$d - \beta_1^{-1} \alpha_1 d_1 = \alpha w + \sum_{i=2}^{m} (\alpha_i - \beta_1^{-1} \beta_i \alpha_1) w_i + b - \beta_1^{-1} \alpha_1 b_1,$$

which diminishes m.

The coproduct of G-monomials has the form $\Delta(w) = w \otimes h + g \otimes w + \beta$, $\Delta(w_i) = w_i \otimes h_i + g_i \otimes w_i + \beta_i$, where $\beta, \beta_i \in H_\gamma^- \otimes H_\gamma^-$, and g, h, g_i, h_i are group-like elements.

Therefore, we may write

$$\Delta(d) - d \otimes h - g \otimes d \equiv \sum_{i=1}^{m} \alpha_i w_i \otimes (h_i - h) + \sum_{i=1}^{m} \alpha_i (g_i - g) \otimes w_i \qquad (1.72)$$

modulo $H_\gamma^- \otimes H_\gamma^-$. Because d belongs to J, this statement implies

$$\sum_{i=1}^{m} \alpha_i w_i \otimes (h_i - h) + \sum_{i=1}^{m} \alpha_i (g_i - g) \otimes w_i \in H_\gamma^- \otimes H_\gamma^- + J \otimes H + H \otimes J. \qquad (1.73)$$

Consider the canonical linear projections $\varphi : H \to H/(J + H_\gamma^-)$ and $\psi : H \to H/J$. Applying the map $\varphi \otimes \psi$ to relation (1.73), we obtain

$$\sum_{i=1}^{m} \alpha_i \varphi(w_i) \otimes \psi(h_i - h) = 0. \qquad (1.74)$$

Since $\varphi(w_i)$ are linearly independent in $H/(J + H_\gamma^-)$ and $\psi(h_i - h)$ are nonzero in H/J, provided that $h_i \neq h$, we have that all coefficients α_i with $h_i \neq h$ are equal to zero. Likewise applying the map $\psi \otimes \varphi$ to (1.73), it follows that $\alpha_i = 0$, provided that $g_i \neq g$. With the help of these equalities, the relation (1.72) implies

$$\Delta(d) - d \otimes h - g \otimes d \in (H_\gamma^- \otimes H_\gamma^-) \cap (J \otimes H + H \otimes J) = 0,$$

for $J \cap H_\gamma^- = 0$. Thus d is skew-primitive, which contradicts the choice of d. □

Remark 1.7 In Theorem 1.6, we do not suppose that H is D-homogeneous.

Although not every character Hopf algebra is defined by skew-primitive relations, Theorem 1.6 provides a way to define any Hopf algebra H step-by step using skew-primitive relations.

Denote by J the kernel of $\xi : G\langle X\rangle \to H$. By Theorem 1.6 the Hopf ideal J has nonzero skew primitive elements. Let J_1 be an ideal generated by all skew primitive elements of J. Clearly, J_1 is a Hopf ideal. Now, consider the Hopf ideal J/J_1 in the quotient Hopf algebra $G\langle X\rangle/J_1$. Again, by Theorem 1.6, either $J_1 = J$ or J/J_1 has nonzero skew primitive elements. Denote by J_2/J_1 an ideal generated by all skew primitive elements of J/J_1, and by J_2, denote its pre-image with respect to the natural homomorphism $G\langle X\rangle \to G\langle X\rangle/J_1$. In continuation of this process, we will find a strictly increasing, finite or infinite chain of Hopf ideals of $G\langle X\rangle$

$$0 = J_0 \subset J_1 \subset J_2 \subset \ldots \subset J_s \subset \ldots. \qquad (1.75)$$

In this chain, the ideal J_s/J_{s-1} of $G\langle X\rangle/J_{s-1}$ is generated by skew-primitive elements.

Lemma 1.24

$$\bigcup_{s=1}^{\infty} J_s = J.$$

Proof Given $\gamma = \{n_i \mid i \in I\} \in D$, define $|\gamma| = \sum_{i \in I} n_i$, and set $G\langle X\rangle_s = \sum_{|\gamma| \leq s} G\langle X\rangle_\gamma$. Because $\bigcup_{s=0}^{\infty}(G\langle X\rangle_s \cap J) = J$, it suffices to demonstrate that for all s we have

$$G\langle X\rangle_s \cap J \subseteq J_{s+1}. \tag{1.76}$$

We perform induction on s. If $s = 0$, then $G\langle X\rangle_s = \mathbf{k}[G]$. Lemma 1.22 applied to the coideal $J \cap \mathbf{k}[G]$ implies (1.76) with $s = 0$, for J_1 contains all skew-primitive elements from $J \cap \mathbf{k}[G]$.

Assume that (1.76) holds for a given s. Let $H' = G\langle X\rangle/J_{s+1}$. In H' consider the coideal $J' = J/J_{s+1}$. If $|\gamma| = s+1$, then $(H')^-_\gamma \subseteq G\langle X\rangle_s/J_{s+1}$, which implies

$$(H')^-_\gamma \cap J' \subseteq (G\langle X\rangle_s \cap J) + J_{s+1}/J_{s+1} = 0.$$

In other words, according to Definition 1.13, the space $(H')_\gamma \cap J'$ consists of D-minimal elements with respect to J'. With Lemma 1.23 applied to the Hopf algebra H' and coideal J', we obtain that $H'_\gamma \cap J'$ is contained in the space spanned by skew-primitive elements of J'. The skew-primitive elements of J' belong to J_{s+2}/J_{s+1}, hence $H'_\gamma \cap J' \subseteq J_{s+2}/J_{s+1}$. Considering that $H'_\gamma = G\langle X\rangle_\gamma/J_{s+1}$, we have $G\langle X\rangle_\gamma \cap J \subseteq J_{s+2}$. Because γ is an arbitrary element with $|\gamma| = s+1$, this statement yields $G\langle X\rangle_{s+1} \cap J \subseteq J_{s+2}$. □

Definition 1.14 The length $\kappa(H)$ of chain (1.75) is called a *combinatorial rank* of the character Hopf algebra H with respect to the generators a_i, $i \in I$.

Remark 1.8 The arguments of Lemma 1.23 remain valid even if the skew-primitive generators are not semi-invariants. Hence, Theorem 1.6 is valid for (and Definition 1.14 may be applied to) an arbitrary Hopf algebra generated by the skew-primitive elements.

1.5.7 Noncommutative Differential Calculi

A *differential calculus* on an associative unitary algebra R is defined by a linear map d from R into a (R, R)-bimodule M such that the Leibniz formula

$$d(uv) = du \cdot v + u \cdot dv$$

is valid. The map d is called a *differential*. If R has a distinguished (finite or infinite) set of generators $x_i,\ i \in I$ such that M is freely generated by the differentials $dx_i,\ i \in I$ as a right module then the calculus is called *right coordinate*, originally a *first order differential calculus with right partial derivatives*, whereas the generators may be considered coordinate "noncommutative functions". In this case, we may define on R the *right partial derivatives* $\partial_i : R \to R,\ i \in I$ according to the formula $du = \sum_i dx_i \cdot \partial_i(u)$. The bimodule structure defines commutation rules

$$v \cdot dx_i = \sum_s dx_s \cdot A(v)_i^s, \quad v \in \mathbf{k}\langle X\rangle,\ i \in I. \tag{1.77}$$

Applying this relation to the product of two elements, we have

$$\sum_t dx_t A(uv)_i^t = (uv)dx_i = u\sum_s dx_s \cdot A(v)_i^s = \sum_s\sum_t dx_t A(u)_s^t A(v)_i^s.$$

Thus, the linear maps $A_i^t : R \to R$ satisfy the following comultiplication formula

$$A(uv)_i^t = \sum_s A(u)_s^t A(v)_i^s, \quad u, v \in \mathbf{k}\langle X\rangle,\ t, i \in I. \tag{1.78}$$

Since $d(uv) = \sum_i dx_i\, \partial_i(uv)$, it follows that the Leibniz formula yields

$$d(uv) = \sum_i dx_i\, \partial_i(u)\cdot v + u\sum_k dx_k\, \partial_k(v) = \sum_i dx_i\, \partial_i(u)\cdot v + \sum_i\sum_k dx_i A(u)_k^i\, \partial_k(v);$$

that is, the Leibniz formula reduces to the following relations for the partial derivatives:

$$\partial_i(uv) = \partial_i(u)v + \sum_k A(u)_k^i\, \partial_k(v), \quad u, v \in \mathbf{k}\langle X\rangle,\ i \in I. \tag{1.79}$$

Conversely, if maps A_i^t that satisfy (1.78) are given, then commutation rules (1.77) uniquely define the structure of a left module on the free right R-module generated by symbols $dx_i,\ i \in I$, whereas maps $\partial_i : R \to R,\ i \in I$ define a right coordinate differential calculus provided that (1.79) holds.

The free algebra $\mathbf{k}\langle X\rangle$ considered as a subalgebra of the free character Hopf algebra $G\langle Y\rangle$, has a right coordinate differential calculus related to the coproduct, where, as above X, are the normalized skew-primitive generators $x_i = h_i^{-1}y_i,\ i \in I$.

Proposition 1.8 *A right coordinate calculus set up by commutation rules* (1.77) *with* $A(u)_i^s = \delta_i^s\, \chi^u(g_s)u$ *is well-defined on* $\mathbf{k}\langle X\rangle$. *The partial derivatives connect this calculus with the coproduct on* $G\langle X\rangle$ *via*

$$\Delta(u) \equiv u \otimes 1 + \sum_i g_i\partial_i(u) \otimes x_i \quad (mod\ G\langle X\rangle \otimes \Lambda^2), \tag{1.80}$$

where Λ is an ideal of $\mathbf{k}\langle X\rangle$ generated by $x_i,\ i \in I$. The Leibniz formula for right partial derivatives takes the form

$$\partial_i(uv) = \partial_i(u)\cdot v + \chi^u(g_i)u\cdot\partial_i(v) \ \textit{with}\ \partial_i(x_k) = \delta_i^k. \tag{1.81}$$

Proof Because $\Delta(x_i) = x_i \otimes 1 + g_i \otimes x_i$ and Δ is a homomorphism of algebras, we have $\Delta(\mathbf{k}\langle X\rangle) \subseteq G\langle X\rangle \otimes \mathbf{k}\langle X\rangle$. The space $G\langle X\rangle \otimes \Lambda^2$ is an ideal of the algebra $G\langle X\rangle \otimes \mathbf{k}\langle X\rangle$, so that (1.80) is an equality in the related quotient algebra. Equation (1.80) uniquely defines linear maps $\partial_i : \mathbf{k}\langle X\rangle \to \mathbf{k}\langle X\rangle$. Let us verify that these linear maps satisfy (1.81). We have, first, $\partial_i(x_k) = \delta_i^k$, and then,

$$\begin{aligned}\Delta(uv) = \Delta(u)\Delta(v) &\equiv (u\otimes 1 + \sum_i g_i\partial_i(u)\otimes x_i)(v\otimes 1 + \sum_i g_i\partial_i(v)\otimes x_i)\\ &\equiv uv\otimes 1 + \sum_i ug_i\partial_i(v)\otimes x_i + \sum_i g_i\partial_i(u)v\otimes x_i\\ &= uv\otimes 1 + \sum_i g_i(\chi^u(g_i)u\partial_i(v) + \partial_i(u)v)\otimes x_i,\end{aligned}$$

which demonstrates (1.81).

The maps A_i^t defined in the proposition satisfy (1.78):

$$A(uv)_i^t = \delta_t^i\,\chi^{uv}(g_t) = \delta_t^i\,\chi^u(g_t)\chi^v(g_t) = \sum_s \delta_t^s\,\chi^u(g_t)\delta_s^i\,\chi^v(g_s) = \sum_s A(u)_s^t A(v)_i^s.$$

The relations (1.79) for the commutation rules (1.77) with $A(u)_i^s = \delta_i^s\,\chi^u(g_s)u$ take the form (1.81). Hence, $d(u) = \sum_i \partial_i(u)dx_i$ is a required differential calculus. □

If the module of differentials M is freely generated by $dx_i,\ i\in I$ as a left module, then the calculus is called *left coordinate*. In this case, we may define on R *left partial derivatives* $\partial_i^* : R\to R,\ i\in I$ according to the formula $du = \sum_i \partial_i^*(u)\cdot dx_i$. The bimodule structure defines commutation rules

$$dx_i\cdot u = \sum_s B(u)_i^s\cdot dx_s,\quad u\in\mathbf{k}\langle X\rangle,\ i\in I. \tag{1.82}$$

Applying this relation to the product of two elements uv, we see that B_i^s satisfy the same comultiplication formula

$$B(uv)_i^s = \sum_k B(u)_i^k B(v)_k^s,\quad u,v\in\mathbf{k}\langle X\rangle,\ i,s\in I. \tag{1.83}$$

The Leibniz formula reduces to relations for the left partial derivatives

$$\partial_i^*(uv) = \sum_s \partial_s^*(u)\,B(v)_s^i + u\,\partial_i^*(v),\quad u,v\in\mathbf{k}\langle X\rangle,\ i\in I. \tag{1.84}$$

Proposition 1.9 *A left coordinate calculus d^* set up by commutation rules* (1.82) *with $B(u)_i^s = \delta_i^s\,\chi^s(g_u)\,u$ is well-defined on $\mathbf{k}\langle X\rangle$. The left partial derivatives connect the calculus d^* with the coproduct on $G\langle X\rangle$ via*

$$\Delta(u) \equiv g_u \otimes u + \sum_i g_u g_i^{-1} x_i \otimes \partial_i^*(u) \quad (\mathit{mod}\ G\Lambda^2 \otimes \mathbf{k}\langle X\rangle), \tag{1.85}$$

where u is a homogeneous polynomial in each x_i, $i \in I$, and as above Λ is the ideal of $\mathbf{k}\langle X\rangle$ generated by x_i, $i \in I$. The Leibniz formula in terms of the left partial derivatives takes the form

$$\partial_i^*(uv) = \chi^i(g_v)\partial_i^*(u)\cdot v + u\cdot\partial_i^*(v) \ \ \mathit{with}\ \ \partial_i^*(x_s) = \delta_i^s. \tag{1.86}$$

Proof The space $G\Lambda^2 \otimes \mathbf{k}\langle X\rangle$ is an ideal of the algebra $G\langle X\rangle \otimes \mathbf{k}\langle X\rangle$, whereas (1.85) is an equality in the related quotient algebra. Equality (1.85) uniquely defines linear maps $\partial_i^* : \mathbf{k}\langle X\rangle \to \mathbf{k}\langle X\rangle$, $i \in I$. These linear maps satisfy (1.86). Indeed, for homogeneous polynomials $u, v \in \mathbf{k}\langle X\rangle$, we have

$$\begin{aligned}\Delta(uv) = \Delta(u)\Delta(v) &\equiv (g_u \otimes u + \sum_i g_u g_i^{-1} x_i \otimes \partial_i^*(u))\\ &\quad\times (g_v \otimes v + \sum_i g_v g_i^{-1} x_i \otimes \partial_i^*(v))\\ &\equiv g_u g_v \otimes uv + \sum_i g_u g_v g_i^{-1} x_i \otimes u\partial_i^*(v) + \sum_i g_u g_i^{-1} x_i\, g_v \otimes \partial_i^*(u) v\\ &= g_{uv} \otimes uv + \sum_i g_{uv} g_i^{-1} x_i \otimes (\chi^i(g_v)\partial_i^*(u)\, v + u\,\partial_i^*(v)),\end{aligned}$$

which demonstrates (1.86), as $\partial_i^*(x_k) = \delta_i^k$ clearly holds.

The maps B_i^s defined in the proposition satisfy (1.83):

$$B(uv)_i^s = \delta_i^s\,\chi^s(g_{uv})\,uv = \sum_k \delta_i^k\,\chi^i(g_u)u\,\delta_k^s\,\chi^k(g_v)v = \sum_k B(u)_i^k\,B(v)_k^s.$$

The relations (1.84) for the commutation rules (1.82) with $B(u)_i^s = \delta_i^s\,\chi^s(g_u)\,u$ take the form (1.86). Hence, $d^*(u) = \sum_i \partial_i^*(u) d^* x_i$ is a required differential calculus. □

Remark 1.9 We must stress that, in general, d and d^* are different differentials defined on $\mathbf{k}\langle X\rangle$. They may be identified only if $p_{is}p_{si} = 1$ for all $i, s \in I$.

Let $\varphi : R \to S$ be an epimorphism of algebras, and assume that R has a right (left) coordinate calculus d with respect to generators x_i, $i \in I$. We would like to understand when the homomorphism φ induces a right (left) coordinate calculus with respect to the generators $a_i = \varphi(x_i) \in S$, so that the homomorphism φ becomes a homomorphism of differential algebras:

$$\varphi(\partial_i(u)) = \partial_i(\varphi(u)). \tag{1.87}$$

Definition 1.15 An ideal J of an algebra R with right (left) coordinate differential calculus is said to be *differential* if $\partial_i(J) \subseteq J,\ i \in I$.

Lemma 1.25 *An epimorphism $\varphi : R \to S$ induces a right (left) coordinate calculus with respect to the generators $a_i = \varphi(x_i) \in S$ so that the homomorphism φ becomes a homomorphism of differential algebras if and only if $J = \ker\varphi$ is a differential ideal.*

Proof If we take $u \in \ker\varphi$, then (1.87) implies $\partial_i(u) \in \ker\varphi,\ i \in I$.

Conversely, if J is a differential ideal, then the formula $\bar{\partial}_i(\varphi(u)) = \varphi(\partial_i(u))$ correctly defines the linear maps $\bar{\partial}_i : S \to S$ because an equality $\varphi(u) = \varphi(v)$ implies $u - v \in J$ and $\partial_i(u) - \partial_i(v) \in \partial_i(J) \subseteq J$, which yields $\varphi(\partial_i(u)) = \varphi(\partial_i(v))$. In particular, we have $\bar{\partial}_i(a_k) = \varphi(\partial_i(x_k)) = \delta_i^k$.

The Leibniz formula (1.79) implies $\partial_i(ux_s) = A(u)_i^s$, and hence $A(J)_s^i \subseteq J$. For this reason, the formula $\bar{A}(\varphi(u))_s^i = \varphi(A(u)_s^i)$ correctly defines the linear maps $\bar{A}_s^i : S \to S$. Both the comultiplication formula (1.78) and the Leibniz formula (1.79) with $\partial \leftarrow \bar{\partial},\ A_s^i \leftarrow \bar{A}_s^i$ remain valid, and they define the required differential calculus on S.

Similarly, in the case of a left coordinate calculus, the Leibniz formula (1.84) implies $\partial_i^*(x_s u) = B(u)_i^s$ and $B(J)_i^s \subseteq J$. Therefore, the formula $\bar{B}(\varphi(u))_i^s = \varphi(B(u)_i^s)$ correctly defines the linear maps $\bar{B}_i^s : S \to S$. Both the comultiplication formula (1.83) and the Leibniz formula (1.84) with $\partial^* \leftarrow \bar{\partial},\ B_i^s \leftarrow \bar{B}_i^s$ remain valid, and they define the required differential calculus on S. □

If H is an arbitrary character Hopf algebra generated by normalized skew-primitive semi-invariants $a_i,\ i \in I$ and group G, then there exists a Hopf algebra homomorphism

$$\xi : G\langle X\rangle \to H,\ \ \xi(x_i) = a_i,\ \ \xi(g) = g,\ \ i \in I,\ \ g \in G. \tag{1.88}$$

Let A be a subalgebra of H generated by the elements $a_i,\ i \in I$.

Proposition 1.10 *If $\ker\xi \subseteq G\Lambda^2$, then ξ induces a right coordinate calculus of A with respect to the generators a_i so that the restriction of ξ on $\mathbf{k}\langle X\rangle$ becomes a homomorphism of differential algebras. In particular, the partial derivatives connect the induced calculus with the coproduct on H via*

$$\Delta(u) \equiv u \otimes 1 + \sum_i g_i\, \partial_i(u) \otimes a_i \quad (mod\ H \otimes \Lambda_a^2), \tag{1.89}$$

where $\Lambda_a = \xi(\Lambda)$ is an ideal of A generated by $a_i,\ i \in I$. The Leibniz formula (1.81) *also remains valid for A.*

Proof By Lemma 1.25, it suffices to check that $\ker \xi \cap \mathbf{k}\langle X\rangle$ is a differential ideal. If $u \in \ker \xi \cap \mathbf{k}\langle X\rangle$, then by (1.80) there exists $W \in G\langle X\rangle \otimes \Lambda^2$ such that

$$\Delta(u) = u \otimes 1 + \sum_i g_i\, \partial_i(u) \otimes x_i + W \in \ker \xi \otimes G\langle X\rangle + G\langle X\rangle \otimes \ker \xi. \quad (1.90)$$

Consider a projection $\eta : G\langle X\rangle \to G + G\cdot X$ that acts identically on $G + G\cdot X$ and $\eta(G\Lambda^2) = 0$. Applying $\mathrm{id} \otimes \eta$ to (1.90), we obtain

$$u \otimes 1 + \sum_i g_i\, \partial_i(u) \otimes x_i \in \ker \xi \otimes (G + G\cdot X). \quad (1.91)$$

Let η_i be a linear map $\eta_i : G + G\cdot X \to \mathbf{k}$ such that $\eta_i(x_k) = \delta_k^s,\ k, s \in I$, and $\eta_i(g) = 0,\ g \in G,\ \eta_i(hx_s) = 0,\ h \in G,\ h \neq 1,\ s \in I$. Applying $\mathrm{id} \otimes \eta_i$ to (1.91), we have $g_i\partial_i(u) \in \ker \xi$, which implies $\partial_i(u) \in \ker \xi$. □

Given $h \in G$, consider a linear space $G\langle X\rangle_h$ spanned by all monomials $gu,\ g \in G$ such that $gg_u = h$, where g_u is a group-like element that appears from the word u under the substitutions $x_i \leftarrow g_i,\ i \in I$. In this way the free Hopf algebra $G\langle X\rangle$ becomes a G-graded algebra:

$$G\langle X\rangle = \bigoplus_{h\in G} G\langle X\rangle_h. \quad (1.92)$$

In particular, to define a linear map, it suffices to consider only homogeneous elements with respect to the above grading.

Lemma 1.26 *If* $\ker \xi \subseteq G\Lambda$, *then* H *maintains the* G*-grading* (1.92). *Here* $G\Lambda$ *denotes the ideal of* $G\langle X\rangle$ *generated by* $x_i,\ i \in I$.

Proof We must demonstrate that $\ker \xi$ is a homogeneous ideal with respect to (1.92). Let $U = \sum_{k=1}^m \alpha_k f_k u_k \in \ker \xi$, where $h_k = f_k g_{u_k},\ 1 \le k \le m$ are different group-like elements, $f_k \in G$. We have,

$$\Delta(U) \in \ker \xi \otimes G\langle X\rangle + G\langle X\rangle \otimes \ker \xi. \quad (1.93)$$

Consider a projection $\eta : G\langle X\rangle \to \mathbf{k}[G]$ that acts identically on G and $\eta(G\Lambda) = 0$. Applying $\eta \otimes \mathrm{id}$ to (1.93), we obtain

$$\sum_{k=1}^m \alpha_k h_k \otimes f_k u_k \in \mathbf{k}[G] \otimes \ker \xi \quad (1.94)$$

because for each G-monomial fu we have $\Delta(fu) \in fg_u \otimes fu + G\Lambda \otimes G\langle X\rangle$. Considering that $h_k,\ 1 \le k \le m$ are linearly independent, there exist linear maps $\eta_k : \mathbf{k}[G] \to \mathbf{k}$, such that $\eta_k(h_s) = \delta_k^s,\ 1 \le k, s \le m$. The application of $\eta_i \otimes \mathrm{id}$ to (1.94) yields $\alpha_i f_i u_i \in \ker \xi,\ 1 \le i \le m$, as required. □

Proposition 1.11 *If* $\ker \xi \subseteq G\Lambda^2$, *then* ξ *induces a left coordinate calculus of* A *with respect to the generators* a_i, *so that the restriction of* ξ *on* $\mathbf{k}\langle X\rangle$ *becomes a homomorphism of differential algebras. In particular, the partial derivatives connect the induced calculus with the coproduct on* H *via*

$$\Delta(u) \equiv g_u \otimes u + \sum_i g_u g_i^{-1} x_i \otimes \partial_i^*(u) \quad (mod\, G\Lambda_a^2 \otimes A), \tag{1.95}$$

and the Leibniz formula (1.86) *remains valid for* A, *where* u *is a homogeneous element with respect to grading* (1.92).

Proof According to Lemma 1.25, we must verify that $\ker \xi \cap \mathbf{k}\langle X\rangle$ is a differential ideal. If $u \in \ker \xi \cap \mathbf{k}\langle X\rangle$ is a homogeneous element with respect to (1.92), then by (1.85) there exists $W \in G\Lambda^2 \otimes \mathrm{k}\langle X\rangle$ such that

$$\Delta(u) = g_u \otimes u + \sum_i g_u g_i^{-1} x_i \otimes \partial_i^*(u) + W \in \ker \xi \otimes G\langle X\rangle + G\langle X\rangle \otimes \ker \xi. \tag{1.96}$$

Applying $\mathrm{id} \otimes \eta$ to (1.96), where η is defined in the proof of Proposition 1.10, we obtain

$$g_u \otimes u + \sum_i g_u g_i^{-1} x_i \otimes \partial_i^*(u) \in (G + G\cdot X) \otimes \ker \xi. \tag{1.97}$$

Consider a linear map $\eta_i : G + G\cdot X \to \mathbf{k}$, such that $\eta_i(g_u g_i^{-1} x_i) = 1$, and $\eta_i(G) = \eta_i(hx_s) = 0$ if $h \neq g_u g_i^{-1}$ or $s \neq i$. Applying $\eta_i \otimes$ id to (1.97), we have $\partial_i^*(u) \in \ker \xi$. □

Definition 1.16 An element $a \in A$ is called a *constant* with respect to differential calculus $d : A \to M$ if $d(a) = 0$. Of course, if d is a right (left) coordinate calculus, then this statement is equivalent to the equalities $\partial_i(u) = 0,\ i \in I$ (respectively $\partial_i^*(u) = 0,\ i \in I$).

Lemma 1.27 *If* $\ker \xi \subseteq G\Lambda^2$, *then each skew-primitive element* $u \in \Lambda_a^2$ *is a constant with respect to both calculi.*

Proof It follows from (1.89) and (1.95). □

The converse statement is not valid. Indeed, the Leibniz formula implies that the set $C(H)$ of all constants with respect to the right coordinate calculus d is a subalgebra. Similarly, the set of all constants $C^*(H)$ with respect to d^* is a subalgebra too. However, sets of skew-primitive elements almost never form subalgebras.

Lemma 1.28 *In any Hopf algebra, a product of two nonzero normalized skew primitive elements* u, v *with* $g_u \neq g_v$ *is not skew-primitive.*

Proof Let $\Delta(u) = u \otimes 1 + g_u \otimes u,\ \Delta(v) = v \otimes 1 + g_v \otimes v$ If uv is skew-primitive, then for suitable $f, h \in G$, we have

$$0 = \Delta(uv) - uv \otimes h - f \otimes uv = uv \otimes (1-h) + g_u v \otimes u + ug_v \otimes v + (g_u g_v - f) \otimes uv.$$

If $g_u g_v \neq f$, then the element uv is a linear combination of three skew-primitive elements $(1-h), u, v$. As all of them are normalized, uv is also normalized; that is, $h = 1$. In this case, $uv = \alpha u + \beta v$ and we have

$$(g_u v + \alpha(g_u g_v - f)) \otimes u + (ug_v + \beta(g_u g_v - f)) \otimes v = 0.$$

By (1.59) the elements u, v are linearly independent. Hence $g_u v + \alpha(g_u g_v - f) = 0$, $ug_v + \beta(g_u g_v - f) = 0$. In particular $\alpha \neq 0,\ \beta \neq 0$. However in this case (1.59) implies that $\alpha u + \beta v$ is skew primitive only if $g_u = g_v$.

If $g_u g_v = f$, then $u, v, 1-h$ must be linearly dependent, say, $\alpha u + \beta v + \gamma(1-h) = 0$. Decomposition (1.59) implies that this requirement can be satisfied only if either $\alpha = 0$ or $\beta = 0$. Let, for example, $\alpha \neq 0$ and $\beta = 0$. In this case we obtain

$$(uv - \alpha^{-1}\gamma g_u v) \otimes (1-h) + ug_v \otimes v = 0,$$

which is impossible due to decomposition (1.59). □

1.6 Filtrations

Let R be a linear space. Consider a basis B of R. A subspace $S \subseteq R$ is said to be *admissible* with respect to B if S is spanned by a subset $b(S)$ of B. A sum of admissible subspaces is admissible, $b(\sum_\lambda S_\lambda) = \cup_\lambda b(S_\lambda)$, and an intersection of admissible subspaces is also admissible: $b(\cap_\lambda S_\lambda) = \cap_\lambda b(S_\lambda)$.

Below, a linear space R is said to be *filtered* if there is established an increasing chain of linear subspaces

$$\{0\} = R_{(-\infty)} \subseteq R_{(0)} \subseteq R_{(1)} \subseteq \ldots \subseteq R_{(n)} \subseteq \ldots,$$

such that $\bigcup_i R_{(i)} = R$. The filtration defines a degree function $d : R \to N \cup \{-\infty\}$ by

$$d(x) = \min\{i \mid x \in R_{(i)}\}.$$

We may choose a basis B^R of R so that all spaces $R_{(n)}$ are admissible with respect to B^R. To this end, we choose a basis $B^R_{(0)}$ of $R_{(0)}$, then extend it to a basis $B^R_{(1)}$ of $R_{(1)}$ and so on.

With every filtered space R, a graded space $\mathrm{gr}\, R$ is associated as follows:

$$\mathrm{gr}\, R = \bigoplus_{i=0}^{\infty} \mathrm{gr}_i\, R,$$

where $\mathrm{gr}_i\, R,\ i \geq 0$ is the quotient space $R_{(i)}/R_{(i-1)}$. A linear map $\varphi : R \to T$ of filtered spaces is a *homomorphism of filtered spaces* or a *filtered linear map* if

$$\varphi(R_{(i)}) \subseteq T_{(i)}, \quad i \geq 0. \tag{1.98}$$

A homomorphism induces a linear map $\mathrm{gr}\, \varphi$ of associated graded spaces:

$$\mathrm{gr}\, \varphi : u + R_{(n-1)} \mapsto \varphi(u) + T_{(n-1)}, \quad u \in R_{(n)},$$

or, equivalently, in terms of operators acting on $R_{(n)}$,

$$\pi_n \mathrm{gr}\, \varphi = \varphi \pi_n, \tag{1.99}$$

where π_n is the natural linear map $R_{(n)} \to R_{(n)}/R_{(n-1)} = \mathrm{gr}_n\, R$, while the operators act from the left to the right: $u \cdot \varphi\pi_n = \pi_n(\varphi(u))$. In this way, the map $\mathrm{gr}\, \varphi$ is well-defined because $u \equiv v \ (\mathrm{mod}\, R_{(n-1)})$ implies

$$\varphi(u) - \varphi(v) = \varphi(u - v) \in \varphi(R_{(n-1)}) \subseteq T_{(n-1)}.$$

The following lemma shows that the operator gr is a functor from filtered spaces to graded spaces.

Lemma 1.29 *If $R,\ T,\ S$ are filtered spaces and $\varphi : R \to T,\ \xi : T \to S$ are filtered linear maps, then the superposition $\varphi \cdot \xi$ is a filtered map and* $\mathrm{gr}\, (\varphi \cdot \xi) = \mathrm{gr}\, \varphi \cdot \mathrm{gr}\, \xi$.

Proof We have $\varphi \cdot \xi(R_{(n)}) = \xi(\varphi(R_{(n)})) \subseteq \xi(T_{(n)}) \subseteq S_{(n)}$. Using (1.99), we obtain

$$\pi_n(\mathrm{gr}\, \varphi \cdot \mathrm{gr}\, \xi) = \varphi \pi_n \mathrm{gr}\, \xi = (\varphi\xi)\pi_n.$$

Therefore, $\mathrm{gr}\, (\varphi \cdot \xi) = \mathrm{gr}\, \varphi \cdot \mathrm{gr}\, \xi$. □

We stress that functor gr is not exact. The simplest example is as follows. Let $R = \mathbf{k}$ be a filtered space with the filtration $R_{(0)} = 0,\ R_{(n)} = \mathbf{k},\ n \geq 1$, while $T = \mathbf{k}$ is a filtered space with the filtration $T_{(n)} = \mathbf{k},\ n \geq 0$. Then, the identical map $\varphi : R \to T$ is a filtered map, and $\mathrm{gr}\, \varphi = 0$. Due to this example, the following simple lemma is of interest.

Lemma 1.30 *If $\varphi : R \to T$ is a filtered linear map such that $\varphi(R_{(n)}) = T_{(n)},\ n \geq 0$, then $\mathrm{gr}\, \varphi$ is an epimorphism. If φ is an embedding and $\varphi(R_{(n)}) = T_{(n)} \cap \varphi(R),\ n \geq 0$, then $\mathrm{gr}\, \varphi$ is an embedding too.*

Proof If we have $T_{(n)} = \varphi(R_{(n)})$, then each element $u + T_{(n-1)}$, $u \in T_{(n)}$ has at least one preimage $v + R_{(n-1)}$, where $u = \varphi(v)$.

If φ is an embedding and $u\pi_n^R \in \ker \mathrm{gr}\,\varphi$, then $0 = u\pi_n^R \mathrm{gr}\,\varphi = \varphi(u)\pi_n^T$, and hence $\varphi(u) \in \ker \pi_n^T \cap \varphi(R) = \varphi(R_{(n-1)})$. The fact that φ is an embedding implies that $u \in R_{(n-1)}$ and $u\pi_n^R = 0$. □

If R, T are filtered spaces, then $R\otimes T$ has an induced filtration, where by definition

$$(R \otimes T)_{(n)} = \sum_{i+s=n} R_{(i)} \otimes T_{(s)}, \quad n \geq 0. \tag{1.100}$$

Respectively, the degree function on $R \otimes T$ is defined by $d(u \otimes v) = d(u) + d(v)$. Each subspace $R_{(i)} \otimes T_{(s)}$, $i, s \geq 0$ is admissible with respect to the basis $B^R \otimes B^T$ of $R \otimes T$ because it is spanned by the tensors $u \otimes v$, $u \in b(R_{(i)})$, $v \in b(T_{(s)})$.

Given $n \geq 0$, let $\overline{\pi}_i^R$, $0 \leq i \leq n$ denote the natural maps $\overline{\pi}_i^R : R_{(n)} \to R_{(n)}/R_{(i-1)}$. Of course, the quotient space $R_{(n)}/R_{(i-1)}$ contains $\mathrm{gr}_i R = R_{(i)}/R_{(i-1)}$ as a subspace. Let $\pi_i^R \otimes \pi_s^T$, $i + s = n$ be the linear map $\overline{\pi}_i^R \otimes \overline{\pi}_s^T$ restricted on $(R \otimes T)_{(n)}$.

Lemma 1.31 *The image of* $\pi_i^R \otimes \pi_s^T$ *equals* $\mathrm{gr}_i R \otimes \mathrm{gr}_s T$.

Proof If $k > i$ and $k + m = n$, then $m < s$; therefore, we have $\overline{\pi}_s^T(T_{(m)}) = 0$. If $k < i$, then $\overline{\pi}_i^R(R_{(k)}) = 0$, which implies

$$\mathrm{im}\,(\pi_i^R \otimes \pi_s^T) = (\sum_{k+m=n} R_{(k)} \otimes T_{(m)})\overline{\pi}_i \otimes \overline{\pi}_s = (R_{(i)} \otimes T_{(s)})(\pi_i^R \otimes \pi_s^T) = \mathrm{gr}_i R \otimes \mathrm{gr}_s T.$$

as required. □

Lemma 1.32 *If* R, T *are filtered spaces, then* $\mathrm{gr}(R \otimes T) = \mathrm{gr}R \otimes \mathrm{gr}T$ *with the decomposition of homogeneous components,*

$$\mathrm{gr}_n(R \otimes T) = \bigoplus_{i+s=n} \mathrm{gr}_i R \otimes \mathrm{gr}_s T,$$

via the identification

$$\pi_n^{R\otimes T} = \bigoplus_{i+s=n} \pi_i^R \otimes \pi_s^T. \tag{1.101}$$

Proof By definition, the operator $\bigoplus_{i+s=n}(\pi_i^R \otimes \pi_s^T)$ acts on $(R \otimes T)_{(n)}$ as follows:

$$u \otimes v \mapsto \bigoplus_{i+s=n} (u\pi_i^R \otimes v\pi_s^T) \in \bigoplus_{i+s=n} \mathrm{gr}_i R \otimes \mathrm{gr}_s R.$$

In particular, the kernel of that operator equals the intersection of the kernels of all $\pi_i^R \otimes \pi_s^T$, $i + s = n$. The functorial property of the tensor product, (1.49), yields

$$\ker(\pi_i^R \otimes \pi_s^T) = \ker(\overline{\pi}_i \otimes \overline{\pi}_s) \cap (R \otimes T)_{(n)} = (R_{(i-1)} \otimes T_{(n)} + R_{(n)} \otimes T_{(s-1)}) \cap (R \otimes T)_{(n)}.$$

The latter intersection as an admissible subspace is spanned by the set

$$B_{i,s} \stackrel{\text{df}}{=} \left((B^R_{(i-1)} \otimes B^T_{(n)}) \cup (B^R_{(n)} \otimes B^T_{(s-1)}) \right) \cap \Big(\bigcup_{k+m=n} B^R_{(k)} \otimes B^T_{(m)} \Big).$$

A tensor $u \otimes v$, with $u \in B^R_{(n)}$ $v \in B^T_{(n)}$ belongs to this set if and only if, first, either $d(u) < i$ or $d(v) < s$ and, next, $d(u) + d(v) \leq n$. These conditions are met if $d(u) + d(v) < n$. If both conditions are valid for all pairs (i, s), $i + s = n$, then taking $i = d(u)$, we have $d(v) < s = n - i = n - d(u)$, which implies $d(u) + d(v) < n$. In other words, the intersection of all $B_{i,s}$, $i + s = n$ equals $(B^R \otimes B^T)_{(n-1)}$, which implies that the kernel of the operator in the right-hand side of (1.101) (as an admissible set) equals $(R \otimes T)_{(n-1)}$, the kernel of the left-hand side. Thus, we may identify those operators. Finally,

$$\text{gr}_n(R \otimes T) = (R \otimes T)_{(n)} \pi_n^{R\otimes T} = (R \otimes T)_{(n)} \bigoplus_{i+s=n} \pi_i^R \otimes \pi_s^T = \bigoplus_{i+s=n} \text{gr}_i R \otimes \text{gr}_s T.$$

□

Lemma 1.33 *If $\varphi : R \to R'$ and $\xi : T \to T'$ are filtered linear maps, then so are $\varphi \otimes \xi : R \otimes T \to R' \otimes T'$, and* $\text{gr}\,(\varphi \otimes \xi) = \text{gr}\,\varphi \otimes \text{gr}\,\xi$.

Proof We have

$$(R \otimes T)_{(n)}(\varphi \otimes \xi) = \sum_{i+s=n} \varphi(R_{(i)}) \otimes \xi(T_{(s)}) \subseteq \sum_{i+s=n} R'_{(i)} \otimes \xi(T'_{(s)} = (R' \otimes T')_{(n)}.$$

Using identification (1.101) we have

$$\pi_n^{R\otimes T}(\text{gr}\,\varphi \otimes \text{gr}\,\xi) = \bigoplus_{i+s=n} \pi_i^R \text{gr}\,\varphi \otimes \pi_s^T \text{gr}\,\xi = \bigoplus_{i+s=n} \varphi \pi_i^{R'} \otimes \xi \pi_s^{T'} = (\varphi \otimes \xi) \pi_n^{R'\otimes T'}.$$

Hence, according to definition (1.99), we obtain $\text{gr}\,(\varphi \otimes \xi) = \text{gr}\,\varphi \otimes \text{gr}\,\xi$. □

One may introduce the concepts of filtered algebras, coalgebras, modules, comodules, Hopf algebras and other objects defined on linear spaces by linear maps (operations) and tensor products by requiring that the ground space and all maps (operations) be filtered. In this case, for every such object R, the graded object $\text{gr}\,R$ is associated. If the axioms are given by certain operator relations, then the Lemmas 1.29 and 1.33 proven above demonstrate that the operator relations remain valid on $\text{gr}\,R$. In other words, the associated graded object retains the structure of

R. Moreover, considering that homomorphisms are normally defined in terms of equalities of operators, gr is a functor of related categories, but it is certainly not exact in general. Let us consider somewhat more thoroughly algebras, coalgebras and Hopf algebras.

Definition 1.17 An algebra R is said to be *filtered* if R is a filtered space and the multiplication $\mathbf{m} : R \otimes R \to R$ and the unit map $\alpha \mapsto \alpha \cdot 1$ are filtered linear maps; that is, $R_{(i)}R_{(s)} \subseteq R_{(i+s)},\ i, s \geq 0$ and $1 \in R_{(0)}$.

The operator form of the product $\mathrm{gr}\,\mathbf{m}$ on $\mathrm{gr}\,R$ is

$$(\pi_i \otimes \pi_s)\mathrm{gr}\,\mathbf{m} = \mathbf{m}\,\pi_{i+s}, \quad i, s \geq 0, \tag{1.102}$$

where as above the operators act from the left to the right, $(u \otimes v)\mathbf{m}\pi_{i+s} = \pi_{i+s}(uv)$. The product can be written in the elementary form as follows:

$$(u + R_{(i-1)})(v + R_{(s-1)}) = uv + R_{(i+s-1)}, \quad u \in R_{(i)},\ v \in R_{(s)} \quad i,\ s \geq 0. \tag{1.103}$$

To ascertain that $\mathrm{gr}\,R$ is associative, we must simply apply gr to the associativity of $\mathbf{m}$ written in the operator form, $(\mathbf{m} \otimes \mathrm{id})\mathbf{m} = (\mathrm{id} \otimes \mathbf{m})\mathbf{m}$, and use Lemmas 1.29 and 1.33:

$$(\mathrm{gr}\,\mathbf{m} \otimes \mathrm{id})\mathrm{gr}\,\mathbf{m} = \mathrm{gr}\,((\mathbf{m} \otimes \mathrm{id})\mathbf{m}) = \mathrm{gr}\,((\mathrm{id} \otimes \mathbf{m})\mathbf{m}) = (\mathrm{id} \otimes \mathrm{gr}\,\mathbf{m})\mathrm{gr}\,\mathbf{m}.$$

It is also easy to check the associativity using the elementary form (1.103).

Let R, T be filtered algebras. A linear map $\varphi : R \to T$ is called a *homomorphism of filtered algebras* if it is a homomorphism of algebras and filtered spaces.

Lemma 1.34 *If $\varphi : R \to T$ is a homomorphism of filtered algebras, then the map* $\mathrm{gr}\,\varphi : \mathrm{gr}\,R \to \mathrm{gr}\,T$ *is a homomorphism of associated graded algebras.*

Proof It suffices to apply gr to the equality $\varphi(uv) = \varphi(u)\varphi(v)$ written in the operator form, $\mathbf{m}\,\varphi = (\varphi \otimes \varphi)\,\mathbf{m}$, and use Lemmas 1.29 and 1.33. □

Example 1.4 Every algebra R with a fixed set of generations $A = \{a_i \mid i \in I\}$ has a natural filtration, called a *filtration by formal degree*. Let us assign certain natural degrees to the generators, $d(a_i) = d_i,\ i \in I$. As usual, the degree of a word in the a_i's is the sum of the degrees of its letters. By definition, the space $R_{(n)}$ is spanned by values of all words of degree $\leq n$. In particular, $R_{(0)} = \mathbf{k} \cdot 1$ is spanned by the value of the empty word. R is a filtered algebra with respect to this filtration.

Let $\mathbf{k}\langle X\rangle$ be a free associative algebra freely generated by a set $X = \{x_i \mid i \in I\}$. If $f \in \mathbf{k}\langle X\rangle$, then there is a decomposition $f = f_n + f_{n-1} + \ldots + f_1 + f_0$ of the polynomial f into homogeneous components, where f_i is the linear combination of all monomials of degree i occurring in f. If $f_n \neq 0$, then by definition n is the formal degree of f. The polynomial $\bar{f} \stackrel{\mathrm{df}}{=} f_n$ is called a *leading component* of f.

The quotient space $\mathbf{k}\langle X\rangle_{(n)}/\mathbf{k}\langle X\rangle_{(n-1)}$ is isomorphic to the space spanned by all words of degree n; that is, we may consider $f_i \in \mathrm{gr}_i\, \mathbf{k}\langle X\rangle$. In this case, the map

$$f \mapsto f_n \oplus f_{n-1} \oplus \ldots \oplus f_1 \oplus f_0$$

is an isomorphism between $\mathbf{k}\langle X\rangle$ and $\mathrm{gr}\,\mathbf{k}\langle X\rangle$ that allows us to identify

$$\mathrm{gr}\,\mathbf{k}\langle X\rangle = \mathbf{k}\langle X\rangle.$$

For the algebra R given in Example 1.4, a natural homomorphism $\varphi : \mathbf{k}\langle X\rangle \to R$ defined by $x_i \mapsto a_i$ is filtered because $\varphi(\mathbf{k}\langle X\rangle_{(n)}) = R_{(n)}$. Moreover, Lemma 1.30 implies that $\mathrm{gr}\,\varphi : \mathbf{k}\langle X\rangle \to \mathrm{gr}\,R$ is an epimorphism.

Proposition 1.12 *The ideal* $\ker(\mathrm{gr}\,\varphi)$ *is spanned by all leading components* $\overline{f}$ *when* f *runs through the ideal* $\ker\varphi$.

Proof If $f \in \ker\varphi$, then $f = 0$ in R. Hence $f_n = -f_{n-1} - \ldots - f_1 - f_0 \in R_{(n-1)}$; that is, $\overline{f} = f_n \in \ker \mathrm{gr}\,\varphi$. Conversely, if $f = f_n + \ldots + f_1 + f_0 \in \ker(\mathrm{gr}\,\varphi)$, then all homogeneous components $f_i,\ 0 \le i \le n$ also belong to $\ker(\mathrm{gr}\,\varphi)$. By definition, for each f_i, there exists $g_i \in \mathbf{k}\langle X\rangle_{(i-1)}$ such that $f_i = g_i$ in R, which implies $f_i - g_i \in \ker\varphi$ and $f_i = \overline{f_i - g_i}$. □

Suppose that the algebra R is defined by relations $F_i = 0,\ i \in I$. Then, the above proposition states that all leading components $\overline{F_i}$ are relations of the associated graded algebra. In other words, there exists a natural epimorphism

$$\overline{\varphi} : \overline{R} \stackrel{\mathrm{df}}{=} \langle X \,||\, \overline{F_i} = 0,\ i \in I\rangle \longrightarrow \mathrm{gr}\,R. \tag{1.104}$$

This epimorphism is not always an isomorphism. For example, consider an algebra R with a generator x and two relations $x^2 = 1,\ x^2 = x$. Then, of course, $R \cong \mathbf{k} \cong \mathrm{gr}\,R$, whereas $\overline{R} \cong \mathbf{k}[x \,|\, x^2 = 0]$ is a two-dimensional algebra. The situation is changed if the set $\{F_i = 0\}$ is a Gröbner-Shirshov system of relations; see Sect. 1.2.3.

Theorem 1.7 *If* $\prec$ *is a Deg-Lex ordering, and the set of relations* $\{F_i = 0,\ i \in I\}$ *is closed with respect to the compositions, then* (1.104) *is an isomorphism.*

Proof If $F \in \ker(\mathrm{gr}\,\varphi)$, then by Proposition 1.12, we have $F = \overline{f},\ f \in \ker\varphi$. By the definition of the Deg-Lex ordering, the leading word w of f has the maximal formal degree. Therefore, w is one of the monomials of $\overline{F}$. If $\overline{\varphi}$ is not an isomorphism, then we may choose $F \notin \ker\overline{\varphi}$ with the minimal possible w. By Theorem 1.2, the word w contains one of the leading words w_i of $F_i = 0,\ i \in I$ as a subword, $w = uw_iv$. In this case, $f' = f - uF_iv$ still belongs to $\ker\varphi$, and the leading word of f' is less than w. Therefore, $f'_n \in \ker\overline{\varphi}$. We have $\overline{f} = f'_n + u\overline{F_i}v$ because w_i is one of the monomials of $\overline{F_i}$, which implies $F = \overline{f} \in \ker\overline{\varphi}$ in view of the fact that $\overline{F_i} \in \ker\overline{\varphi}$. □

If U is a subalgebra of $\mathbf{k}\langle X\rangle$, then U has a filtration $U_{(n)} = \{u \in U \mid d(u) \leq n\}$ induced by formal degree. By Lemma 1.30 the associated graded algebra $\mathrm{gr}\, U$ is a subalgebra of $\mathrm{gr}\, \mathbf{k}\langle X\rangle = \mathbf{k}\langle X\rangle$.

Proposition 1.13 *The algebra* $\mathrm{gr}\, U$ *is spanned by all leading components* $\overline{f}$ *when* f *runs through the algebra* U.

Proof If $f \in U_{(n)}$, then $f + U_{(n-1)}$ is identified with $f + \mathbf{k}\langle X\rangle_{(n-1)}$, which, in turn, was identified with $\overline{f}$, so that $\overline{f} \in \mathrm{gr}\, U$.

Conversely, if $f = f_n + \ldots + f_1 + f_0 \in \mathrm{gr}\, U$, then each homogeneous component f_i, $0 \leq i \leq n$ belongs to $\mathrm{gr}_i\, U$. By definition, for each f_i, there exists $g_i \in U$ such that $f_i \equiv g_i$ modulo $\mathbf{k}\langle X\rangle_{(i-1)}$. In this case, the homogeneous element f_i equals $\overline{f_i + (g_i - f_i)} = \overline{g_i}$. □

Assume that on a filtered space R, a coalgebra structure is given with a coproduct Δ and a counit ε.

Definition 1.18 The coalgebra R is said to be *filtered* if $\Delta : R \otimes R \to R$ is a filtered linear map; that is,

$$\Delta(R_{(n)}) \subseteq \sum_{s+i=n} R_{(i)} \otimes R_{(s)}, \quad n \geq 0, \tag{1.105}$$

or, equivalently, the degree function satisfies $d(x) \geq d(x^{(1)}) + d(x^{(2)})$. Note that the counit $\varepsilon : R \to \mathbf{k}$ is always a filtered map because $\mathbf{k}_{(0)} = \mathbf{k}$.

Lemma 1.35 *If* R *is a filtered coalgebra, then the associated graded space* $\mathrm{gr}\, R$ *is a coalgebra with the coproduct* $\mathrm{gr}\, \Delta$ *and the counit* $\mathrm{gr}\, \varepsilon$.

Proof The coalgebra axioms may be written in the form of equalities of operators defined on tensor products:

1. Coassociativity: $\Delta \cdot (\Delta \otimes \mathrm{id}) = \Delta \cdot (\mathrm{id} \otimes \Delta)$;
2. Counit: $\Delta \cdot (\varepsilon \otimes \mathrm{id}) = \mathrm{id} = \Delta \cdot (\mathrm{id} \otimes \varepsilon)$.

Applying gr to those equalities and using Lemmas 1.29 and 1.33, we see that $\mathrm{gr}\, \Delta$ and $\mathrm{gr}\, \varepsilon$ also satisfy the axioms. □

In the elementary form, the coproduct $\mathrm{gr}\, \Delta$ and counit $\mathrm{gr}\, \varepsilon$ behave as follows:

$$\mathrm{gr}\, \Delta : u + R_{(n-1)} \mapsto \bigoplus_{i+s=n} (u^{(1)} + R_{(i-1)}) \otimes (u^{(2)} + R_{(s-1)}), \quad u \in R_{(n)}, \tag{1.106}$$

$$\mathrm{gr}\, \varepsilon : u + R_{(n-1)} \mapsto \begin{cases} \varepsilon(u), \text{ if } n = 0, \ u \in R_{(0)}; \\ 0, \quad \text{if } n > 0, \ u \in R_{(n)}. \end{cases} \tag{1.107}$$

Example 1.5 Every coalgebra C has a natural filtration, called a *coradical filtration*. By definition, $C_{(0)}$ is the coradical (a sum of all simple subcoalgebras of C), and $C_{(n)}$

is defined inductively as follows:

$$C_{(n)} = \Delta^{-1}(C \otimes C_{(n-1)} + C_{(0)} \otimes C), \quad n \geq 1.$$

The coalgebra C is a filtered coalgebra with respect to the coradical filtration. One may find the basic properties of the coradical filtration in the books [1, 176, 220].

Let R, T be filtered coalgebras. A linear map $\varphi : R \to T$ is called a *homomorphism of filtered coalgebras* if it is a homomorphism of coalgebras and filtered spaces.

Lemma 1.36 *If* $\varphi : R \to T$ *is a homomorphism of filtered coalgebras, then the map* $\operatorname{gr}\varphi : \operatorname{gr} R \to \operatorname{gr} T$ *is a coalgebra homomorphism.*

Proof The definition of a coalgebra homomorphism in the operator form reads as follows: $\Delta \cdot (\varphi \otimes \varphi) = \varphi \cdot \Delta$. It remains to apply Lemmas 1.29 and 1.33. □

Definition 1.19 A Hopf algebra H is said to be *filtered* if on the space H, a filtration is fixed so that the product $\mathbf{m}$, the unit map $\alpha \to \alpha \cdot 1$, the coproduct Δ, and the antipode σ are filtered linear maps; that is,

1. $R_{(i)}R_{(s)} \subseteq R_{(i+s)}, \quad 1 \in R_{(0)}, \qquad i, s \geq 0;$
2. $\Delta(R_{(n)}) \subseteq \sum_{s+i=n} R_{(i)} \otimes R_{(s)}, \quad n \geq 0;$
3. $\sigma(R_{(n)}) \subseteq R_{(n)}, \qquad n \geq 0.$

Theorem 1.8 *If H is a filtered Hopf algebra, then the associated graded space* $\operatorname{gr} H$ *is a Hopf algebra with the product* $\operatorname{gr}\mathbf{m}$, *the coproduct* $\operatorname{gr}\Delta$, *the counit* $\operatorname{gr}\varepsilon$, *and the antipode* $\operatorname{gr}\sigma$.

Proof By the above two lemmas $\operatorname{gr} H$ is both an algebra and a coalgebra. The rest of the Hopf algebra axioms may be written in the form of equalities of operators as follows:

1. Antipode: $\Delta \cdot (\sigma \otimes \mathrm{id})\,\mathbf{m} = \Delta \cdot (\mathrm{id} \otimes \sigma)\,\mathbf{m} = \varepsilon \cdot 1;$
2. The coproduct is a homomorphism: $\mathbf{m}\,\Delta = (\Delta \otimes \Delta)(\mathrm{id} \otimes \theta \otimes \mathrm{id})(\mathbf{m} \otimes \mathbf{m}),$

   where θ is the flip map, $\theta : u \otimes v \mapsto v \otimes u$. The flip map is filtered, and $\operatorname{gr}\theta = \theta$.

Applying gr to those equalities and using Lemmas 1.29 and 1.33, we determine that the associated graded maps satisfy these axioms too. □

In the elementary form, the graded antipode has a very clear representation

$$\operatorname{gr}\sigma : u + R_{(n-1)} \mapsto \sigma(u) + R_{(n-1)}, \quad u \in R_{(n)}. \tag{1.108}$$

Let R, T be filtered Hopf algebras. A linear map $\varphi : R \to T$ is called a *homomorphism of filtered Hopf algebras* if it is a homomorphism of algebras, coalgebras, and filtered spaces.

Lemma 1.37 *If* $\varphi : R \to T$ *is a homomorphism of filtered Hopf algebras, then* $\operatorname{gr}\varphi : \operatorname{gr} R \to \operatorname{gr} T$ *is also a Hopf algebra homomorphism.*

Proof This statement follows from Lemmas 1.34 and 1.36. □

Example 1.6 The free character Hopf algebra (as well as each of its homomorphic images) has two natural filtrations, the coradical filtration and the filtration defined by the formal degree, $d(x_i) = 1$, $d(G) = 0$. In general, these two filtrations are different. The zero components are the same, $G\langle X\rangle_{(0)} = \mathbf{k}[G]$, but the first component of the coradical filtration contains all skew-primitive elements that may have a formal degree greater than 1. For example, if $p_{is}p_{si} = 1$, then $x_ix_s - p_{is}x_sx_i$ is such an element. Thus, at least two graded Hopf algebras may be associated with each character Hopf algebra.

1.7 Certain Concepts of P.M. Cohn's Theory

A filtered space R is said to be *connected* if $R_{(0)} = \mathbf{k}$. Each subspace of R has the induced connected filtration. For example, an algebra with a distinguished set of generators is connected with respect to a filtration defined by the formal degree; see Example 1.4.

Let R be a connected filtered algebra, and let d be the degree function $d(u) = \min\{n \mid u \in R_{(n)}\}$; we shall say that the family $\{a_i | 1 \leq i \leq n\}$ of elements of R is *right d-dependent* if there exist elements $b_i \in R$, such that

$$d(\sum a_ib_i) < \max\{d(a_i) + d(b_i)\},$$

or if some $a_i = 0$. Otherwise, the family $\{a_i\}$ is *right d-independent*. An element a of R is said to be *right d-dependent* on a family $\{a_i\}$ if $a = 0$ or if there exist $b_i \in R$ such that

$$d(a - \sum a_ib_i) < d(a), \quad \text{and } d(a_i) + d(b_i) = d(a) \text{ for all } i.$$

In the contrary case, a is said to be right *d-independent* of $\{a_i\}$.

Definition 1.20 A set X in R is called a *weak algebra basis* if all words in X (including the empty one) span R, and no element of X is right d-dependent on the rest.

Lemma 1.38 *Every connected filtered algebra R has a weak algebra basis.*

Proof For each $n > 0$ denote by $R'_{(n)}$ the subspace of $R_{(n)}$ spanned by the products ab, where $a, b \in R_{(n-1)}$ and $d(a)+d(b) \leq n$. Now choose a minimal set X_n spanning $R_{(n)}$ (mod$R'_{(n)}$) over $\mathbf{k}$, that is a set of representatives for a basis of $R_{(n)}/R'_{(n)}$, and set $X = \cup X_n$.

To show that X is a weak algebra basis, suppose that an element $x \in X$ is right d-dependent on other elements $x_1, \ldots, x_m$ of X.

$$x \equiv \sum x_i b_i \pmod{R_{(n-1)}}, \tag{1.109}$$

where $n = d(x)$. Any terms $x_i b_i$ with $d(x_i) < n$ belong to R'_n, so (1.109) implies that

$$x \equiv \sum \alpha_i x_i \pmod{R'_{(n)}},$$

where $\alpha_i \in \mathbf{k}$ and $d(x_i) = n$ whenever $\alpha_i \neq 0$. However, this statement contradicts the construction of X; thus, no element of X is right d-dependent on the rest. An easy induction on the degree shows that the monomials in X span R; more precisely, the monomials with a formal degree of, at most, n span $R_{(n)}$. □

Definition 1.21 The algebra R is said to satisfy the *weak algorithm* relative to d, if in any right d-dependent family, say $a_1, \ldots, a_m$, where

$$d(a_1) \leq \ldots \leq d(a_m),$$

some a_i is right d-dependent on $a_1, \ldots, a_{i-1}$.

Theorem 1.9 *Let R be a connected filtered algebra with the degree function d. Then, R satisfies the weak algorithm relative to d if and only if R is the free associative algebra on a set X such that the filtration is defined by the formal degree induced from $d : X \to N_+$.*

Proof Let $X = \{x_i \mid i \in I\}$ be a weak algebra basis constructed in Lemma 1.38. By induction on the length, we shall prove that all monomials in X are linearly independent. The monomials of length zero certainly are. Assume that all monomials in X of length $< n$ are linearly independent as elements of R, and let $\sum_s \alpha_s w_s = 0$, where w_s are monomials of length $\leq n$. By splitting off the left-hand factor from X in each monomial w_s, we can write $\alpha + \sum_{i\in I} x_i f_i = 0$, $\alpha \in \mathbf{k}$, $f_i \in R$, which implies

$$d(\sum_{i\in I} x_i f_i) = d(-\alpha) = 0 < \max\{d(x_i) + d(f_i)\}.$$

Therefore, X is a right d-dependent family. By the weak algorithm, either one of the x_i's is d-dependent on the rest or each f_i equals zero in R. The former option contradicts the choice of X. Thus, each $f_i = 0$ in R, and so $\alpha = 0$. By the induction assumption, all f_i are zero polynomials. Hence, the given relation $f = 0$ was trivial, which completes the induction.

To demonstrate that $d(f) = d'(f)$, where $f = \sum_s \alpha_s w_s$ is a linear combination of monomials, we perform induction on the length of the monomials. It is clear that d coincides with d' on $R_{(0)} = \mathbf{k}$. Assume that $d(f) = d'(f)$ whenever all monomials of f are of length $< n$. The formal degree d' with $d'(x_i) = d(x_i),\ i \in I$ is defined by the equality

$$d'(\sum_i x_i f_i) = \max\{d(x_i) + d'(f_i)\},$$

where f_i are polynomials in X. If all monomials of f are of length $\leq n$, then $f = \sum_i x_i f_i$ where all monomials of f_i are of length $< n$, which implies

$$d'(f) = \max\{d(x_i) + d'(w_i)\} = \max\{d(x_i) + d(w_i)\}.$$

At the same time, if $d(f) < \max\{d(x_i) + d(w_i)\}$, then the weak algorithm states that one of the x_s is d-dependent on the rest, which contradicts the choice of X. Hence $d(f) = \max\{d(x_i) + d(w_i)\} = d'(f)$.

We complete the proof by showing that the free algebra $\mathbf{k}\langle X\rangle$ filtered by the formal degree $d(x_i) = d_i > 0$ satisfies the weak algorithm.

Let us fix a monomial $x_1 \dots x_h$ of degree r, and define the *transduction* for this monomial as the linear map $b \mapsto b^*$ of $\mathbf{k}\langle X\rangle$ into itself, which sends any monomial of the form $ax_1 \dots x_h$ to a and all other monomials to zero. Thus, b^* is the "left cofactor" of $x_1 \dots x_h$ in the canonical expression for b. Clearly, for any $b \in \mathbf{k}\langle X\rangle$ we have $d(b^*) \leq d(b) - r$. Further, if $a, b \in \mathbf{k}\langle X\rangle$, then

$$(ab)^* \equiv ab^* \pmod{\mathbf{k}\langle X\rangle_{(d(a)-1)}}.$$

This statement is clear if b is a monomial term of degree at least r; in fact, we then have equality. If b is a monomial term of degree less than r, the right-hand side is zero, and the congruence holds. The general case follows by linearity.

Assume now that $a_1, \dots, a_n$ is a d-dependent family:

$$d(\sum_i a_i b_i) < m = \max_i\{d(a_i) + d(b_i)\}.$$

Taking the a_i's as ordered so that $d(a_1) \leq \dots \leq d(a_n)$, we must show that some a_i is d-dependent on those that precede. By omitting terms if necessary, we may assume that $d(a_i) + d(b_i) = m$ for all i and, hence, that $d(b_1) \geq \dots \geq d(b_n)$.

Let $x_1 \dots x_h$ be a product of maximal degree $r = d(b_n)$ occurring in b_n with a nonzero coefficient α, and denote the transduction for $x_1 \dots x_h$ by *. Consider now $\sum a_i b_i^*$; the ith term differs from $(a_i b_i)^*$ by a term of degree less than $d(a_i) \leq d(a_n)$. Hence the sum will differ by a term of degree less than $d(a_n)$ from $(\sum a_i b_i)^*$, which has degree $\leq d(\sum_i a_i b_i) - r < m - r = d(a_n)$. Therefore, $d(\sum a_i b_i^*) < d(a_n)$, which gives a relation of left d-dependence of a_n on a_i, $i < n$, as $b_n^* = \alpha \in \mathbf{k}$, $\alpha \neq 0$. □

Remark 1.10 One can easily define left d-dependent families and a left weak algorithm. The above theorem implies that the concept of a weak algorithm for connected filtered algebras is right-left symmetric because the free algebra R is isomorphic to the opposite algebra $R^{\mathrm{op}} = \langle R, *\rangle$ with the product $u * v = vu$.

We conclude this section by describing a condition for a subalgebra to inherit the weak algorithm. If U is a subalgebra of a filtered algebra R, then U has an induced filtration $U_{(n)} = R_{(n)} \cap U$ with the same degree function. By Lemma 1.30 the associated graded algebra $\mathrm{gr}\, U$ is a subalgebra of $\mathrm{gr}\, R$ by means of the identification $u + U_{(n-1)} = u + R_{(n-1)},\ u \in U_{(n)}$.

Definition 1.22 We call homogeneous elements $u_1, \dots, u_n \in \mathrm{gr}\, U$ *right linearly independent* over $\mathrm{gr}\, R$ if $\sum u_i r_i = 0$ implies $r_i = 0$ for arbitrary $r_i \in \mathrm{gr}\, R$. The subalgebra $\mathrm{gr}\, U$ is said to be *right closed* in $\mathrm{gr}\, R$ if for any homogeneous $u_1, \dots, u_n \in \mathrm{gr}\, U$, which are right linearly independent over $\mathrm{gr}\, R$, if $\sum u_i r_i \in \mathrm{gr}\, U$, then $r_i \in \mathrm{gr}\, U$.

Similarly, $u_1, \dots, u_n \in \mathrm{gr}\, U$ are *left linearly independent* over $\mathrm{gr}\, R$ if $\sum r_i u_i = 0$ implies $r_i = 0$ for arbitrary $r_i \in \mathrm{gr}\, R$. The subalgebra $\mathrm{gr}\, U$ is said to be *left closed* in $\mathrm{gr}\, R$ if for any homogeneous $u_1, \dots, u_n \in \mathrm{gr}\, U$, which are left linearly independent over $\mathrm{gr}\, R$, if $\sum r_i u_i \in \mathrm{gr}\, U$, then $r_i \in \mathrm{gr}\, U$.

In other words, $\mathrm{gr}\, U$ is left closed in $\mathrm{gr}\, R$ if $\mathrm{gr}\, U^{\mathrm{op}}$ is right closed in $\mathrm{gr}\, R^{\mathrm{op}}$.

Proposition 1.14 *Let R be a connected filtered algebra with the weak algorithm, and let U be a subalgebra such that* $\mathrm{gr}\, U$ *with respect to the induced filtration is right or left closed in* $\mathrm{gr}\, R$. *In this case, U also has the weak algorithm.*

Proof Assume that $\mathrm{gr}\, U$ is right closed in $\mathrm{gr}\, R$, and $A = \{a_1, \dots, a_n\}$ is a d-dependent set of U. This set is also d-dependent in R:

$$d(\sum_i a_i b_i) < \max_i\{d(a_i) + d(b_i)\}, \quad a_i \in U,\ b_i \in R.$$

We may suppose that all proper subsets of A are d-independent in R, as otherwise one may diminish the number n in the above relation.

Taking the a_i's as ordered so that $d(a_1) \leq \dots \leq d(a_n)$, we claim that $\overline{a_1}, \dots, \overline{a_{n-1}}$, where $\overline{a_i} = a_i + R_{(i)} \in \mathrm{gr}\, U$ are right linearly independent over $\mathrm{gr}\, R$. Indeed, if $\sum_{i<n} \overline{a_i}\, \overline{r_i} = 0$, then $\{a_1, \dots, a_{n-1}\}$ is a proper d-dependent in R subset of A.

Since R has the weak algorithm, it follows that some a_i is d-dependent on the rest:

$$d(a_i - \sum_{k=1}^{i-1} a_k r_k) < d(a_i), \quad d(a_k) + d(r_k) = d(a_i).$$

In the homogeneous component $\mathrm{gr}_h R$, $h = d(a_i)$ this relation reduces to $\sum_{k<i} \overline{a_k}\, \overline{r_k} = \overline{a_i} \in \mathrm{gr}\, U$, which implies $\overline{r_k} \in \mathrm{gr}\, U$, $1 \leq k < i$ because $\mathrm{gr}\, U$ is right closed in $\mathrm{gr}\, R$. In terms of the algebra R, this statement reads as follows: $r_k - u_k \in R_{(d(r_k)-1)}$, $u_k \in U$. Finally, we have

$$d(a_i - \sum_{k=1}^{i-1} a_k u_k) = d(a_i - \sum_{k=1}^{i-1} a_k r_k + \sum_{k=1}^{i-1} a_k(r_k - u_k)) = d(a_i - \sum_{k=1}^{i-1} a_k r_k) < d(a_i);$$

that is, a_i is d-dependent on $a_1, \ldots, a_{i-1}$ in U.

If $\operatorname{gr} U$ is left closed in $\operatorname{gr} R$, then $\operatorname{gr} U^{\mathrm{op}}$ is right closed in $\operatorname{gr} R^{\mathrm{op}}$, where R^{op}, U^{op} are opposite algebra and subalgebra. By the above arguments, U^{op} has the weak algorithm. Since the concept of weak algorithm is left-right symmetric (see Remark 1.10), it follows that U also has the weak algorithm. □

1.8 Representation Theory and Crossed Products

The word "representation" in Representation Theory originally stands for a concrete representation of an abstract algebra (abstract group) by matrices or, equivalently, by linear transformations of a linear space V. Certainly, such a representation is equivalent to a consideration of a right module over the given algebra (correspondingly, over the group algebra of the given group). The representation theory of the symmetric group (more generally, of all finite groups) starts with the highly important classical theorem of Maschke.

Theorem 1.10 *Let G be a finite group of order n and let $\mathbf{k}$ be a field of characteristic 0 or of characteristic p where p is not a divisor of n. Then, the group algebra $\mathbf{k}[G]$ is semisimple.*

Proof See, for example, [57, Theorem 10.8], or [98, Theorem 1.4.1], or [200, Theorem 1.5.3]. □

Recall that by definition, a finite-dimensional algebra R is *semisimple* if each right (left) module over R is a direct sum of simple submodules (a nonzero module N is *simple* or *irreducible* if it has no proper submodules other than N and $\{0\}$). In terms of representations, this case is equivalent to the condition that each invariant subspace $W \subseteq V$, $WR \subseteq W$ has an invariant complement $V = W \oplus W'$, $W'R \subseteq W'$. In terms of structural theory, this case is equivalent to the radical of R being zero; that is, R has no nilpotent ideals: $I^n = 0 \Longrightarrow I = 0$.

Lemma 1.39 *Let R be a finite-dimensional semisimple algebra, and let $\rho \neq 0$ be a right (left) ideal of R. Then, $\rho = eR$ (respectively, $\rho = Re$) for some idempotent $e = e^2$ in R.*

Proof See [98, Theorem 1.4.2]. □

Corollary 1.3 *If ρ is a right (left) ideal of the group algebra $\mathbf{k}[S_n]$, then $rl(\rho) = \rho$ (respectively, $lr(\rho) = \rho$) provided that* $\operatorname{char} \mathbf{k} = 0$. *Here $l(\rho) = \{x \in R \mid x\rho = 0\}$ is the left annihilator, and $r(\rho) = \{x \in R \mid \rho x = 0\}$ is the right annihilator.*

Proof By Maschke theorem $\mathbf{k}[S_n]$ is a semisimple algebra. Lemma 1.39 states that $\rho = eR$, $e^2 = e$. The equality $xeR = 0$ implies $xe = 0$ and hence $x = x(1-e)$. In other words $l(\rho) = R(1-e)$. Similarly, the right annihilator $r(\rho)$ of a left ideal $\rho = Re$ equals $(1-e)R$. Since $1-e$ is also idempotent, it follows that $rl(eR) = eR$ and $lr(Re) = Re$. □

Interestingly, the above statement remains true even if the characteristic p is a divisor of $n!$, see [57, Theorems 61.3, 62.1].

The fundamental theorem of Wedderburn connects abstract semisimple algebras with matrix algebras.

Theorem 1.11 *A finite-dimensional algebra is semisimple if and only if it is the direct sum of simple algebras, whereas a finite-dimensional algebra is simple if and only if it is isomorphic to the algebra of all $n \times n$ matrices over a division ring.*

Proof See, [98, Theorem 1.4.4] and [98, Theorem 2.1.6]. □

Corollary 1.4 *Let M be the algebra of all $n \times n$ matrices over a field $\mathbf{F}$. Then, each right M-module is a direct sum of simple submodules, whereas all simple submodules are isomorphic to the n-rows module over $\mathbf{F}$.*

Proof By Wedderburn's theorem M is a simple algebra. Since each finite-dimensional simple algebra is semisimple, it follows that each right M-module is a direct sum of simple submodules.

If N is a simple right M-module, then the annihilator $r(N) = \{x \in M \mid Nx = 0\}$ is a two-sided ideal of M. Therefore, $r(N) = 0$. Let $e_{11} = \mathrm{diag}\,(1, 0, 0, \dots, 0)$ be the matrix with only one nonzero entire. Of course, $e_{11}M$ is precisely the n-rows module over $\mathbf{F}$. We have $Ne_{11} \neq 0$. Let us fix $n \in N$ such that $ne_{11} \neq 0$. Then, $ne_{11}M$ is a nonzero submodule of N, which implies $ne_{11}M = N$; that is, each element x of N has a representation $x = ne_{11}m,\ m \in M$. Now it is easy to check that the map $ne_{11}m \mapsto e_{11}m$ is the required isomorphism between N and $e_{11}M$. □

If, in the construction of the skew group ring given in Sect. 1.3.3, the algebra R is a field, and G is a finite group of its automorphisms of order m, then $R * G$ has a special name: a *trivial crossed product* or a *crossed product with trivial factor set*. The theory of crossed products is an important part of the theory of central simple algebras and has many of applications in modern algebra, beginning with the following beautiful result.

Theorem 1.12 *The trivial crossed product $R * G$ is isomorphic to the full algebra of $m \times m$ matrices over a Galois subfield $R^G \stackrel{df}{=} \{a \in R \mid a^g = a \text{ for all } g \in G\}$.*

Proof See, [98, Lemma 4.4.2]. □

In turn, the theory of central simple algebras starts with the following fundamental statement.

Theorem 1.13 *Let M be an algebra and M_1 be a finite-dimensional simple subalgebra with center* $\mathbf{k} \cdot 1$. *Then,* $M = M_1 \otimes Z_1$, *where*

$$Z_1 = \{m \in M \mid am = ma \text{ for all } a \in M_1\}$$

is a centralizer of M_1 in M.

Proof See, for example, [98, Theorem 4.4.2]. We are reminded that by definition, the multiplication in the tensor product of algebras, $M_1 \otimes Z_1$, is given as follows: $(m \otimes z) \cdot (m' \otimes z') = (mm' \otimes zz')$. □

1.9 Chapter Notes

The second chapter of the book [142] by Klimyk and Schmüdgen contains all basic formulas of what is commonly referred to as q-calculus, along with detailed proofs. Among the considered topics are the following: q-numbers, q-factorials, q-differentiation, basic hypergeometric functions, and q-orthogonal polynomials.

The associative standard words first appeared in an article published by Lyndon [154] in 1954 during the investigation of the Burnside problem for groups, and then in an article published by Shirshov [211] in 1958 while studying Lie algebras. The famous theorem concerning standard words, Theorem 1.1, firstly appeared in an explicit form in a paper by Schützenberger and Sherman [206, Lemma 2, p. 486], though the authors credited this result to Shirshov [211]. In fact, the combinatorics of words has arisen independently within several branches of mathematics, including number theory, group theory and probability, and appears frequently in problems related to theoretical computer science. The unified treatment of the area was specified in Lothaire's "Combinatorics on Words" [148] and again in two other books: "Algebraic Combinatorics on Words" [149] and "Applied Combinatorics on Words" [150].

What is now referred to as the Gröbner–Shirshov "bases" theory was articulated independently by Shirshov [212] in 1962 for Lie algebras explicitly and for associative algebras implicitly, by Hironaka [99] in 1964 for formal and convergent infinite series algebras, and by Buchberger [43, 44] in 1965 for commutative algebras. Buchberger named this theory in honor of his supervisor W. Gröbner. The Gröbner theory introduced by Buchberger provides a solution to the reduction problem for commutative algebras and is now included in the standard algebra curriculum in many universities. Given the ubiquity of scientific problems modeled by polynomial equations, this subject is of interest to not only mathematicians but also an increasing number of scientists and engineers, see [218, 219].

In [30], Bergman generalized the Gröbner theory to associative algebras by proving the Diamond Lemma. In [32], Bokut' noted that the parallel theory developed early for Lie algebras by Shirshov can be applied to associative algebras as well, and thus, he adapted the proofs for undergraduate students in [33]. The key component of the theory is precisely the Composition Lemma given in Theorem 1.2. We refer the reader to a survey by Bokut' and Kolesnikov [36] and to a survey by Bokut' and Chen [35] for detailed information on the modern development of this theory.

Notably, the Gröbner–Shirshov theory also applies to braid monoids. In the case of monoids (groups), the Gröbner–Shirshov method and the method of rewriting systems are equivalent. Both specify a method of constructing the normal form for words of a monoid and are a powerful tool to solve many combinatorial problems, see [34, 37, 38].

The symmetric group is undoubtedly the most important object of modern mathematics. It arises not only in algebra and combinatorics but also in physics [31], probability and statistics [61], topological graphs theory [227], the theory of partially ordered sets [216] and many other branches of modern science. As an introductory course, we recommend the book [200] written by Sagan for graduate students. The monograph [45] by Cameron provides the general method for investigating abstract groups represented by permutations, see also [57, 64, 228].

The concept of Hopf algebra appeared long before that of quantum groups in a paper by Hopf [101] on algebraic topology. This concept was rediscovered in a pure algebraic context by Kac [111]. Additionally to the basic early monographs [1, 220], a modern treatment of the Hopf algebra theory is provided in the most recent book by Radford "Hopf Algebras" [191] and in notes by Montgomery "Hopf Algebras and Their Actions on Rings" [176]. The initial chapters of the books on quantum group theory, Kassel, "Quantum Groups" [120], Joseph, "Quantum Groups and Their Primitive Ideals" [107], and Klimik, Schmüdgen, "Quantum Groups and Their Representations" [142], also contain the foundations of the Hopf algebra theory. Books written for physicists, Shneider, Sternberg, "Quantum Groups" [213], and Chaichian, Demichev, "Introduction to Quantum Groups" [46], provide a specific perspective on the subject.

The notion of a combinatorial rank appeared in [127, 128]. In the context of braided Hopf algebras it was investigated by Ardizzoni [7, 8]. In [62, 137, 138], Alvarez, Díaz Sosa, and the author determine the combinatorial rank of the Frobenius–Lusztig kernels of types A_n, B_n, and D_n to be $\lfloor \log_2 n \rfloor + 1$, $\lfloor \log_2(n-1) \rfloor + 2$, and $\lfloor \log_2(2n-3) \rfloor + 1$, respectively. There exists an infinitely generated character Hopf algebra of an infinite combinatorial rank. Whether there exists a finitely generated character Hopf algebra of infinite combinatorial rank remains unknown.

Fox [79] applied a special type of noncommutative differential calculus to various problems in topology and combinatorial group theory. The general notion of noncommutative differential calculus appeared later in the famous paper by Woronowicz [230]. The fourth part of the book by Klimyk and Schmüdgen [142] focuses on this topic. Numerous studies have focused on noncommutative differential calculi. The notion of a coordinate calculus, inspired by Wess and Zumino [226], was introduced by Borowiec, Oziewicz, and the author [39–41]. In [81, 82], Frønsdal and Galindo considered coordinate calculi with diagonal commutation rules, whereas the author in [129] generalized their results to commutation rules defined by Yang–Baxter operators.

Cohn theory [53] is one of the greatest algebra achievements of the twentieth century. This theory was inspired by Maltcev's (negative) solution of the Van der Waerden problem of embedding of a ring with no zero divisors in a skew field [161], followed by discovery of conditions when a monoid can be embedded into a group. The necessary and sufficient conditions have been revealed to be so complicated that they may not be able to be expressed as a finite number of elementary axioms [162, 163]. In turn, P.M. Cohn has found necessary and sufficient conditions for a ring to be embeddable in a skew field [52, 53]. His theory of matrix-inverting homomorphisms and matrix ideals is devised as if created for the solution of the problem concerning embedding of a bialgebra into a Hopf algebra.

Chapter 2
Poincaré-Birkhoff-Witt Basis

Abstract In this chapter, we demonstrate that every character Hopf algebra has a PBW basis. A Hopf algebra H is referred to as a character Hopf algebra if the group G of all group-like elements is commutative and H is generated over $\mathbf{k}[G]$ by skew-primitive semi-invariants, whereas a well-ordered subset $V \subseteq H$ is a set of PBW generators of H if there exists a function $h : V \to \mathbf{Z}^+ \cup \{\infty\}$, called the height function, such that the set of all products

$$gv_1^{n_1}v_2^{n_2}\cdots v_k^{n_k},$$

where $g \in G,\ v_1 < v_2 < \ldots < v_k \in V,\ n_i < h(v_i), 1 \leq i \leq k$ is a basis of H.

In this chapter, we demonstrate that every character Hopf algebra has a PBW basis. According to Definition 1.11, a Hopf algebra H is referred to as a character Hopf algebra if the group G of all group-like elements is commutative and H is generated over $\mathbf{k}[G]$ by skew-primitive semi-invariants.

Definition 2.1 A well-ordered subset V of a character Hopf algebra H is considered a set of *PBW generators* of H if there exists a function $h : V \to \mathbf{Z}^+ \cup \{\infty\}$, called the *height function*, such that the set of all products

$$gv_1^{n_1}v_2^{n_2}\cdots v_k^{n_k}, \tag{2.1}$$

where $g \in G,\ v_1 < v_2 < \ldots < v_k \in V,\ n_i < h(v_i), 1 \leq i \leq k$ is a basis of H. The value $h(v)$ is referred to as the *height* of v in V.

For example, the standard words, due to Theorem 1.1, form a set of PBW generators with infinite heights of the free character Hopf algebra $G\langle X\rangle$. This fact provides an idea concerning how to find the PBW basis of an arbitrary character Hopf algebra.

We establish a homomorphism $G\langle X\rangle \to H$ of the character Hopf algebras. The values of elements (2.1) in H span all of H but may be linearly dependent. If the value of a standard word v is a linear combination of the monomials (2.1) with $v_i < v$, then the values of elements (2.1), where $v_i \neq v$, continue to span H. Hence, the set of all standard words may be reduced to the set of "hard" standard words,

V. Kharchenko, *Quantum Lie Theory*, Lecture Notes in Mathematics 2150,
DOI 10.1007/978-3-319-22704-7_2

i.e., standard words v whose values in H are not linear combinations of (2.1) with $v_i < v$.

Then, one must demonstrate that the increasing products of "hard" standard words are linearly independent in H. For this task, we must use the coproduct. If U is such a linear combination, then we may (somehow) find its coproduct in the free character Hopf algebra

$$\Delta(U) = U \otimes 1 + \sum U_i' \otimes U_i'' + g \otimes U, \quad g \in G.$$

If $U = 0$ in H, then in $H \otimes H$ we have the equality

$$\sum U_i' \otimes U_i'' = 0. \tag{2.2}$$

This equality of tensors provides one equation corresponding to each basis element of the space spanned by all U_i''. Because the U_i'''s have degrees less than that of U, we may theoretically decompose them in linear combinations of increasing products of "hard" standard words that are already linearly independent in H (by induction). This amount of information is sufficient for obtaining the required contradiction.

Because of technical reasons, it was impossible to realize these considerations directly for "hard" standard words; Instead, developing the above logic for nonassociative standard words seemed possible, interpreting the bracket as the skew commutator of polynomials. Surprisingly, after this logic was developed, demonstrating that the "hard" standard words are indeed the PBW generators became straightforward.

The equality (2.2) is not equivalent to setting U to be zero but does indicate that U is skew-primitive. In other words, while solving the above system of equations, we will obtain information on the skew-primitive elements of character Hopf algebras. This information is given in Theorem 2.3.

2.1 PBW Bases of the Free Character Hopf Algebra

Let $G\langle Y\rangle = G\langle X\rangle$ be the free character Hopf algebra, see Sect. 1.5.3. Recall that x_i, $i \in I$ are free variables with the coproduct given by

$$\Delta(x_i) = x_i \otimes 1 + g_i \otimes x_i, \quad \Delta(g_i) = g_i \otimes g_i, \tag{2.3}$$

whereas associated with each variable x_i is a character $\chi^i : G \to \mathbf{k}^*$ such that $g^{-1}x_i g = \chi_i(g) g x_i$, for all $g \in G$, see (1.66).

For every word u in X let g_u denotes a group-like element that appears from u by replacing each x_i with g_i. Similarly, χ^u is a character that appears from u by replacing each x_i with χ^i. Because both the group G and the group of characters are commutative, the values g_u, χ^u are defined on the set of all homogeneous elements in each $x_i \in X$. For a pair u, v of homogeneous polynomials in X put

$$p_{u,v} = \chi^u(g_v). \tag{2.4}$$

Obviously, the following equalities hold:

$$p_{uv,w} = p_{u,w}p_{v,w}, \quad p_{u,vw} = p_{u,v}p_{u,w}. \tag{2.5}$$

Sometimes it is more convenient to denote this bimultiplicative operator by $p(u, v)$. Of course, the operator $p(\text{-}, \text{-})$ is completely defined by the parameters $p_{ik} = \chi^i(g_k)$.

In terms of this operator, the brackets (1.67) take the form

$$[u, v] = uv - p_{u,v}vu, \quad [u, v]^* = uv - p_{v,u}^{-1}vu. \tag{2.6}$$

Lemma 2.1 *The brackets* $[,]$ *satisfy the following "Jacobi identity":*

$$[[u, v], w] = [u, [v, w]] + p_{w,v}^{-1}[[u, w], v] + (p_{v,w} - p_{w,v}^{-1})[u, w] \cdot v, \tag{2.7}$$

where $\cdot$ *stands for usual multiplication in the free algebra.*

Proof We have

$$[[u, v], w] = [uv - p_{u,v}vu, w] = uvw - p_{uv,w}wuv - p_{u,v}vuw + p_{u,v}p_{vu,w}wvu.$$

Under the substitution $w \leftrightarrow v$, this equality becomes

$$[[u, w], v] = uwv - p_{uw,v}vuw - p_{u,w}wuv + p_{u,w}p_{wu,v}vwu.$$

Similarly,

$$[u, [v, w]] = [u, vw - p_{v,w}wv] = uvw - p_{u,wv}vwu - p_{v,w}uwv + p_{v,w}p_{u,wv}wvu,$$

and

$$[u, w] \cdot v = uwv - p_{u,w}wuv.$$

It remains to compare the coefficients at all six permutations of uvw in (2.7).

$$uvw: \quad 1 = 1;$$

$$wuv: \quad -p_{uv,w} = -p_{w,v}^{-1}p_{u,w} + (p_{v,w} - p_{w,v}^{-1})p_{u,w};$$

$$vuw: \quad -p_{u,v} = -p_{w,v}^{-1}p_{uw,v};$$

$$wvu: \quad p_{u,v}p_{vu,w} = p_{v,w}p_{u,wv};$$

$$uwv: \quad 0 = p_{w,v}^{-1} - p_{v,w} + (p_{v,w} - p_{w,v}^{-1});$$
$$vwu: \quad 0 = p_{w,v}^{-1} p_{u,w} p_{wu,v} - p_{u,vw}.$$

□

Lemma 2.2 *The following formulas link the brackets to multiplication:*

$$[u, v \cdot w] = [u, v] \cdot w + p_{u,v} v \cdot [u, w], \tag{2.8}$$

$$[u \cdot v, w] = p_{v,w} [u, w] \cdot v + u \cdot [v, w]. \tag{2.9}$$

Proof We have, $[u, v \cdot w] = uvw - p_{u,vw} vwu = uvw - p_{u,v} vuw + p_{u,v} vuw - p_{u,v} p_{u,w} vwu = [u, v] \cdot w + p_{u,v} v \cdot [u, w]$. Similarly, $[u \cdot v, w] = uvw - p_{uv,w} wuv = uvw - p_{v,w} uwv + p_{v,w} uwv - p_{uv,w} wuv = u \cdot [v, w] + p_{v,w} [u, w] \cdot v$. □

Definition 2.2 A *super-letter* is a polynomial that equals a standard nonassociative word where the brackets [,] are defined in (2.6).

Every noncommutative polynomial f in X is a linear combination of different words $f = \sum \alpha_i u_i$. Recall that a leading word of f is the maximal word u_i that occurs in this decomposition with nonzero coefficient.

Lemma 2.3 *A leading word of a super-letter* $[u]$ *with respect to the lexicographical order is the word* u*, and it occurs in the decomposition of* $[u]$ *with coefficient 1.*

Proof We use induction on length. If $[u] = [[v][w]]$ then the super-letter $[u]$ equals $[v][w] - p_{u,w}[w][v]$. By the inductive hypothesis, $[v]$ and $[w]$ are homogeneous polynomials with the leading words v and w, respectively. The leading word with respect to the lexicographical order of a product of two homogeneous polynomials equals the product of leading words of the factors. Therefore, the leading word of $[v][w]$ equals vw and has coefficient 1; the leading word of $[w][v]$ equals wv and is less than vw because $vw = u$ is a standard word. □

The proven Lemma demonstrates that different standard words u and v define distinct super-letters $[u]$ and $[v]$. We define the order on the set of all super-letters thus:

$$[u] > [v] \iff u > v. \tag{2.10}$$

Definition 2.3 A word in super-letters is called a *super-word*. A super-word is said to be *increasing* if it has the form

$$W = [u_1]^{k_1} [u_2]^{k_2} \cdots [u_m]^{k_m}, \tag{2.11}$$

where $u_1 < u_2 < \ldots < u_m$. On the set of all super-words, we fix the lexicographic order defined by the ordering of super-letters in (2.10).

Lemma 2.4 *An increasing super-word* $W = [w_1]^{k_1}[w_2]^{k_2}\cdots[w_m]^{k_m}$ *is greater than an increasing super-word* $V = [v_1]^{m_1}[v_2]^{m_2}\cdots[v_k]^{m_k}$ *if and only if the word* $w = w_1^{k_1}w_2^{k_2}\cdots w_m^{k_m}$ *is greater than the word* $v = v_1^{m_1}v_2^{m_2}\cdots v_k^{m_k}$. *Moreover, the leading word of the polynomial* W, *when decomposed into a linear combinations of words, equals* w *and has coefficient 1.*

Proof Let $W > V$. Then $w_1 \geq v_1$ in view of the ordering of super-letters. If $w_1 = v_1$, we can remove one factor from the left of both V and W, and then proceed by induction. Therefore, we will put $w_1 > v_1$. If w_1 is not the beginning of v_1, then the inequality $w_1 > v_1$ can be multiplied from the right by suitable distinct elements, which yields $w > v$, as required.

Let $v_1 = w_1T$, $T = (w_1^{k_1-1}w_2^{k_2}\cdots w_{s-1}^{k_{s-1}})w_s^l \cdot v_1'$, where $0 \leq l < k_s$. Here w_s is not a beginning of v_1', whereas the term between the parentheses may be missing (in this case $s = 1, l > 0$).

If v_1' is a nonempty word, then $v_1' < v_1 < w_1 \leq w_s$ because v_1 is standard. The inequality $v_1' < w_s$ implies $av_1'b < aw_sc$ for all words a, b, c because w_s is not a beginning of v_1'. Taking $a = (w_1^{k_1}w_2^{k_2}\cdots w_{s-1}^{k_{s-1}})w_s^l$ and suitable b, c, we obtain $v < w$.

Let v_1' is the empty word. If $l > 0$, then the word v_1 should be greater than its end w_s. Therefore, $w_1 > v_1 > w_s$, which contradicts the fact that $w_1 \leq w_s$ is valid for all $s \geq 1$. If $l = 0$, then $s > 1$ because v_1 begins with w_1. It follows that v_1 is greater than its end w_{s-1}, which is again a contradiction with $w_1 > v_1 > w_{s-1}$.

The second part of the lemma follows from Lemma 2.3 and the fact that the leading word of a product of homogeneous polynomials equals the product of leading words of the factors. □

Remark 2.1 We stress that the above lemma cannot be extended to all super-words, for example if $x_1 > x_2 > x_3$, then $[x_1]\cdot[x_3] > [x_1x_2]$ and $x_1x_3 < x_1x_2$.

Lemma 2.5 *Let* u, u_1 *be standard words and* $u > u_1$. *The polynomial* $[[u],[u_1]]$ *is a linear combination of super-words in the super-letters* $[w]$ *such that* $uu_1 \geq w > u_1$, *in which case the constitution of the super-words equals the constitution of* uu_1.

Proof If the nonassociative word $[[u][u_1]]$ is standard then it defines a super-letter $[w]$ and $uu_1 = w > u_1$ by Lemma 1.4. In particular, the lemma is valid if u and u_1 are letters. We can therefore proceed by induction on the length of uu_1.

Suppose that the lemma is true if the length of uu_1 is less than m. Choose a pair u, u_1 with a greatest word u, so that the polynomial $[[u],[u_1]]$ does not enjoy the required decomposition and the length of uu_1 equals m. Then the nonassociative word $[[u][u_1]]$ is not standard. By Lemma 1.10, we have $[u] = [[u_3][u_2]]$ with $u_2 > u_1$.

We fix the notation for super-letters $U_i = [u_i]$, $i = 1, 2, 3$. By Jacobi identity (2.1), we can write

$$[[U_3, U_2], U_1] = [U_3, [U_2, U_1]] + p_{u_1,u_2}^{-1}[[U_3, U_1], U_2] \\ +(p_{u_2,u_1} - p_{u_1,u_2}^{-1})[U_3, U_1] \cdot U_2. \qquad (2.12)$$

We have $u_3 > u > u_2 > u_1$. By the inductive hypothesis, $[U_3, U_1]$ can be represented as $\sum_i \alpha_i \prod_k [w_{ik}]$, where $u_3 > u_3u_1 \geq w_{ik} > u_1$. Using Lemma 1.7, we obtain $u > uu_1 > u_3u_1 \geq w_{ik}$; that is, all super-letters $[w_{ik}]$ satisfy the requirements of the present lemma. Furthermore, the word u cannot be the beginning of u_2, and so $u > u_2$ implies $uu_1 > u_2$. Thus, the super-letter U_2, too, satisfies the requirements. Consequently, the second [in view of (2.6)] and third summands of (2.12) have the required decomposition.

Using the inductive hypothesis, for the first summand we obtain

$$[U_2, U_1] = \sum_i \beta_i \prod_k [v_{ik}], \qquad (2.13)$$

where $u_2u_1 \geq v_{ik} > u_1$. By Lemma 1.7, $uu_1 > u_2u_1 \geq v_{ik}$; that is, the super-letters $[v_{ik}]$ satisfy the conditions of the lemma. Rewrite the first summand using skew-derivation formula (2.8), with the first factor replaced by (2.13). In this way, the first summand turns into a linear combination of words in the super-letters $[v_{ik}]$ and skew commutators $[[u_3], [v_{ik}]]$. Because $u_3 > u > u_2 > v_{ik}$ and the length of v_{ik} does not exceed that of u_2u_1, the inductive hypothesis applies to yield

$$[[u_3], [v_{ik}]] = \sum_j \gamma_j \prod_t [w_{jt}], \qquad (2.14)$$

where $u_3 > u_3v_{ik} \geq w_{jt} > v_{ik}$. In this case $u_2u_1 \geq v_{ik}$ implies

$$uu_1 = u_3u_2u_1 \geq u_3v_{ik} \geq w_{jt};$$

in addition, $w_{jt} > v_{ik} > u_1$, i.e., the super-letters $[w_{jt}]$ also satisfy the conditions. □

Lemma 2.6 *Every nonincreasing super-word W is a linear combination of lesser increasing super-words of the same constitution whose super-letters all lie (not strictly) between the greatest and the least super-letters of W.*

Proof We proceed by induction on the length of the super-word. Assume that the lemma is true for super-words of length $\leq t$, and let $W = UU_1 \cdots U_t$ be a least super-word of length $t + 1$ for which our lemma fails.

If the super-word $U_1 \cdots U_t$ is not increasing, then by the inductive hypothesis it is a linear combination of lesser increasing super-words W_i. In this case $UW_i < W$, and according to the choice of W, all super-words UW_i have the required representation.

Let

$$W = UU_1^{k_1} \cdots U_t^{k_t}, \quad U_1 < U_2 < \ldots < U_t. \tag{2.15}$$

If $U \leq U_1$, then W is increasing, and there is nothing to prove. Let $U > U_1$. Then

$$W = [U, U_1]U_1^{k_1-1} \cdots U_t^{k_t} + p_{u,u_1} U_1 U U_1^{k_1-1} \cdots U_t^{k_t}. \tag{2.16}$$

The second summand is less than W as a super-word, and so we can write it in the required form. By Lemma 2.5, the factor $[U, U_1]$ in the first term can be represented as $\sum_i \alpha_i \prod_s [w_{is}]$, where the super-letters $[w_{is}]$ are less than U. Consequently, the super-words $\prod_s [w_{is}] U_1^{k_1-1} \cdots U_t^{k_t}$ are less than W; that is, the first term has the required representation too. □

Theorem 2.1 *The set of all super-words*

$$[u_1]^{n_1} [u_2]^{n_2} \cdots [u_k]^{n_k}, \tag{2.17}$$

where $u_1 < u_2 < \ldots < u_k$ are standard words, forms a basis of $\mathbf{k}\langle X \rangle$.

Proof Since by definition all words of length one are standard, the letters $x_i = [x_i]$ are super-letters. Hence, by Lemma 2.6, every polynomial is a linear combination of increasing super-words. It remains to prove that the set of all increasing super-words is linearly independent. Let

$$\sum_i \alpha_i W_i = 0 \tag{2.18}$$

and assume that $W = [w_1]^{k_1} [w_2]^{k_2} \cdots [w_m]^{k_m}$ is a leading super-word in (2.18). By Lemma 2.4, the leading word of W equals $w = w_1^{k_1} w_2^{k_2} \cdots w_m^{k_m}$. This word occurs exactly once in (2.18). Suppose, to the contrary, that W does also occur in the decomposition of $V = [v_1]^{m_1} [v_2]^{m_2} \cdots [v_k]^{m_k}$. Then the word w is less than or equal to the leading word $v = v_1^{m_1} v_2^{m_2} \cdots v_k^{m_k}$ in the decomposition of V, which contradicts the fact that $W > V$ by Lemma 2.4. □

2.2 Coproduct on Super-Letters

Theorem 2.1 demonstrates that the super-letters are PBW generators of infinite height for the free character Hopf algebra $G\langle X \rangle$. Our next goal is to describe properties of the coproduct of these PBW generators.

Lemma 2.7 *The coproduct of a super-letter $W = [w]$ has a representation*

$$\Delta([w]) = [w] \otimes 1 + g_w \otimes [w] + \sum_i \alpha_i g(W_i'')W_i' \otimes W_i'', \tag{2.19}$$

where W_i' are nonempty words in less super-letters than is $[w]$. Moreover, the sum of constitutions of W_i' and W_i'' equals the constitution of V. Here $g(u)$ denotes the group-like element g_u.

Proof We use induction on the length of a word w. For letters, there is nothing to prove. Let $W = [U, V]$, $U = [u]$, and $V = [v]$. Assume that the decompositions

$$\Delta(U) = U \otimes 1 + g_u \otimes U + \sum_i \alpha_i g(U_i'')U_i' \otimes U_i'', \tag{2.20}$$

and

$$\Delta(V) = V \otimes 1 + g_v \otimes V + \sum_j \beta_j g(V_j'')V_j' \otimes V_j'' \tag{2.21}$$

satisfy the requirements of the lemma. Using (2.6) and properties of p, we can write

$$\begin{aligned}\Delta(W) = \Delta(U)\Delta(V) - p_{u,v}\Delta(V)\Delta(U) &= W \otimes 1 + g_w \otimes W \\ &+ (1 - p_{u,v}p_{v,u})g_u V \otimes U + \sum \beta_j p(U, V_j'')g(V_j'')[U, V_j'] \otimes V_j'' \\ &+ \sum \beta_j g_u g(V_j'')V_j' \otimes (UV_j'' - p_{u,v}p(V_j', U)V_j''U) \\ &+ \sum \alpha_i g(U_i'')(U_i' \cdot V - p_{u,v}p(V, U_i'')V \cdot U_i') \otimes U_i'' \\ &+ \sum \alpha_i p(U_i', V)g_v g(U_i'')U_i' \otimes [U_i'', V] \\ &+ \sum \alpha_i \beta_j g(U_i''V_j'')(p(U_i', V_j'')U_i'V_j' \otimes U_i''V_j'' \\ &- p_{u,v}p(V_j', U_i'')V_j'U_i' \otimes V_j''U_i'').\end{aligned} \tag{2.22}$$

Collecting similar terms in this formula was result in the canceling of terms of the form $g_v U \otimes V$ only. We claim that all left parts of the remaining tensors in (2.22) admit the required decomposition. First, in view of the inductive hypothesis, all super-letters of all super-words V_j' are less than V, which are in turn less than W because v is the end of a standard word w. Moreover, by the inductive hypothesis again, u cannot be the beginning of any word u' such that the super-letter $[u']$ would occur in super-words U_i'. Therefore, $u > u'$ implies $uv > u'$ and $W > [u']$. Thus, all but the first and fourth super-words on the left-hand sides of all tensors depend only on super-letters which are less than W.

We want to apply Lemma 2.5 to the fourth tensor. Let $V'_j = \prod_k V_{ik}$, where $V_{ik} = [v_{ik}]$ are less than V. By Eq. (2.8), the polynomial $[U, V'_j]$ is a linear combination of words in the super-letters V_{ik} and skew commutators $[U, V_{ik}]$. By Lemma 2.5, each of these commutators is a linear combination of words in the super-letters $[v']$ such that $v' \leq uv_{ik}$. In view of $v_{ik} < v$, we obtain $v' < uv = w$.

The statement concerning the constitutions follows immediately from formula (2.22) and the inductive hypothesis. □

Lemma 2.8 *The coproduct of a super-word W has a decomposition*

$$\Delta(W) = W \otimes 1 + g(W) \otimes W + \sum_i \alpha_i g(W''_i) W'_i \otimes W''_i, \tag{2.23}$$

where the sum of constitutions of W'_i and W''_i equals the constitution of W.

Proof It suffices to observe that Δ is an homomorphism of algebras. Here, we can no longer assert that $W'_i < W$. □

Lemma 2.9 *If $[w]$ is a super-letter, then*

$$\Delta([w]^m) = \sum_{j=0}^{m} \begin{bmatrix} m \\ j \end{bmatrix}_q g_w^{m-j} [w]^j \otimes [w]^{m-j} + \sum_i \alpha_i g(V_i) U_i \otimes V_i, \tag{2.24}$$

where $\begin{bmatrix} m \\ j \end{bmatrix}_q$ are the Gauss polynomials considered in Sect. 1.1 with $q = p(w, w)$, whereas the super-words U_i are less than $[w]^m$ with respect to the lexicographical ordering of words in super-letters.

Proof After developing of the product, the mth power of the right hand side of (2.19) takes the form (2.24), where each of U_i is a product of m super-words some of whom equal to $[w]$ (but not all of them!) and others equal to some of the W'_i's. By Lemma 2.7, all super-letters that occur in W_i are less than $[w]$. Hence, the super-word U_i is less than $[w]^m$ with respect to the lexicographical ordering of words in super-letters. □

2.3 Hard Super-Letters

Consider a character Hopf algebra H. By definition H is generated over $\mathbf{k}[G]$ by skew-primitive semi-invariants b_i, $i \in I$:

$$\Delta(b_i) = b_i \otimes h_i + f_i \otimes b_i, \ h_i, f_i \in G, \quad b_i g = \chi^{b_i}(g) \cdot g b_i, \ g \in G, \ i \in I. \tag{2.25}$$

As the skew-primitive elements are closed with respect to the multiplication by group-like elements, we may normalize the generators, $a_i = h_i^{-1} b_i$, diminishing the number of group-like elements related to them:

$$\Delta(h_i^{-1}b_i) = h_i^{-1}u \otimes 1 + h_i^{-1}f_i \otimes h_i^{-1}b_i.$$

In what follows, we fix a set of *normalized skew-primitive* generators $\{a_i\}$, so that

$$\Delta(a_i) = a_i \otimes 1 + g_i \otimes a_i, \;\; \Delta(g_i) = g_i \otimes g_i, \;\; a_i g = \chi^{a_i}(g) \cdot g a_i, \;\; g \in G, \; i \in I. \tag{2.26}$$

Let $G\langle X\rangle$, $X = \{x_i \mid i \in I\}$ be the free character Hopf algebra such that $\chi^i = \chi^{a_i}$ and $g_i = g_{a_i}$, $i \in I$. Then there exists a natural homomorphism of Hopf algebras

$$\varphi : G\langle X\rangle \to H, \tag{2.27}$$

which maps x_i to a_i, $i \in I$.

Definition 2.4 Let Γ be a well-ordered additive (commutative) monoid. With each x_i, $i \in I$ we associate a nonzero element $d_i \in \Gamma$. The *D-degree* of a word, a super-letter, a super-word, or more generally, a homogeneous polynomial f in X of a constitution $\{m_i \mid i \in I\}$ is

$$D(f) = \sum_i m_i d_i = \sum_i d_i \deg_i(f). \tag{2.28}$$

In what follows, we fix a well-ordered monoid Γ and elements $d_i = D(x_i)$. For example, Γ may be the monoid related to the constitution given in the construction after Definition 1.3. For the first reading, one may suppose that $\Gamma = Z^+$ is the monoid of nonnegative integer numbers, whereas $d_i = 1$. However, we should stress that the resulting set of PBW generators and its properties essentially depend on the chosen D-degree function.

Lemma 2.10 *The set X_m^* of all words of a fixed D-degree m is well-ordered with respect to the lexicographical order.*

Proof We note, first, that Γ has no negative elements: if $a < 0$, then there appears an infinite descending chain $0 > a > 2a > 3a > \dots$. Additionally, Γ has the cancelation property, $a + x = a + y$ implies $x = y$: if $x > y$, then $a + x > a + y$.

Let F be a subset of X_m^*. As $\langle X, <\rangle$ is well-ordered, the set A of all first letters of words from F has a least element, say, $x_1 \in X$. If $x_1 u$, $x_1 v \in F$, then $D(x_1) + D(u) = D(x_1) + D(v) = m$. Hence, $D(u) = D(v) < m$ because $D(v) \geq m$ and $D(x_1) > 0$ would imply $D(x_1) + D(v) > m$. By these reasons, we may apply the induction supposition to the set $B = \{u \in X^* \mid x_1 u \in F\}$. If u_0 is a least element of B, then $x_1 u_0$ is a least element of F. □

Definition 2.5 A *G-super-word* is a product of the form gW, where $g \in G$ and W is a super-word. The degree, constitution, length, and other concepts which apply with G-super-words are defined by the super-word W. In other words, we assume that the D-degree and the constitution of $g \in G$ are equal to zero. In view of (2.26), every product of super-letters and group-like elements equals a linear combination of G-super-words of the same constitution.

Definition 2.6 A super-letter $[u]$ is said to be *hard* if its value $\varphi([u])$ in H is not a linear combination of values of words of the same D-degree in less super-letters than is $[u]$ and of G-super-words of a lesser D-degree.

We are remanded that a primitive tth root of 1 is an element $\alpha \in \mathbf{k}$ such that $\alpha^t = 1$ and $\alpha^r \neq 1$ for all r, $1 \leq r < t$. In particular, 1 is the 1st primitive root of 1.

Definition 2.7 We say that the *height* of a super-letter $[u]$ of D-degree $d \in \Gamma$ equals $h = h([u])$ if h is the smallest natural number such that:

(1) $p_{u,u}$ is a primitive tth root of 1 and either $h = t$ or $h = tl^r$, where l is the characteristic of $\mathbf{k}$.
(2) the value in H of $[u]^h$ is a linear combination of values of super-words of D-degree hd in less super-letters than is $[u]$ and of G-super-words of a lesser D-degree.

If, for the super-letter $[u]$, the number h with the above properties does not exist, then we say that the height of $[u]$ is infinite.

Theorem 2.2 *The set of values in H of all G-super-words W in the hard super-letters $[u_i]$,*

$$W = g[u_1]^{n_1}[u_2]^{n_2}\cdots[u_k]^{n_k}, \tag{2.29}$$

where $g \in G$, $u_1 < u_2 < \ldots < u_k$, $n_i < h([u_i])$, forms a basis of H.

The proof will proceed through a number of lemmas. For brevity, we call a G-super-word (2.29) *restricted* if each of the numbers n_i is less than the height of $[u_i]$. A super-word (a G-super-word) is said to be *admissible* if it is increasing restricted and is a word in hard super-letters only.

First of all, we have to demonstrate that every element of H is a linear combination of values of admissible G-super-words. Clearly, every element is a linear combination of values of not necessarily admissible G-super-words because each variable x_i is a super-letter, $x_i = [x_i]$. In fact, there exist a natural diminishing procedure, based on Lemma 2.5 and on the definitions of hard super-letters and their heights, that allows one to find the required linear combination.

Lemma 2.11 *The value of each non-admissible super-word of D-degree d is a linear combination of values of lesser admissible super-words of D-degree d and of admissible G-super-words of a lesser D-degree. Also, all super-letters occurring in the super-words of D-degree d of this linear combination are less than or equal to a greatest super-letter of the super-word given.*

Proof Assume that the lemma is valid for super-words of D-degree $< m$. Let W be a least super-word of D-degree m for which the required representation fails. By Lemma 2.6, the super-word W is increasing. If it has a non-hard super-letter, by definition, we can replace it with a linear combination of G-super-words of a lesser D-degree and of words in less super-letters of the same D-degree. Developing the product turns W into a linear combination of G-super-words of a lesser D-degree and of lesser super-words of the same D-degree, a contradiction with the choice of W. If W contains a subword $[u]^k$, where k equals the height of $[u]$, then we can replace it as is specified above, which gives us a contradiction again. Thus the W is itself increasing restricted and is a word in hard super-letters only. □

In order to prove Theorem 2.2, it remains to show that admissible G-super-words are linearly independent. Consider an arbitrary linear combination $\mathbf{T}$ of admissible G-super-words and let $U = V_1^{n_1} V_2^{n_2} \cdots V_k^{n_k}$ be its leading (maximal) super-word of D-degree m. Multiplying, if necessary, that combination by a group-like element, we can assume that U occurs once without a group-like element:

$$\mathbf{T} = U + \sum_{j=1}^{r} \alpha_j g_j U + \sum_{\mathbf{i}=(i1,i2,\ldots,is)} \alpha_{\mathbf{i}}\, g_{\mathbf{i}}\, W_{\mathbf{i}}, \quad W_{\mathbf{i}} = V_{i1}^{n_{i1}} V_{i2}^{n_{i2}} \cdots V_{is}^{n_{is}}. \tag{2.30}$$

In the next three lemmas, we accept the following assumptions on m, U and r:

1. The admissible G-super-words of D-degree $< m$ are linearly independent;
2. The admissible G-super-words of D-degree m which are less than U are linearly independent modulo the space spanned by G-super-words mentioned in 1;
   and, if $r > 0$, then
3. The super-words $g_j U$, $1 \le j \le r$ are linearly independent modulo the space spanned by G-super-words mentioned in 1 and 2.

In view of these assumptions and Lemma 2.11, every super-word of D-degree m which is less than U, and every super-word of D-degree $< m$, can be *uniquely* decomposed into a linear combination of admissible G-super-words. For brevity, such will be referred to as a *basis* decomposition.

Lemma 2.12 *Under the assumptions 1, 2, 3, if the value of* $\mathbf{T}$ *in* H *is a skew-primitive element, then* $r = 0$ *and* $g_{\mathbf{i}} = 1$ *for all* $\mathbf{i}$ *such that* $D(W_{\mathbf{i}}) = m$.

Proof Rewrite the linear combination $\mathbf{T}$ as follows:

$$\mathbf{T} = U + \sum_{\mathbf{i} \in I} \alpha_{\mathbf{i}} g_{\mathbf{i}} W_{\mathbf{i}} + W', \tag{2.31}$$

where $g_{\mathbf{i}} W_{\mathbf{i}}$ are distinct G-super-words of D-degree m in (2.30) (including $\alpha_j\, g_j\, U$) and W' is a linear combination of G-super-words of D-degree $< m$. In the expression

$$\Delta(\mathbf{T}) - \mathbf{T} \otimes h_t - f_t \otimes \mathbf{T}, \quad h_t, f_t \in G \tag{2.32}$$

consider all tensors of the form $gW \otimes \dots$, where $D(W) = m$. By Lemma 2.8, the sum of all such tensors equals

$$\sum_{\mathbf{i}\in I} \alpha_{\mathbf{i}} g_{\mathbf{i}} W_{\mathbf{i}} \otimes g_{\mathbf{i}} - \sum_{\mathbf{i}\in I} \alpha_{\mathbf{i}} g_{\mathbf{i}} W_{\mathbf{i}} \otimes 1 = \sum_{\mathbf{i}\in I} \alpha_{\mathbf{i}} g_{\mathbf{i}} W_{\mathbf{i}} \otimes (g_i - 1). \quad (2.33)$$

By assumptions 1, 2, 3, the elements $g_{\mathbf{i}} W_{\mathbf{i}}$, $\mathbf{i} \in I$ are linearly independent modulo all left parts of tensors of D-degree $< m$ in (2.32). Therefore, if (2.32) vanishes in H, then either $\alpha_{\mathbf{i}} = 0$ or $g_{\mathbf{i}} = 1$ for every $\mathbf{i} \in I$, as required. □

Lemma 2.13 *Under the assumptions 1, 2, 3, if* $\mathbf{T}$ *is a skew-primitive element, then* $U = [u]^h$ *and all super-words of D-degree m except U are words in less super-letters than* $[u]$ *is.*

Proof By the preceding lemma, we can assume that

$$\mathbf{T} = \sum_{\mathbf{i}=(i1,i2,\dots,is)} \alpha_{\mathbf{i}} g_{\mathbf{i}} W_{\mathbf{i}}, \; W_{\mathbf{i}} = V_{i1}^{n_{i1}} V_{i2}^{n_{i2}} \cdots V_{is}^{n_{is}}, \quad (2.34)$$

where one of the $W_{\mathbf{i}}$'s is U, whereas $V_{ij} = [v_{ij}]$ are hard super-letters, $\alpha_{\mathbf{i}}$ are nonzero coefficients, and $g_{\mathbf{i}} = 1$ if $W_{\mathbf{i}}$ is of D-degree m. By Lemma 2.7, we have

$$\Delta(g_{\mathbf{i}} W_{\mathbf{i}}) = (g_{\mathbf{i}} \otimes g_{\mathbf{i}}) \prod_{j=1}^{s} (V_{ij} \otimes 1 + g_{ij} \otimes V_{ij} + \sum_{\theta} g_{ij\theta} V'_{ij\theta} \otimes V''_{ij\theta})^{n_{ij}}, \quad (2.35)$$

where $V'_{ij\theta} < V_{ij}$ and $\deg V'_{ij\theta} + \deg V''_{ij\theta} = \deg V_{ij}$.

Let $[u]$ be the greatest super-letter occurring in super-words of D-degree m in (2.34). Because all super-words of (2.34) are increasing, this super-letter stands at the end of some super-words $W_{\mathbf{i}}$, i.e., $[u] = V_{is}$. If one of these super-words depends only on $[u]$; that is, $W_{\mathbf{i}} = [u]^h$, then $W_{\mathbf{i}}$ is a leading super-word, $W_{\mathbf{i}} = U$ as required. Therefore, we assume that every super-word of D-degree m ending with $[u]$ is a word in more than one different super-letters.

Let $h = n_{is}$ be the largest exponent of $[u]$ in (2.34). Consider all tensors of the form $g[u]^k \otimes \dots$ obtained in (2.35) by removing the parentheses and applying the basis decomposition to all left parts of tensors in all terms except $\mathbf{T} \otimes 1$ (all of these terms are of D-degree $< m$).

All left parts of tensors which appear in (2.35) removing the parentheses arise from the G-super-word $g_i V_{i1}^{n_{i1}} V_{i2}^{n_{i2}} \cdots V_{is}^{n_{is}}$ by replacing some of the super-letters V_{ij} either with group-like element g_{ij} or with G-super-word $g_{ij\theta} V'_{ij\theta}$ of a lesser D-degree in less super-letters. The right parts are, respectively, products obtained by replacing super-letters V_{ij} with super-words $V''_{ij\theta}$ multiplied from the left by g_i.

Let $gR \otimes g'S$ be a resulting tensor under the replacements above and followed then basis decomposition.

If $D(R) < hD(u)$, then its basis decomposition may give rise to terms of the form $g[u]^k \otimes \ldots$. In this case, however, $D(S) < (m-h)D(u)$ because the sum of D-degrees of both parts of the tensors either remains equal to m or decreases.

If $D(R) < hD(u)$, or R is itself less than $[u]^h$ as a super-word, then the basis decomposition of R have no terms of the form $g[u]^h$; see Lemma 2.9.

If $D(R) = hD(u)$, while $D(W_{\mathbf{i}}) < m$, then R can be greater than or equal to $[u]^h$, but in this case $D(S) < (m-h)D(u)$ because $D(R) + D(S) \leq D(W_{\mathbf{i}}) < m$.

If $D(R) = hD(u)$, while $D(W_{\mathbf{i}})$ does not end with $[u]^h$; that is, $W_{\mathbf{i}} = W'_{\mathbf{i}}[u]^r$, $0 \leq r < h$ and $W'_{\mathbf{i}}$ ends with a lesser than $[u]$ supper-letter, then S is less than $[u]^h$ because, due to Lemma 2.7, its first super-letter is less than $[u]$: if all super-letters of $W'_{\mathbf{i}}$ are replaced with group-like elements, then $D(R) \leq D([u]^r) < hD(u)$.

Finally, if $W_{\mathbf{i}} = W'_{\mathbf{i}}[u]^h$, then a super-word R of D-degree $hD(u)$, which is greater than or equal to $[u]^h$, may appear only if all super-letters of the super-word $W'_{\mathbf{i}}$ are replaced with group-like elements, but $[u]$ is not. Here, the resulting tensor is of the form $g_{\mathbf{i}}\, g(W'_{\mathbf{i}})[u]^h \otimes g_{\mathbf{i}}\, W'_{\mathbf{i}}$.

We fix an index $\mathbf{i}$ such that $W_{\mathbf{i}}$ ends with $[u]^h$. Then the sum of all tensors of the form $g_{\mathbf{i}}\, g(W'_{\mathbf{i}})[u]^h \otimes \ldots$ in $\Delta(\mathbf{T}) - \mathbf{T} \otimes h_t$ is equal to

$$g_{\mathbf{i}} g(W'_{\mathbf{i}})[u]^h \otimes \Big(\sum_{\mathbf{j}} \alpha_{\mathbf{j}}\, g_{\mathbf{j}}\, W'_{\mathbf{j}} + \mathbf{W}''\Big), \tag{2.36}$$

where $\mathbf{W}''$ is a linear combination of basis elements of D-degree less than $(m-h)D(u)$, and $\mathbf{j}$ runs through the set of all indices such that $W_{\mathbf{j}} = W'_{\mathbf{j}}[u]^h$, $g_{\mathbf{j}}\, g(W'_{\mathbf{j}}) = g_{\mathbf{i}}\, g(W'_{\mathbf{i}})$, and $D(W_{\mathbf{j}}) = (m-h)D([u])$.

Because $W'_{\mathbf{j}}$ are distinct nonempty basis super-words of D-degree $(m-h)D(u)$, the value of tensor (2.36) in H is nonzero. A contradiction. □

Lemma 2.14 *Under the conditions of the above lemma, $p_{u,u}$ is a tth primitive root of* 1 *with $t \geq 1$ and $h = t$, or the characteristic of* $\mathbf{k}$ *equals $l > 0$ and $h = tl^k$.*

Proof By Lemma 2.13, the linear combination $\mathbf{T}$ can be written in the form

$$\mathbf{T} = [u]^h + \sum_{\mathbf{i}=(i1,i2,\ldots,is)} \alpha_{\mathbf{i}} g_{\mathbf{i}} W_{\mathbf{i}}, \quad W_{\mathbf{i}} = V_{i1}^{n_{i1}} V_{i2}^{n_{i2}} \cdots V_{is}^{n_{is}}, \tag{2.37}$$

where $[u]$ is greater than all super-letters V_{ij} for $W_{\mathbf{i}}$ of D-degree m. First let $\xi = 1 + p_{uu} + p_{uu}^2 + \ldots + p_{uu}^{h-1} \neq 0$ and assume $h > 1$.

In the basis decomposition of $\Delta(\mathbf{T}) - \mathbf{T} \otimes 1$, consider tensors of the form $[u]^{h-1} \otimes \ldots$. All super-letters V_{ij} in super-words of D-degree m are less than or equal to $[u]$; therefore, tensors of this form may appear under the basis decomposition of a tensor of $\Delta(W_{\mathbf{i}}) - W_{\mathbf{i}} \otimes 1$, $V_i = V_{i1}^{n_{i1}} V_{i2}^{n_{i2}} \cdots V_{is}^{n_{is}}$, only if either the left part of that tensor is of D-degree greater than $(h-1)D(u)$ or $W_{\mathbf{i}}$ is of D-degree less than m. In either case, the right part is of less D-degree than is $[u]$. As above, if we remove the parentheses in

$$\Delta([u]^h) = ([u] \otimes 1 + g_u \otimes [u] + \sum_\tau g_\tau U'_\tau \otimes U''_\tau)^h, \tag{2.38}$$

we see that the left parts of the resulting tensors arise from the super-word $[u]^h$ by replacing some super-letters $[u]$ either with g_u or with G-super-words $g_\tau U'_\tau$ of a lesser D-degree in less super-letters than is $[u]$. It follows that a super-word of D-degree $(h-1)D(u)$ which is greater than or equal to $[u]^{h-1}$ appears only if exactly one super-letter is replaced with a group element. Using the commutation rule $[u]^s g_u = p_{u,u}^s g_u [u]^s$, we see that the sum of all tensors of the form $g_u[u]^{k-1} \otimes \ldots$ equals

$$g_u[u]^{k-1} \otimes (\xi[u] + F + \mathbf{W}), \tag{2.39}$$

where F is a linear combination of super-words in less than $[u]$ super-letters, and $\mathbf{W}$ is a linear combination of basis G-super-words of D-degree less than $D(u)$. Consequently, (2.32) is nonzero provided that $\xi \neq 0$.

Now let $\xi = 0$. In this case $p_{u,u}^h = 1$. Therefore, $p_{u,u}$ is a tth primitive root of 1, and $h = t \cdot q$ or, if $\mathbf{k}$ has a characteristic $l > 0$, then $h = tl^r \cdot q$ with $q, t \neq 0 \pmod{l}$. Our aim is to demonstrate that $q = 1$. Let $h' = h/q$.

The commutation rule $([u] \otimes 1) \cdot (g_u \otimes [u]) = p_{u,u}(g_u \otimes [u]) \cdot ([u] \otimes 1)$ implies

$$([u] \otimes 1 + g_u \otimes [u])^{h'} = [u]^{h'} \otimes 1 + g_u^{h'} \otimes [u]^{h'}. \tag{2.40}$$

If we remove the parentheses in

$$\Delta([u]^{h'}) = (([u] \otimes 1 + g_u \otimes [u]) + \sum_i g(U''_i) U'_i \otimes U''_i)^{h'}, \tag{2.41}$$

then Lemma 2.9 implies

$$\Delta([u]^{h'}) = [u]^{h'} \otimes 1 + g_u^{h'} \otimes [u]^{h'} + \sum_\theta g(U''_\theta) U'_\theta \otimes U''_\theta, \tag{2.42}$$

where all super-words U'_θ are less than $[u]^{h'}$ (in particular, $U'_\theta \neq [u]^d$, $d < h'$) and $D(U'_\theta) < h' \cdot D(u)$.

This allows us to treat $[u]^{h'}$ in (2.37) as a single block, or as a new formal super-letter $\{[u]^{h'}\}$ such that $\{[u]^{h'}\} < [u]$, and $\{[u]^{h'}\} > [v_{ij}]$ if $u^{h'} > v_{ij}$ (the latter inequality is equivalent to $u > v_{ij}$ by Lemma 1.5):

$$\mathbf{T} = \{[u]^{h'}\}^q + \sum_i \alpha_i g_i V_{i1}^{n_{i1}} V_{i2}^{n_{i2}} \cdots V_{is}^{n_{is}}. \tag{2.43}$$

Considering that $p([u]^{h'},[u]^{h'}) = p_{u,u}^{h'\cdot h'} = 1$, we have

$$\xi_1 = 1 + p([u]^{h'},[u]^{h'}) + \ldots + p([u]^{h'},[u]^{h'})^{q-1} = q \neq 0 \ (\mathrm{mod}\, l).$$

As in the case above, assuming that $\{[u]^{h'}\}$ is a single block, we can compute the sum of all tensors of the form $g_u^{h'}\{[u]^{h'}\}^{q-1} \otimes \ldots$ in the basis decomposition of $\Delta(\mathbf{T}) - \mathbf{T}\otimes 1$ (provided that $q > 1$):

$$g_u^{h'}\{[u]^{h'}\}^{q-1} \otimes (q \cdot \{[u]^{h'}\} + F + \mathbf{W}), \tag{2.44}$$

where F is a linear combination of super-words in less than $[u]^{h'}$ super-letters, and $\mathbf{W}$ is a linear combination of basis G-super-words of less D-degree than is $[u]^{h'}$. By the induction hypothesis, tensor (2.44) is nonzero in $H \otimes H$, and so is (2.32). □

Now we are ready to complete the proof of Theorem 2.2 by induction on m, U, and r. The least super-word of the minimal D-degree is a least variable x_i with minimal d_i. In (2.30), the minimal value of r is zero. For these values of the induction parameters, we have $\mathbf{T} = x_i$. If $x_i = 0$ in H then $U = [x_i]$ is not a hard super-letter.

If under the induction assumptions 1, 2, 3, we have $\mathbf{T} = 0$ in H, then value of $\mathbf{T}$ is a skew-primitive element. By Lemmas 2.13, 2.14, the equality $\mathbf{T} = 0$ takes the form

$$[u]^h = - \sum_{\mathbf{i}=(i1,i2,\ldots,is)} \alpha_{\mathbf{i}}\, g_{\mathbf{i}}\, W_{\mathbf{i}}, \quad W_{\mathbf{i}} = V_{i1}^{n_{i1}} V_{i2}^{n_{i2}} \cdots V_{is}^{n_{is}},$$

where $V_{ij} < [v]$ if $D(W_{\mathbf{i}}) = D([u]^h)$, whereas for h there are just the following options: $h = 1$; or $p_{u,u}$ is a primitive tth root of 1, and either $h = t$ or, in case when the characteristic l is positive, $h = tl^k$

If $h = 1$, then Definition 2.6 implies that $[u]$ is not hard. In other cases, Definition 2.7 implies that the height of $[u]$ is less than h. Theorem 2.2 is proved.

The skew-primitive elements in character Hopf algebras have a special form in the basis decomposition related to hard super-letters. We are remanded that if $a \in \mathbf{k}[G]$ is a skew-primitive element, then a is proportional to $h - f$, see Lemma 1.19.

Theorem 2.3 *If $a \notin \mathbf{k}[G]$ is a skew-primitive element, then $a = \alpha g\, \varphi(\mathbf{T})$, where $0 \neq \alpha \in \mathbf{k}$, $g \in G$, and $\mathbf{T}$ has the following expansion:*

$$\mathbf{T} = [u]^h + \sum \alpha_i W_i + \sum \beta_j\, g_j\, W_j'. \tag{2.45}$$

Here, $[u]$ is a hard super-letter, W_i are basis super-words in super-letters less than $[u]$, $D(W_i) = hD([u])$, and $D(W_j') < hD([u])$. Moreover, if $p_{u,u}$ is not a root of 1, then $h = 1$; if $p_{u,u}$ is a primitive tth root of 1, then $h = 1$, or $h = t$, or (in case of characteristic $l > 0$) $h = tl^k$.

Proof By Theorem 2.2, the element a is a linear combination of values of increasing restricted G-super-words, $a = \varphi(\mathbf{T}')$,

$$\mathbf{T}' = \alpha g U + \sum_{i=1}^{k} \gamma_i g_i W_i + W', \ \alpha \neq 0, \tag{2.46}$$

where gU, g_iW_i are admissible G-super-words of maximal degree, and either $U > W_i$ or $U = W_i$ but $g_i \neq g$. Considering that, due to Theorem 2.2, assumptions 1, 2, 3 are universally true, we may apply Lemmas 2.12–2.14 to $\mathbf{T} = \alpha^{-1}g^{-1}\mathbf{T}'$. □

2.4 Monomial PBW Basis

In this section, we prove that values of standard words corresponding to hard super-letters form a set of *PBW* generators for H also. Additionally we find some criterion for a super-letter $[u]$ to be hard in terms of the values of monomials. This criterion allows one to forget about skew brackets while computing the hard super-letters.

We keep the notations of the above section. In particular, H is a Hopf algebra generated by an Abelian group G of all group-like elements and by skew-primitive semi-invariants $a_1, \dots, a_n$ with which degrees $d_1, \dots, d_n$ are associated. We fix the homomorphism of Hopf algebras $\varphi : G\langle X\rangle \to H, \ x_i \mapsto a_i, \ 1 \leq i \leq n.$

Let w be an arbitrary word. By Theorem 1.1, there exists a unique decomposition of the word w in the product: $w = w_1^{n_1} \cdot w_2^{n_2} \cdot \dots \cdot w_m^{n_m}$, where w_i, $1 \leq i \leq m$ are standard words such that $w_1 < w_2 < \dots < w_m$. Let $W = [w_1]^{n_1} \cdot [w_2]^{n_2} \cdot \dots \cdot [w_m]^{n_m}$.

Lemma 2.15 *If the super-word W is admissible, then the leading super-word of the basis decomposition of $\varphi(w)$ is precisely W and it occurs with the coefficient 1 only. If W is not admissible, then each super-word of the basis decomposition of $\varphi(w)$ either is less than W or is of a lesser D-degree.*

Proof Lemma 2.4 implies that the leading word of the polynomial W is precisely w. Hence, $W - w$ is a linear combination of words that are less than w.

If W is admissible, then the decomposition $w = W + (w - W)$ allows one to perform the evident induction.

If W is not admissible, then by Lemma 2.9, there is a decomposition $\varphi(W) = \sum_j \alpha_j g_j \varphi(W_j)$, where W_j are admissible super-words and for each j either $W_j < W$ or $D(W_j) < D(w)$. Let $W_j = [w_{1j}]^{n_1} \cdot [w_{2j}]^{n_2} \cdot \dots \cdot [w_{mj}]^{n_{mj}}$ and $w_j = w_{1j}^{n_1} \cdot \dots \cdot w_{mj}^{n_{mj}}$. Lemma 2.4 implies that $w_j < w$ provided that $D(w_j) = D(w)$. Thus, we have a representation of $\varphi(w)$ as a linear combination of lesser words of the same D-degree and G-words of lesser D-degree:

$$\varphi(w) = \varphi(w - W) + \sum_j \alpha_j g_j \varphi(w_j) - \sum_j \alpha_j g_j \varphi(W_j - w_j). \tag{2.47}$$

The induction applies. □

Theorem 2.4 *The set of values in H of all G-words*

$$gu_1^{n_1} \cdot u_2^{n_2} \cdot \ldots \cdot u_k^{n_k}, \tag{2.48}$$

where $g \in G$, $u_1 < u_2 < \ldots < u_k$ are standard words such that $[u_i]$ are hard super-letters, $n_i < h([u_i])$ forms a basis of H.

Proof Suppose that values of all words of degree $< m$ belong to the space H_0 spanned by (2.48). Among the words of D-degree m, let w be the minimal one with respect to the lexicographic order, such that $\varphi(w) \notin H_0$. If W is admissible, then w itself has the form (2.48). If W is not admissible, than by induction (2.47) implies that $\varphi(w) \in H_0$. Hence, $H_0 = H$.

Let w_j, $j \in J$ be different words of the type (2.48); that is, $w_j = w_{1j}^{n_1} \cdot \ldots \cdot w_{mj}^{n_{mj}}$, whereas $W_j = [w_{1j}]^{n_1} \cdot [w_{2j}]^{n_2} \cdot \ldots \cdot [w_{mj}]^{n_{mj}}$ are admissible super-words. By Lemma 2.15, the super-word W_j is a leading super-word of the PBW decomposition $w_j = W_j + \sum_i \alpha_{ij} W_{ij}$. Let W_k is the maximal super-word among the W_j's of maximal D-degree. Considering that different W_j, W_{ij}, $j \in J$ are linearly independent in H, we obtain that a linear dependence

$$\sum_{j\in J,\ t\in T} \alpha_{jt} h_{jt} \varphi(w_j) = 0,\ 0 \neq \alpha_{jt} \in \mathbf{k},\ h_{jt} \in G, \tag{2.49}$$

would imply $\sum_{t\in T} \alpha_{kt} g_{kt} \varphi(W_k) = 0$. This contradicts to Theorem 2.2. □

Corollary 2.1 *A super-letter $[u]$ is hard if and only if the value of u is not a linear combination of values of lesser words of D-degree $D(u)$ and of G-words of a lesser D-degree.*

Proof Let $\varphi(u) = \sum_i \alpha_i \varphi(w_i) + u_0$, $\alpha_i \in \mathbf{k}$, where $w_i < u$, $D(w_i) = D(u)$ and $D(u_0) < D(u)$. By Lemma 2.15, we obtain $u = [u] + \sum_j \beta_j U_j$ where the super-words U_j are less than $[u]$.

Let $w_i = w_{1i}^{n_1} \cdot w_{2i}^{n_2} \cdot \ldots \cdot w_{mi}^{n_{mi}}$, where w_{ki}, $1 \leq k \leq mi$ are standard words such that $w_{1i} < w_{2i} < \ldots < w_{mi}$, and let $W_i = [w_{1i}]^{n_1} \cdot [w_{2i}]^{n_2} \cdot \ldots \cdot [w_{mi}]^{n_{mi}}$. Lemma 2.15 demonstrates that all super-words V of the basis decomposition of w_i are less than or equal to W_i unless $D(V) < D(w_i)$. Because $u > w_i$, by Lemma 2.4, we have $[u] > W_i$, for all i.

Therefore $[u]$ is greater than all super-words of degree $D(u)$ in the basis decomposition of $\sum_i \alpha_i \varphi(w_i)$. Thus, Theorem 2.2 implies that $\varphi(u) \neq \sum_i \alpha_i \varphi(w_i) + u_0$.

Conversely, if $\varphi([u]) = \sum \alpha_i \varphi(W_i) + U_0$, where W_i depends on super-letters less than $[u]$ only, and $D(U_0) < D(u)$, then

$$\varphi(u) = \varphi([u]) + \varphi(u - [u]) = \sum \alpha_i \varphi(W_i) + U_0 + \varphi(u - [u]).$$

Due to Lemma 2.4, the latter polynomial has no one monomial whose D-degree equals $D(u)$ and which is greater than or equal to u. □

2.5 Serre Skew-Primitive Polynomials

In this section, using Theorem 2.3, we shall describe all skew-primitive polynomials in two variables linear in one of them. We keep notation of Sect. 1.5.3:

$$\Delta(y_i) = y_i \otimes h_i + f_i \otimes y_i, \ \ y_i g = \chi^i(g) g y_i, \quad h_i, f_i, g \in G, \ \ i = 1, 2.$$

We know that $G\langle y_1, y_2\rangle$ as a Hopf algebra with group G of group-like elements is completely defined by the following four parameters

$$p_{ik} = q_{ik}^{-1} q'_{ik} = \chi^i(h_k^{-1} f_k), \quad 1 \leq i, k \leq 2 \tag{2.50}$$

related to the normalized skew-primitive generators $x_1 = h_1^{-1} y_1, \ x_2 = h_2^{-1} y_2$ because $G\langle y_1, y_2\rangle = G\langle x_1, x_2\rangle$.

Theorem 2.5 *There exists a nonzero linear in y_1 skew-primitive element W of degree n in y_2 if and only if either*

$$p_{12} p_{21} = p_{22}^{1-n} \tag{2.51}$$

or p_{22} is a primitive mth root of 1, $m|n$, *and*

$$p_{12}^m p_{21}^m = 1. \tag{2.52}$$

If one (or both) of these conditions is satisfied, then

$$W = \alpha g\,[\ldots[[y_1, y_2], y_2], \ldots, y_2], \quad \alpha \in \mathbf{k}, \ g \in G, \tag{2.53}$$

where the brackets are defined in (1.67).

Proof Let W be a skew-primitive element of constitution $(1, n)$. By Theorem 2.3 the element W has a representation (2.45) up to a factor αg. Considering that the free character Hopf algebra is homogeneous in each variable, there are no terms W'_s in that representation. There exist only one standard word of constitution $(1, n)$: this is $x_1 x_2^n$. The standard alignment of brackets is precisely $[x_1 x_2^n] = [[\ldots[[x_1 x_2] x_2], \ldots] x_2]$. Hence, (2.45) reduces to $W = \alpha g\,[x_1 x_2^n]$. Due to Lemma 1.21, the G-super-word $h_1 h_2^n [x_1 x_2^n]$ becomes $[y_1 y_2^n]$ up to a scalar factor if we distribute the group-like factors among the variables using the commutation rules (1.62):

$$h_1 h_2^n [[\ldots[[x_1, x_2], x_2], \ldots], x_2] \sim [[\ldots[[y_1, y_2], y_2], \ldots], y_2]. \tag{2.54}$$

This proportion proves (2.53).

It remains to analyze when $[x_1x_2^n]$ is skew-primitive. By induction on n we shall prove the following explicit coproduct formula

$$\Delta([x_1x_2^n]) = [x_1x_2^n] \otimes 1 + \sum_{k=0}^{n} \alpha_k^{(n)} g_1 g_2^{n-k} x_2^k \otimes [x_1 x_2^{n-k}], \tag{2.55}$$

$$\alpha_k^{(n)} = \begin{bmatrix} n \\ k \end{bmatrix}_{p_{22}} \cdot \prod_{s=n-k}^{n-1} (1 - p_{12}p_{21}p_{22}^s). \tag{2.56}$$

If $n = 0$, then the equality reduces to $\Delta(x_1) = x_1 \otimes 1 + g_1 \otimes x_1$, whereas $\alpha_0^{(0)} = 1$. Moreover, it is clear that $\alpha_0^{(n)} = 1$ for all n. We have,

$$\Delta([x_1x_2^n]) \cdot (x_2 \otimes 1) = [x_1x_2^n]x_2 \otimes 1 + \sum_{k=0}^{n} \alpha_k^{(n)} g_1 g_2^{n-k} x_2^{k+1} \otimes [x_1x_2^{n-k}], \tag{2.57}$$

$$\Delta([x_1x_2^n]) \cdot (g_2 \otimes x_2) = [x_1x_2^n]\, g_2 \otimes x_2 + \sum_{k=0}^{n} \alpha_k^{(n)} g_1 g_2^{n-k} x_2^k g_2 \otimes [x_1x_2^{n-k}]x_2, \tag{2.58}$$

$$(x_2 \otimes 1) \cdot \Delta([x_1x_2^n]) = x_2[x_1x_2^n] \otimes 1 + \sum_{k=0}^{n} \alpha_k^{(n)} x_2\, g_1 g_2^{n-k} x_2^k \otimes [x_1x_2^{n-k}], \tag{2.59}$$

$$(g_2 \otimes x_2) \cdot \Delta([x_1x_2^n]) = g_2[x_1x_2^n] \otimes x_2 + \sum_{k=0}^{n} \alpha_k^{(n)} g_1 g_2^{n-k+1} x_2^k \otimes x_2[x_1x_2^{n-k}]. \tag{2.60}$$

In the second and third relations we may move the group-like factors to the left:

$$[x_1x_2^n]g_2 = p_{12}p_{22}^n\, g_2[x_1x_2^n], \quad x_2^k\, g_2 = p_{22}^k\, g_2\, x_2^k, \quad x_2\, g_1 g_2^{n-k}\, x_2^k$$
$$= p_{21}p_{22}^{n-k}\, g_1 g_2^{n-k+1}\, x_2^{k+1}.$$

Using all that relations, we develop the coproduct of

$$[x_1x_2^{n+1}] = [x_1x_2^n]x_2 - p_{12}p_{22}^n\, x_2[x_1x_2^n]$$

taking into account that $\Delta(x_2) = x_2 \otimes 1 + g_2 \otimes x_2$. The sums of (2.57) and (2.59) provide the tensors

$$\sum_{k=0}^{n} \alpha_k^{(n)} (1 - p_{12}p_{21}p_{22}^{2n-k}) g_1 g_2^{n-k}\, x_2^{k+1} \otimes [x_1x_2^{n-k}],$$

whereas the sums of (2.57) and (2.59) produce the following ones:

$$\sum_{k=0}^{n} \alpha_k^{(n)} p_{22}^k g_1 g_2^{n-k+1} x_2^k \otimes [x_1 x_2^{n-k+1}].$$

The first term of (2.58) cancels with the first term of (2.60). Finally, we arrive to a formula (2.55) with $n \leftarrow n+1$ and coefficients

$$\alpha_k^{(n+1)} = \alpha_{k-1}^{(n)} (1 - p_{12}p_{21}p_{22}^{2n-k+1}) + \alpha_k^{(n)} p_{22}^k, \ \ k \geq 1, \ \ \alpha_0^{(n+1)} = 1. \qquad (2.61)$$

To prove the coproduct formula (2.55), it remains to check that values (2.56) satisfy the above recurrence relations. To this end, we shall check the equality of the following two polynomials in commutative variables λ, q :

$$\begin{bmatrix} n+1 \\ k \end{bmatrix}_q \cdot (1 - \lambda q^n) = \begin{bmatrix} n \\ k-1 \end{bmatrix}_q \cdot (1 - \lambda q^{2n-k+1}) + \begin{bmatrix} n \\ k \end{bmatrix}_q \cdot (1 - \lambda q^{n-k}) \cdot q^k. \qquad (2.62)$$

If $\lambda = 0$, then the equality reduces to the first q-Pascal identity (1.2). Let us compare the coefficients at λ,

$$\begin{bmatrix} n+1 \\ k \end{bmatrix}_q \cdot q^n = \begin{bmatrix} n \\ k-1 \end{bmatrix}_q \cdot q^{2n-k+1} + \begin{bmatrix} n \\ k \end{bmatrix}_q \cdot q^{n-k} \cdot q^k.$$

This equality differs from the second q-Pascal identity (1.3) just by a common factor q^n. Hence, the equality (2.62) is valid.

If we multiply both sides of (2.62) by $\prod_{s=n-k+1}^{n-1}(1-\lambda q^s)$ and next replace the variables $q \leftarrow p_{22}$, $\lambda \leftarrow p_{12}p_{21}$, then we obtain precisely (2.61) for values (2.56). The proof of (2.55) is complete.

Each $\alpha_k^{(n)}$, $1 \leq k \leq n$ defined by (2.56) has a factor $1 - p_{12}p_{21}p_{22}^{n-1}$. In particular, if $p_{12}p_{21} = p_{22}^{1-n}$, then all of these coefficients are zero, whence $[x_1x_2^n]$ is a skew-primitive polynomial.

If p_{22} is a primitive mth root of 1, $m|n$, and $p_{12}^m p_{21}^m = 1$, then $p_{12}p_{21}$ is a power of p_{22}; that is, $p_{12}p_{21}p_{22}^s = 1$ for some s, $0 \leq s < m$. This implies that the product $\prod_{s=n-k}^{n-1}(1 - p_{12}p_{21}p_{22}^s)$ equals zero provided that $k \geq m$. If $k < m$, then Lemma 1.1 applies.

Conversely, suppose that all coefficients $\alpha_k^{(n)}$, $1 \leq k \leq n$ are zero. In particular, $\alpha_1^{(n)} = (1 - p_{12}p_{21}p_{22}^{n-1})p_{22}^{[n]} = 0$. Therefore, if $p_{12}p_{21} \neq p_{22}^{1-n}$, then $p_{22}^{[n]} = 0$. This implies $p_{22}^n = 1$; that is, p_{22} is a primitive mth root of 1 and $m|n$. In this case, the equality $\alpha_n^{(n)} = \prod_{s=0}^{n-1}(1 - p_{12}p_{21}p_{22}^s) = 0$ implies that $1 - p_{12}p_{21}p_{22}^s = 0$ for some s, $0 \leq s < n$. Hence, $(p_{12}p_{21})^m = p_{22}^{-sm} = 1$ which is required. □

Corollary 2.2 *If one of the existence conditions of the above theorem holds then*

$$[\dots[[y_1, y_2], y_2], \dots, y_2] \sim [y_2, [y_2, \dots, [y_2, y_1]\dots]]. \qquad (2.63)$$

Proof By Lemma 1.21, we have

$$[y_2, [y_2, \dots, [y_2, y_1]\dots]] \sim h_1 h_2^n [x_2, [x_2, \dots, [x_2, x_1]\dots]]. \qquad (2.64)$$

This lemma and (2.54) imply that it suffices to demonstrate (2.63) under the substitution $y_i \leftarrow x_i$.

Let us introduce the opposite order, $x_2 > x_1$. There exist only one standard word of constitution $(1, n)$ with respect to this ordering of variables, $x_2^n x_1$, whereas the standard alignment of brackets is $[x_2[x_2 \dots [x_2, x_1]\dots]]$. As $[\dots[[x_1, x_2], x_2]\dots, x_2]$ is skew-primitive, it has a representation (2.45) where all summands have the same constitution, $(1, n)$. By definition of the lexicographical order $x_2 > x_2^n x_1$. Hence, x_2 does not occur in (2.45) as a super-letter. Since every addend has degree 1 in x_1, it follows that (2.45) reduces to $\mathbf{T} = \alpha[x_2^n x_1]$. □

2.5.1 Examples

In this subsection, we consider in more detail the above-described binary skew-primitive polynomials with $n \leq 3$ and study the Hopf algebras set up by those polynomials (as defining relations).

We fix two normalized skew-primitive variables x_1, x_2 such that

$$\Delta(x_i) = x_i \otimes 1 + g_i \otimes x_i, \quad i = 1, 2.$$

Respectively, we put $p_{is} = \chi^i(g_s),\ i, s = 1, 2$ so that

$$x_1 g_1 = p_{11} g_1 x_1, \quad x_1 g_2 = p_{12} g_2 x_1, \quad x_2 g_1 = p_{21} g_1 x_2, \quad x_2 g_2 = p_{22} g_2 x_2.$$

We always suppose that the variables are ordered so that $x_1 > x_2$.

Example 2.1 If $n = 1$, then the existence condition of Theorem 2.53 reduces to $p_{12} p_{21} = 1$. Under that condition the skew commutator $[x_1, x_2] = x_1 x_2 - p_{12} x_2 x_1$ is a skew primitive element. We have $[x_1, x_2] = -p_{12}[x_2, x_1]$, which is the particular case of the general formula (2.63). The Hopf algebra H defined by the relation $[x_1, x_2] = 0$ is the skew group ring $R * G$, where G is the group generated by $g_1,\ g_2$ and R is the so-called algebra of quantum polynomials

$$R = \{\sum_{m,n} \alpha_{m,n} x_2^m x_1^n \mid x_1 x_2 = p_{12} x_2 x_1\}.$$

Obviously, x_1 and x_2 are the PBW-generators of H. To see this formally, we may apply Composition Lemma (Theorem 1.2). Indeed, $[x_1, x_2] = 0$ is a Gröbner-Shirshov system of relations because there are no compositions at all. Hence, by Composition Lemma, the set Σ of all words without subword x_1x_2 is a basis of R. Of course, $\Sigma = \{x_2^m x_1^n \mid m, n \geq 0\}$.

Example 2.2 If $n = 2$, then the existence condition of Theorem 2.53 reduces to

$$(p_{12}p_{21} = p_{22}^{-1}) \vee (p_{12}p_{21} = 1 \ \& \ p_{22} = -1). \tag{2.65}$$

Under that condition, the polynomial

$$[[x_1, x_2], x_2] = x_1x_2^2 - p_{12}(1 + p_{22})x_2x_1x_2 + p_{12}^2 p_{22} x_2^2 x_1$$

is a skew primitive element. In this case, the general formula (2.63) takes the form $[x_2, [x_2, x_1]] = p_{12}^2 p_{22}[[x_1, x_2], x_2]$. Similarly, condition

$$(p_{12}p_{21} = p_{11}^{-1}) \vee (p_{12}p_{21} = 1 \ \& \ p_{11} = -1) \tag{2.66}$$

implies that

$$[x_1, [x_1, x_2]] = x_1^2 x_2 - p_{12}(1 + p_{11})x_1x_2x_1 + p_{12}^2 p_{11} x_2 x_1^2$$

is a skew-primitive element and $[x_1, [x_1, x_2]] = p_{12}^2 p_{11}[[x_2, x_1], x_1]$.

If both polynomials are skew-primitive, then we may consider the Hopf algebra H defined by relations $[[x_1, x_2], x_2] = 0$ and $[x_1, [x_1, x_2]] = 0$. Of course, $H = R * G$, where R is the algebra defined by the same relations, and G, as above, is the group generated by g_1, g_2.

If $p_{11} = p_{22}$, then the algebra R is precisely the algebra A_2^+ considered in Example 1.1, where $\alpha = -p_{12}(1 + p_{22})$, $\beta = p_{12}^2 p_{22}$. In Example 1.1, we have seen that the system of relations

$$[[x_1, x_2], x_2] = 0, \quad [x_1, [x_1, x_2]] = 0$$

is closed with respect to the compositions, and

$$\Sigma = \{x_2^m (x_1x_2)^n x_1^k \mid m, n, k \geq 0\}$$

is a basis of R. In other words, the elements x_2, x_1x_2, x_1 form a set of PBW generators for H over G. Corollary 2.1 implies that all hard super-letters are precisely $x_2, [x_1x_2], x_1$, and they form a set of PBW generators for H over G as well.

We stress that the existence conditions (2.65), (2.66) imply $p_{11} = p_{22}$ unless $p_{22} = p_{12}p_{21} = 1, p_{11} = -1$ or $p_{22} = -1, p_{12}p_{21} = p_{11} = 1$.

Example 2.3 Note that $[[[x_1, x_2], x_2], x_2]$ is precisely the Lyndon–Shirshov standard word $[x_1x_2^3]$ with the standard alignment of brackets. Due to Theorem 2.53 the polynomial $[x_1x_2^3]$ is skew-primitive if either $p_{12}p_{21} = p_{22}^{-2}$ or $p_{22} = \zeta$ is a primitive third root of 1 and $p_{12}p_{21} \in \{1, \zeta^2\}$. Under that condition the polynomial

$$[x_1x_2^3] = x_1x_2^3 - p(1+q+q^2)x_2x_1x_2^2 + p^2(q+q^2+q^3)x_2^2x_1x_2 - p^3q^3x_2^3x_1,$$

where $p = p_{12}$, $q = p_{22}$ is skew-primitive, and (2.63) takes the form

$$[x_1x_2^3] = -p^3q^3[x_2, [x_2, [x_2, x_1]]].$$

If $p_{11}^{-1} = p_{12}p_{21} = p_{22}^{-2}$, then both $[x_1x_2^3]$ and $[x_1^2x_2]$ are skew-primitive polynomials. Consider the Hopf algebra H defined by two relations: $[x_1x_2^3] = 0$, and $[x_1^2x_2] = 0$. These relations have the form (1.22) considered in Example 1.2 with

$$\alpha = -p(1+q^2),\ \beta = p^2q^2,\ \gamma = -p(1+q+q^2),\ \delta = p^2(q+q^2+q^3),\ \varepsilon = -p^3q^3,$$

whereas before, we put for short $p = p_{12}$, $q = p_{22}$. If we define $\mu = -pq$, then these parameters satisfy the following relations (1.23):

$$\beta = \mu^2,\ \gamma = \alpha + \mu,\ \delta = \gamma\mu,\ \varepsilon = \mu^3.$$

In Example 1.2, we observed that the system of relations $[x_1x_2^3] = 0$ and $[x_1^2x_2] = 0$ becomes closed with respect to the compositions if we add one new relation, (1.27), which is a consequence of the two initial ones. Hence the set

$$\Sigma = \{x_2^m(x_1x_2x_2)^n(x_1x_2)^kx_1^s \mid m, n, k, s \geq 0\}$$

is a basis of R. In other words, the elements x_2, $x_1x_2^2$, x_1x_2, x_1 form a set of PBW-genrators for H over G. Respectively, Corollary 2.1 implies that all hard super-letters are precisely x_2, $[x_1x_2^2]$, $[x_1x_2], x_1$, and they form a set of PBW-generators of H over G also.

Interestingly, by Proposition 1.3 we may replace the very new relation with any other relation with the same leading word. The leading word, $x_1x_2x_1x_2^2$, is standard, and one may show (here we omit the detailed calculations) that $[x_1x_2x_1x_2^2] = 0$ is a relation for R. Therefore the three relations $[x_1x_2^3] = 0$, $[x_1^2x_2] = 0$, and $[x_1x_2x_1x_2^2] = 0$ is a Gröbner–Shirshov system of defining relations for R. Here $[x_1x_2x_1x_2^2] = [[x_1x_2][[x_1x_2]x_2]]$ has the standard alignment of brackets.

There exist five exceptional cases, when $[x_1x_2^3]$, $[x_1^2x_2]$ are still skew-primitive but $p_{11} \neq p_{22}^2$. They are: $p_{11} = p_{12}p_{21} = 1$, $p_{22} = \zeta$; $p_{11} = p_{22} = \zeta$, $p_{12}p_{21} = \zeta^2$; and $p_{11} = -1$, $p_{12}p_{21} = 1$, $p_{22} \in \{1, -1, \zeta\}$; here, ζ is the third primitive root of 1. The analysis of each one of these cases is much easier than that of Example 1.2, and we let the reader find the PBW-generators and Gröbner-Shirshov systems of relations as an exercise.

2.6 Chapter Notes

Examples 2.2 and 2.3 above are particular cases of quantizations of Lie algebras. Gröbner-Shirshov systems of defining relations for quantizations of Lie algebras of infinite series A_n, B_n, C_n, D_n were found by the author [128] using as a basic tool the PBW theorem proved in this chapter. Interestingly, all relations in those systems have the form $[u] = 0$, where $[u]$ is a standard word with standard alignment of brackets. Independently, Chen et al. [48] found the Gröbner-Shirshov systems for quantizations $U_q(\mathfrak{sl}_n)$ of type A_n by means of the specific PBW basis constructed by Rosso [195] and Yamane [234].

There are many publications on the construction of a PBW basis for Hopf algebras. The first PBW-type theorem for *Drinfeld-Jimbo quantizations* (see the next chapter) appeared in the pioneering paper by Jimbo [106], which discusses $U_q(\mathfrak{sl}_2)$ in detail. Rosso [195] and Yamane [234] independently constructed the PBW basis for Drinfeld–Jimbo algebras $U_q(\mathfrak{sl}_n)$ of type A_n, $n > 2$. Thereafter, G. Lusztig, in his fundamental works [151–153], determined the PBW bases for arbitrary Drinfeld-Jimbo and Lusztig quantum enveloping algebras. These bases and their modifications have been considered in a number of subsequent papers, e.g., Kashiwara [119], Concini et al. [58], Berger [28], Towber [224], Bautista [21], Gavarini [84], Chari and Xi [47], Reineke [192], Leclerc [146], Bai and Hu [19]. An original approach based on the Ringel-Hall algebras was also advanced in [59, 60, 194].

The general statement given in Theorem 2.2 can be attributed to the author [124]. This PBW-type theorem was found to be essential in the construction of the *Weyl groupoid* by Heckenberger [91] corresponding to a *Nichols algebra* (see Sect. 6.7 below) of diagonal type. This groupoid was crucial in classifying such Nichols algebras [90]. In turn, knowledge of these Nichols algebras is important to perform the lifting method developed by N. Andruskiewitsch and H.-J. Schneider for classifying pointed Hopf algebras [4] .

Theorem 2.2 was generalized in two different directions by Ufer [225], and by Graña and Heckenberger [87] using similar methods. Instead of character Hopf algebras, S. Ufer considered *braided Hopf algebras* (see Chap. 6 below) with "triangular" braidings, whereas M. Graña and I. Heckenberger replaced the skew-primitive generators with irreducible Yetter–Drinfeld modules and obtained a factorization of the Hilbert series for a wide class of graded Hopf algebras, where the factors are parametrized by Lyndon–Shirshov words in a manner similar to how the PBW generators are parametrized in Theorem 2.2. In [97], I. Heckenberger and H. Yamane modified Theorem 2.2 based on the work of G. Lusztig by using the concept of the Weyl groupoid.

Returning to the main idea of the proof of Theorem 2.2, the right and left sides of the tensors in (2.2) were used differently, although we required detailed information (given in Lemma 2.9) about the left sides only. This information provides a noteworthy idea for applying the method to subalgebras R of H such that $\Delta(R) \subseteq R \otimes H$. A subspace that obeys the latter property is known as a *right coideal*.The author developed this idea in [133] by proving the following statement:

Theorem 2.6 *Every right coideal subalgebra of a character Hopf algebra H that contains all group-like elements of H has a PBW basis that can be extended up to a PBW basis of H.*

One reason that one-sided coideal subalgebras are important is that Hopf algebras do not have a sufficient number of Hopf subalgebras. The straightforward idea to consider Hopf subalgebras as "quantum subgroups" appeared to be inappropriate, whereas the one-sided coideal subalgebras are more precise. The one-sided comodule subalgebras, not the Hopf subalgebras, are found to be the Galois objects in the Galois theory for Hopf algebra actions (Milinski [173, 174], see also a detailed survey by Yanai [235]). In particular, the Galois correspondence theorem for the actions on free algebra establishes a one-to-one correspondence between right coideal subalgebras and intermediate free subalgebras (see Ferreira et al. [73]). In a detailed survey [147], G. Letzter provides a panorama of the use of one-sided coideal subalgebras in constructing quantum symmetric pairs to form Harish-Chandra modules and produce quantum symmetric spaces.

The importance of this concept led to a project to classify one-sided coideal subalgebras of Drinfeld–Jimbo quantizations. In fact, the proof of Theorem 2.6 yields sufficient additional information to try to attempt this classification for the subalgebras containing all group-like elements.

In a series of papers by Lara Sagahón, Garza Rivera and the author [134, 135, 139, 140], using the parallelization technique for supercomputers, this program was developed for a multiparameter version of the Drinfeld–Jimbo and Lusztig quantizations of types A_n and B_n. It was found in [135, 139] that in these cases the number of right coideal subalgebras of the positive Borel part $U_q^+(\mathfrak{g})$ coincides with the order of the Weyl group.

The latter statement was extended to arbitrary quantizations of finite type by Heckenberger and Schneider [96]. The right coideal subalgebras in that case are the well-known spaces $U^+[w]$ defined by the elements w of the Weyl group, which was used by Lusztig [153] to establish a PBW basis for $U_q^+(\mathfrak{g})$. This establishment represents an outstanding achievement of a general theory developed by N. Andruskiewitsch, I. Heckenberger, and H.-J. Schneider in a number of papers [5, 92, 95, 96]. Generally, this theory is a categorical version of the fundamental theory of Lusztig's automorphisms. More precisely, instead of the skew-primitive generators $x_1, \ldots, x_n$ the authors consider irreducible finite-dimensional Yetter–Drinfeld modules $V_1, \ldots, V_n$ over a Hopf algebra H with bijective antipode, and in place of the Weyl group is the Weyl groupoid theorized by I. Heckenberger. The theory includes a PBW theorem for the related Nichols algebras and their right coideal subalgebras.

Using these results as a starting point, Heckenberger and Kolb [94] classified all homogeneous right coideal subalgebras for a quantized enveloping algebra $U_q^+(\mathfrak{g})$ of a complex semisimple Lie algebra $\mathfrak{g}$ with deformation parameter q not a root of unity.

Using the computer algebra program to compute the commutative and non-commutative rings and modules FELIX [6, 72], they determined the number of different right coideal subalgebras when the order $|W|$ of the Weyl group was less than one million, thus confirming results of [139] for the case A_n and reducing the error in the explicit computer calculations for the case B_n presented in [140]. These numbers $|Co|$ are given in the tables below.

Type	A_2	A_3	A_4	A_5	A_6	A_7	A_8	E_6	F_4	G_2
$\|W\|$	6	24	120	720	5040	40,320	362,880	51,840	1152	12
$\|Co\|$	26	252	3368	58,810	1,290,930	34,604,844	1,107,490,596	38,305,190	91,244	68

B_2	B_3	B_4, C_4	B_5, C_5	B_6, C_6	B_7, C_7	D_4	D_5	D_6	D_7
8	48	384	38,400	46,080	645,120	192	1920	23,040	322,560
38	664	17,848	672,004	33,369,560	2,094,849,020	6512	238,720	11,633,624	720,453,984

It is likely that the same numbers remain true for multiparameter and "small" versions of the quantizations. Heckenberger and Kolb [93] recently extended their work on classification problem by considering right coideal subalgebras that do not contain all group-like elements.

Chapter 3
Quantizations of Kac-Moody Algebras

Abstract Numerous books and articles concerning quantizations of Kac-Moody algebras have been published. However, almost all publications have their own modifications in construction and different notations, so it is often unclear whether the results of one work may be applied to the construction of another. Nevertheless all of the constructions are character Hopf algebras. In view of the fact that the number and degrees of relations in all of the constructions related to a given Kac-Moody algebra $\mathfrak{g}$ are identical, we introduce a class of character Hopf algebras defined by the same number of defining relations of the same degrees as the Kac-Moody algebra $\mathfrak{g}$ is. This class contains all possible quantizations of $\mathfrak{g}$ (including multiparameter quantizations), and these Hopf algebras are considered as quantum deformations of the universal enveloping algebra of $\mathfrak{g}$ as well. The unification in the above class provides the potential to understand the differences, if any, between these constructions by comparing the basic invariants inside that class. We demonstrate that if the generalized Cartan matrix A of $\mathfrak{g}$ is connected then the algebraic structure, up to a finite number of exceptional cases, is defined by just one "continuous" parameter q related to a symmetrization of A, and one "discrete" parameter $\mathfrak{m}$ related to the modular symmetrizations of A. The Hopf algebra structure is defined by $n(n-1)/2$ additional "continuous" parameters.

In this chapter, we associate a class of character Hopf algebras $\mathfrak{A}$ with a given Kac-Moody algebra $\mathfrak{g}$. Algebras from $\mathfrak{A}$ are defined by the same number of defining relations and with the same degrees as $\mathfrak{g}$. The class $\mathfrak{A}$ contains all known quantizations of $\mathfrak{g}$ (including multiparameter quantizations). The Hopf algebras from $\mathfrak{A}$ must be considered quantum deformations of the universal enveloping algebra of $\mathfrak{g}$ as well.

In Sect. 3.4, we demonstrate that all Hopf algebras from $\mathfrak{A}$ have the so-called triangular decomposition as coalgebras. In Sect. 3.5, we prove that if the generalized Cartan matrix A of $\mathfrak{g}$ is indecomposable, then the algebraic structure, up to a finite number of exceptional cases, is defined by only one "continuous" parameter q related to a symmetrization of A, and one "discrete" parameter $\mathfrak{m}$ related to the modular symmetrizations of A. In other words, the algebraic variety of parameters that define the algebra structure of the quantizations related to a given $\mathfrak{g}$ has the

V. Kharchenko, *Quantum Lie Theory*, Lecture Notes in Mathematics 2150,
DOI 10.1007/978-3-319-22704-7_3

dimension ≤ 1. The Hopf algebra structure is defined by $n(n-1)/2$ additional "continuous" parameters.

Throughout the chapter, $\bar{\mathbf{k}}$ represents the algebraic closure of the ground field $\mathbf{k}$.

3.1 Kac-Moody Algebras

Recall that due to the Gabber-Kac theorem [83], any Kac-Moody algebra $\mathfrak{g}$ associated with a symmetrizable *generalized Cartan matrix* $A = ||a_{is}||$, $1 \leq i, s \leq n$ (an integral $n \times n$ matrix such that $a_{ii} = 2$, $a_{is} \leq 0$ for $i \neq s$, and $a_{is} = 0$ implies that $a_{si} = 0$) has the following representation by generators and relations. The generators are $3n$ elements $\mathbf{e}_i, \mathbf{f}_i, \mathbf{h}_i$, $1 \leq i \leq n$. The relations are divided into three groups:

$$[\mathbf{h}_i, \mathbf{h}_s] = 0, \quad [\mathbf{h}_i, \mathbf{e}_s] = a_{is}\mathbf{e}_s, \quad [\mathbf{h}_i, \mathbf{f}_s] = -a_{is}\mathbf{f}_s; \tag{3.1}$$

$$[\mathbf{e}_i, \mathbf{f}_s] = 0 \text{ if } i \neq s, \quad [\mathbf{e}_i, \mathbf{f}_i] = \mathbf{h}_i; \tag{3.2}$$

$$(\operatorname{ad} \mathbf{e}_i)^{1-a_{is}}\mathbf{e}_s = 0, \quad (\operatorname{ad} \mathbf{f}_i)^{1-a_{is}}\mathbf{f}_s = 0 \text{ if } i \neq s, \tag{3.3}$$

where by definition $(\operatorname{ad} a)^m b = [\underbrace{\ldots [[b, a], a], \ldots, a]}_{m}$.

A generalized Cartan matrix $A = ||a_{is}||$ is said to be *symmetrizable* if there exist natural numbers $d_1, d_2, \ldots, d_n$ such that $d_i a_{is} = d_s a_{si}$, $\quad 1 \leq i, s \leq n$.

A generalized Cartan matrix A is said to be *indecomposable* if there is no partition of the set $\{1, 2, \ldots, n\}$ into two nonempty subsets such that $a_{is} = 0$ whenever i belongs to the first subset, and s belongs to the second. Evidently, every generalized Cartan matrix is a diagonal sum of its indecomposable components, $A = \bigoplus A_\lambda$, whereas the Kac-Moody algebra $\mathfrak{g}$ defined by A is a direct sum of the Kac-Moody algebras $\mathfrak{g}_\lambda$ defined by A_λ.

A generalized Cartan matrix A is called a *Cartan matrix* if all its indecomposable components belongs to the following list of distinguished matrices:

$$A_n, \quad B_n, \quad C_n, \quad D_n, \quad E_6, \quad E_7, \quad E_8, \quad F_4, \quad G_2. \tag{3.4}$$

The indices coincide with dimensions of the related matrices. The non-diagonal coefficients of the matrix A_n are defined as follows:

$$(A_n) \qquad a_{is} = \begin{cases} -1, & \text{if } |i-s| = 1, \\ 0, & \text{otherwise.} \end{cases}$$

The matrices B_n, C_n, F_4, G_2 differ from the matrix A_n of the same dimension only in one coefficient:

$$(B_n) \ a_{n-1\,n} = -2; \quad (C_n) \ a_{n\,n-1} = -2; \quad (F_4) \ a_{23} = -2; \quad (G_2) \ a_{21} = -3.$$

The matrix D_n differs from A_n in four coefficients:

$$(D_n) \qquad a_{n\,n-1} = a_{n-1\,n} = 0, \ a_{n-2\,n} = a_{n\,n-2} = -1,$$

whereas E_n differs from A_n, $n = 6, 7, 8$ only in the following coefficients:

$$(E_n) \qquad a_{12} = a_{21} = a_{23} = a_{32} = 0, \ a_{13} = a_{31} = a_{24} = a_{42} = -1.$$

The fundamental property of the Cartan matrices is that they define semisimple finite dimensional Kac-Moody algebras provided that the ground field **k** has zero characteristic. Moreover, in this case all semisimple finite dimensional Lie algebras over $\bar{\mathbf{k}}$ are precisely the Kac-Moody algebras defined by the Cartan matrices. Respectively, the list (3.4) corresponds to the finite dimensional simple Lie algebras.

In the theory of Kac-Moody algebras, another 16 types of generalized Cartan matrices, called *affine Cartan matrices*, are important. These matrices also differ from A_n only in a small number of coefficients.

3.2 Quantum Deformations

The universal enveloping algebra $U(\mathfrak{g})$ of a Kac-Moody algebra $\mathfrak{g}$ is the associative algebra defined by the same relations (3.1)–(3.3) when the brackets are replaced by ordinary commutators, $[a, b] = ab - ba$. Of course $U(\mathfrak{g})$ has the structure of a Hopf algebra where all generators are primitive, $\Delta(v) = v \otimes 1 + 1 \otimes v, \ v \in \mathfrak{g}$. We wish to investigate all possible "deformations" H of $U(\mathfrak{g})$ to Hopf algebras which are defined by relations of the same degrees without fixing the coefficients.

We suppose that the Chevalley generators, $\mathbf{e}_i$, $\mathbf{f}_i$ are transformed to skew-primitive generators y_i and y_i^-, $1 \leq i \leq n$, respectively. Relations of the first group demonstrate that the subalgebra generated by the $\mathbf{h}_i$'s (the Cartan subalgebra) is commutative and acts on the Chevalley generators so that they are semi-invariants. Moreover, the second and third relations demonstrate that characters defined by this action on $\mathbf{e}_i$ and $\mathbf{f}_i$ with a fixed i are opposite to each other. For this reason, we suppose that under a deformation the Cartan subalgebra is transformed to a group G with the diagonal action on the skew-primitive generators,

$$g^{-1} y_i g = \chi^i(g) y_i, \quad g^{-1} y_i^- g_i = \chi_-^i(g) y_i^-,$$

and related characters are opposite, $\chi_-^i = (\chi^i)^{-1}$, $1 \leq i \leq n$. We shall identify the group G of group-like elements with diagonal transformations of the space V spanned by y_i, y_i^-, $1 \leq i \leq n$:

$$G = \{\mathrm{diag}\,(\lambda_1, \lambda_2, \ldots, \lambda_n, \lambda_1^{-1}, \lambda_2^{-1}, \ldots, \lambda_n^{-1}) \mid \lambda_i \in \mathbf{k}^*\}. \tag{3.5}$$

The second group of relations demonstrates that y_i and y_s^-, $i \neq s$ must be connected by a bilinear relation:

$$R_{is} \stackrel{\text{df}}{=} \alpha_{is}\, y_i y_s^- + \beta_{is}\, y_s^- y_i = 0, \tag{3.6}$$

whereas for $i = s$, some bilinear combination belongs to the coradical:

$$R_{ii}(y_i, y_i^-) \stackrel{\text{df}}{=} \alpha_{ii}\, y_i y_i^- + \beta_{ii}\, y_i^- y_i - \sum_k \mu_{ik}\, g_k = 0, \;\; g_k \in G. \tag{3.7}$$

Finally, the third group of relations is transformed to relations of the type

$$S_{it}(y_i, y_t) \stackrel{\text{df}}{=} \sum_{k=0}^{1-a_{it}} \gamma_{itk}\, y_i^k y_t y_i^{1-a_{it}-k} = 0, \;\; i \neq t; \tag{3.8}$$

$$S_{it}^-(y_i^-, y_t^-) \stackrel{\text{df}}{=} \sum_{k=0}^{1-a_{it}} \delta_{itk}\, (y_i^-)^k y_t^- (y_i^-)^{1-a_{it}-k} = 0, \;\; i \neq t. \tag{3.9}$$

For arbitrary values of the parameters α_{it}, β_{it}, μ_{ik}, γ_{itk}, δ_{itk}, and arbitrary generalized Cartan matrix A (not necessary symmetrizable) the relations (3.6)–(3.9) define an algebra H, however, this algebra does not always remain a Hopf algebra. Then, our next goal is to understand when H does retain the Hopf algebra structure. As an example we consider the *Drinfeld–Jimbo quantizations*.

Example 3.1 Assume that a generalized Cartan matrix $A = ||a_{is}||$ is symmetrizable; that is, there exist natural numbers $d_1, d_2, \ldots, d_n$ such that

$$d_i a_{is} = d_s a_{si}, \quad 1 \leq i, s \leq n.$$

Traditionally the skew-primitive generators of the Drinfeld–Jimbo algebra $U_q(\mathfrak{g})$ defined by the Kac-Moody algebra $\mathfrak{g}$ and a parameter $q \in \mathbf{k}$ have the designations $y_i = E_i$, $y_i^- = F_i$, $1 \leq i \leq n$. The related group-like elements are $h_i = K_i^{-1}$, $f_i = K_i$ both for E_i and F_i, so that

$$\Delta(E_i) = E_i \otimes K_i^{-1} + K_i \otimes E_i; \;\; \Delta(F_i) = F_i \otimes K_i^{-1} + K_i \otimes F_i, \;\; 1 \leq i \leq n.$$

The characters χ^{E_i} and χ^{F_i} are defined by $\chi^{E_i}(K_s) = q^{-d_i a_{is}}$ and $\chi^{F_i}(K_s) = q^{d_i a_{is}}$:

$$K_i^{-1} E_s K_i = q^{-d_i a_{is}} E_s, \quad K_i^{-1} F_s K_i = q^{d_i a_{is}} F_s,$$

in other words

$$K_i = \text{diag}(q^{-d_i a_{i1}}, q^{-d_i a_{i2}}, \ldots, q^{-d_i a_{in}}, q^{d_i a_{i1}}, q^{d_i a_{i2}}, \ldots, q^{d_i a_{in}}).$$

Relations (3.6) and (3.7) have a symmetric form

$$E_i F_s - F_s E_i = \delta_i^s \left(\frac{K_i^2 - K_i^{-2}}{q^{2d_i} - q^{-2d_i}} \right), \tag{3.10}$$

whereas relations (3.7) are:

$$[\ldots [[E_i, \underbrace{E_s], E_s], \ldots, E_s}_{1-a_{si}}] = 0, \quad 1 \le i \ne s \le n;$$

$$[\ldots [[F_i, \underbrace{F_s], F_s], \ldots, F_s}_{1-a_{si}}] = 0, \quad 1 \le i \ne s \le n, \tag{3.11}$$

where (1.67) defines the brackets, $[u, v] = \chi^v(h_u)uv - \chi^u(f_v)vu$, for example,

$$[E_i, E_s] = \chi^{E_s}(K_i^{-1})E_i E_s - \chi^{E_i}(K_s)E_s E_i = q^{d_i a_{is}}E_i E_s - q^{-d_i a_{is}}E_s E_i,$$

and

$$[F_i, F_s] = \chi^{F_s}(K_i^{-1})F_i F_s - \chi^{F_i}(K_s)F_s F_i = q^{-d_i a_{is}}F_i F_s - q^{d_i a_{is}}F_s F_i.$$

Let us demonstrate that $U_q(\mathfrak{g})$ retain the Hopf algebra structure. By Proposition 1.7 it suffices to check that all defining relations are skew-primitive elements as polynomials of the free character Hopf algebra. Each of the relations is a polynomial in two variables. Therefore we may apply Theorem 2.5. We have

$$[E_i, F_s] = \chi^{F_s}(K_i^{-1})E_i F_s - \chi^{E_i}(K_s)F_s F_i = q^{-d_s a_{si}}E_i F_s - q^{-d_i a_{is}}F_s F_i \sim F_i F_s - F_s F_i.$$

By Theorem 2.5 with $n = 1$, the polynomial $[E_i, F_s]$ is $((K_i K_s)^{-1}, K_i K_s)$-primitive because $p_{12} p_{21} = 1$, where according to (2.50), we have

$$p_{12} = \chi^{E_i}(K_s^2) = q^{-2d_i a_{is}}, \quad p_{21} = \chi^{F_s}(K_i^2) = q^{2d_s a_{si}}.$$

Thus, the left-hand side of (3.10) is always skew-primitive. If $i = s$, then the right-hand side is (K_i^{-2}, K_i^2)-primitive too since it is proportional to a difference of two group-like elements $K_i^{-2} - K_i^2$.

Similarly, by Theorem 2.5 with $n = 1 - a_{si}$, the left-hand sides of (3.11) are skew-primitive provided that $p_{12} p_{21} = p_{22}^{a_{si}}$. For the relations in the E_i's we have,

$$p_{12} \cdot p_{21} = \chi^{E_i}(K_s^2) \cdot \chi^{E_s}(K_i^2) = q^{-2d_i a_{is}} q^{-2d_s a_{si}} = (q^{-4d_s})^{a_{si}} = (\chi^{E_s}(K_s^2))^{a_{si}} = p_{22}^{a_{si}},$$

whereas relations in the F_i's satisfy

$$p_{12} \cdot p_{21} = \chi^{F_i}(K_s^2) \cdot \chi^{F_s}(K_i^2) = q^{2d_i a_{is}} q^{2d_s a_{si}} = (q^{4d_s})^{a_{si}} = (\chi^{F_s}(K_s^2))^{a_{si}} = p_{22}^{a_{si}}.$$

3.3 Defining Relations of the Main Class

To make notations compatible with Sect. 1.5, let us identify $y_k,\ n < k \leq 2n$ with y_{k-n}^-, so that if $k = n + s$ then $y_k = y_s^-,\ h_k = h_s^-,\ f_k = f_s^-$. We keep notation of Sect. 1.5.3:

$$\Delta(y_i) = y_i \otimes h_i + f_i \otimes y_i,\ \ y_i g = \chi^i(g) g y_i,\quad h_i,\ f_i,\ g \in G,\ 1 \leq i \leq 2n.$$

The parameters $q_{ik} = \chi^i(h_k),\ q'_{ik} = \chi^i(f_k),\ 1 \leq i \leq n$ completely define the free character Hopf algebra $G\langle Y\rangle$ over the diagonal group (3.5). Moreover, $G\langle Y\rangle$ as a Hopf algebra is completely defined by parameters

$$p_{ik} = q_{ik}^{-1} q'_{ik},\quad 1 \leq i, k \leq n$$

related to the normalized skew-primitive generators $x_i = h_i^{-1} y_i$ because $G\langle Y\rangle = G\langle X\rangle$.

We fix an algebra H defined by relations (3.6)–(3.9). The following three lemmas demonstrate that if H keeps the structure of Hopf algebra, then values of the parameters $\alpha_{is},\ \beta_{is},\ \mu_{ik},\ \gamma_{isk},\ \delta_{isk}$ are completely defined by the basic parameters $q_{is},\ q'_{is},\ 1 \leq i, s \leq n$. More precisely, we demonstrate that the only option for keeping the coproduct is to replace the Lie operation in (3.2), (3.3) with the brackets (1.67).

If $\{R_1, R_2, \ldots, R_m\}$ is a set of elements from $G\langle Y \cup Y^-\rangle$, then the G-algebra

$$G\langle y_1, y_2, \ldots, y_{2n} \mid\mid R_1, R_2, \ldots, R_m\rangle \stackrel{\text{df}}{=} G\langle Y\rangle / \mathrm{Id}(R_1, R_2, \ldots, R_m)$$

defined by generators $y_1, y_2, \ldots, y_{2n}$ and relations $R_1 = 0,\ R_2 = 0,\ \ldots,\ R_m = 0$ retains the Hopf algebra structure only if the ideal J generated by $R_1, R_2, \ldots, R_m$ is a Hopf ideal: $\Delta(J) \subseteq J \otimes G\langle Y \cup Y^-\rangle + G\langle Y \cup Y^-\rangle \otimes J,\ \sigma(J) \subseteq J,\ \varepsilon(J) = 0$.

On the free character Hopf algebra $G\langle Y \cup Y^-\rangle$ we introduce the following degree function D with integer (not necessary positive) values. We set

$$D(y_i) = 1,\ D(y_i^-) = D(y_{n+i}) = -1,\ 1 \leq i \leq n,\ D(g) = 0,\ g \in G.$$

A degree of a word in $Y \cup Y^- \cup G$ equals the sum of all degrees of its letters, whereas a degree of a linear combination of words equals the maximum degree of the words. The coproduct formula (1.60) is homogeneous with respect to D. Thus, the free character Hopf algebra is homogeneous with respect to D as well. In other words, the free character Hopf algebra is graded by the group of integer numbers,

$$G\langle Y \cup Y^-\rangle = \bigoplus_{i=-\infty}^{\infty} \Gamma_i,\quad \Delta(\Gamma_i) \subseteq \bigoplus_{s+k=i} \Gamma_s \otimes \Gamma_k,$$

where the ith component Γ_i is spanned by all words of degree i.

Below, the relation $a \sim b$ means projective equality: $a = \alpha b,\ \alpha \in \mathbf{k},\ \alpha \neq 0$.

Lemma 3.1 *If the algebra H keeps the Hopf algebra structure then either $y_i = y_s^- = 0$ in H or $R_{is} \sim [y_i, y_s^-]$ and $q'_{ik}q'_{ki} = q_{ik}q_{ki},\ k = n+s$.*

Proof All of the above-defining relations for H are homogeneous with respect to D. Thus, the Hopf algebra H is homogeneous as well; that is, H also has a grading by the integer numbers:

$$H = \bigoplus_{i=-\infty}^{\infty} H_i, \quad \Delta(H_i) \subseteq \bigoplus_{s+k=i} H_s \otimes H_k. \tag{3.12}$$

According to (3.6) in $H \otimes H$, we have

$$0 = \Delta(R_{is}) = R_{is} \otimes h_i h_k + f_i f_k \otimes R_{is} + \alpha_{is}\, y_i f_k \otimes h_i y_k + \alpha_{is} f_i y_k \otimes y_i h_k$$
$$+ \beta_{is}\, y_k f_i \otimes h_k y_i + \alpha_{is} f_k y_i \otimes y_k h_i, \tag{3.13}$$

where $k = n + s$.

Due to the commutation relations (1.62), this equality implies that

$$(\alpha_{is} q'_{ik} + \beta_{is} q_{ki}) f_k y_i \otimes h_i y_k + (\alpha_{is} q_{ik} + \beta_{is} q'_{ki}) f_i y_k \otimes h_k y_i = 0.$$

Note that the elements $f_k y_i$ and $f_i y_k$ are linearly independent in H or $y_i = y_k = 0$ in H. Indeed, because the algebra H is homogeneous with respect to D, a relation $\alpha f_k y_i + \beta f_i y_k = 0$ implies that $f_k y_i = 0,\ f_i y_k = 0$; that is, $y_i = y_k = 0$.

As a consequence of the above note, the tensors $f_k y_i \otimes h_i y_k$ and $f_i y_k \otimes h_k y_i$ are linearly independent in $H \otimes H$. Therefore,

$$\alpha_{is} q'_{ik} + \beta_{is} q_{ki} = 0, \quad \alpha_{is} q_{ik} + \beta_{is} q'_{ki} = 0. \tag{3.14}$$

The above system of equations has a nonzero solution with respect to α_{is}, β_{is} if and only if

$$\det \begin{vmatrix} q'_{ik} & q_{ki} \\ q_{ik} & q'_{ki} \end{vmatrix} = q'_{ik} q'_{ki} - q_{ik} q_{ki} = 0. \tag{3.15}$$

Under that condition, $\alpha_{is} = q_{ki},\ \beta_{is} = -q'_{ik}$ is the only solution up to a scalar factor. Thus,

$$R_{is} \sim q_{ki}\, y_i y_k - q'_{ik}\, y_k y_i = [y_i, y_k] = [y_i, y_s^-].$$

□

Lemma 3.2 *If the algebra H keeps the Hopf algebra structure, then*

$$R_{ii} \sim [y_i, y_i^-] - \alpha_i(h_i h_i^- - f_i f_i^-), \quad \alpha_i \in \mathbf{k}, \tag{3.16}$$

and $q'_{in+i}q'_{n+ii} = q_{in+i}q_{n+ii}$.

Proof Let us denote for short $s = n + i$. Then, the relation (3.7) takes the form

$$R_{ii}^0 \stackrel{\text{df}}{=} \alpha_{ii}\, y_i y_s + \beta_{ii}\, y_s y_i = \sum_k \mu_{ik}\, g_k.$$

Applying the coproduct to this relation, we obtain

$$\begin{aligned} R_{ii}^0 \otimes h_i h_s + f_i f_s \otimes R_{ii}^0 + \alpha_{ii} f_i y_s \otimes y_i h_s + \alpha_{ii}\, y_i f_s \otimes h_i y_s \\ + \beta_{ii} f_s y_i \otimes y_s h_i + \beta_{ii}\, y_s f_i \otimes h_s y_i = \sum_k \mu_{ik}\, g_k \otimes g_k. \end{aligned} \tag{3.17}$$

Due to the commutation relations (1.62) and $R_{ii}^0 = \sum_k \mu_{ik}\, g_k$, this result implies that

$$\begin{aligned} (\alpha_{ii} q'_{is} + \beta_{ii} q_{si}) f_s y_i \otimes h_i y_s + (\alpha_{ii} q_{is} + \beta_{ii} q'_{si}) f_i y_s \otimes h_s y_i \\ = -(\sum_k \mu_{ik}\, g_k) \otimes h_i h_s - f_i f_s \otimes (\sum_k \mu_{ik}\, g_k) + \sum_k \mu_{ik}\, g_k \otimes g_k. \end{aligned}$$

The homogeneous components with respect to (3.12) of this relation are valid too:

$$\alpha_{ii} q'_{is} + \beta_{ii} q_{si} = 0, \quad \alpha_{ii} q_{is} + \beta_{ii} q'_{si} = 0, \tag{3.18}$$

$$(\sum_k \mu_{ik}\, g_k) \otimes h_i h_s + f_i f_s \otimes (\sum_k \mu_{ik}\, g_k) = \sum_k \mu_{ik}\, g_k \otimes g_k. \tag{3.19}$$

The system of equations (3.18) has a nonzero solution α_{ii}, β_{ii} if and only if

$$\det \begin{vmatrix} q'_{is} & q_{ii} \\ q_{is} & q'_{si} \end{vmatrix} = q'_{is} q'_{si} - q_{is} q_{si} = 0. \tag{3.20}$$

Under that condition, $\alpha_{ii} = q_{si}$, $\beta_{ii} = -q'_{is}$ is the only solution up to a scalar factor. Thus, $R_{ii}^0 \sim q_{si}\, y_i y_s - q'_{is}\, y_s y_i = [y_i, y_s] = [y_i, y_i^-]$.

Different group-like elements in a Hopf algebra are always linearly independent. If $g_k \neq h_i h_s$, $g_k \neq f_i f_s$ then the tensor $g_k \otimes g_k$ appears only in the right-hand side of (3.19). This fact implies that $\mu_{ki} = 0$. Thus, the right-hand side of (3.16) has two terms: $g_1 = h_i h_s$, $g_2 = f_i f_s$, in which case, the relation (3.19) implies that $\mu_{2i} = -\mu_{1i}$. Hence, the right-hand side of (3.16) equals $\alpha_i(h_i h_s - f_i f_s)$ with $\alpha_i = \mu_{1i}$. □

Lemma 3.3 *If* $H^+ = G\langle y_1, y_2, \ldots, y_n \mid\mid S_{it}(y_i, y_t), 1 \le i \ne t \le n\rangle$ *maintains the Hopf algebra structure, then* S_{it} *are skew-primitive polynomials. In particular, by Theorem 2.5, we have* $S_{it} \sim [\ldots[[y_t, y_i], y_i], \ldots, y_i]$ *in* $G\langle Y\rangle$.

Proof All defining relations (3.8) are homogeneous with respect to each variable y_i, $1 \le i \le n$, as are the definition formulae (1.60) of the coproduct. Therefore, the Hopf algebra H^+ is also homogeneous with respect to each variable y_i, $1 \le i \le n$; that is, H^+ has a grading by $(Z^+)^{\times n}$:

$$H^+ = \bigoplus_{\mathbf{u}\in(Z^+)^{\times n}} H^+_{\mathbf{u}}, \quad \Delta(H^+_{\mathbf{u}}) = \bigoplus_{\mathbf{u}=\mathbf{v}+\mathbf{w}} H^+_{\mathbf{v}} \otimes H^+_{\mathbf{w}}, \tag{3.21}$$

where $H^+_{\mathbf{u}}$ is spanned by the values of all words in $Y \cup G$ of constitution $\mathbf{u}$.

We observe that the ideal of relations I is generated by elements whose constitutions have precisely two nonzero components, one of which equals 1. In particular, I has no nonzero elements of constitution with just one nonzero component. Moreover, the ideal I has no nonzero elements with a constitution $\mathbf{u} = (d_1, d_2, \ldots, d_n)$ such that $d_t = 1,\ d_i < 1 - a_{it},\ d_k = 0,\ k \ne i, t$. Indeed, only two generators, S_{it} and S_{ti}, have constitutions with $d_k = 0,\ k \ne i, t$. Thus, the ideal I would have a nonzero element with $d_t = 1,\ d_i < 1 - a_{it},\ d_k = 0,\ k \ne i, t$ only if $1 - a_{ti}$, the degree of S_{ti} in y_t, equals 1. In this case, $a_{ti} = 0$, and according to the definition of the generalized Cartan matrix $a_{it} = 0$. Thus, $d_i < 1 - a_{it}$ implies that $d_i = 0$.

The coproduct of the polynomial $S_{it} \in G\langle Y\rangle$ can be decomposed as follows:

$$\Delta(S_{it}) = S_{it} \otimes h_i h_t^{1-a_{it}} + f_i f_t^{1-a_{it}} \otimes S_{it} + \sum_k S_k^{(1)} \otimes S_k^{(2)}, \tag{3.22}$$

where the sum of the constitutions of $S_k^{(1)}$ and $S_k^{(2)}$ equals the constitution of S_{it}. In algebra H^+, we have $S_{it} = 0$; therefore,

$$0 = \sum_k S_k^{(1)} \otimes S_k^{(2)} \quad \text{in } H^+ \otimes H^+. \tag{3.23}$$

One of the polynomials $S_k^{(1)}$, $S_k^{(2)}$ has a constitution with only one nonzero component, whereas another polynomial has a constitution with $d_i = 1,\ d_t < 1 - a_{it},\ d_k = 0,\ k \ne i, t$. We have observed above that the ideal I has no nonzero polynomials with such constitutions. Therefore, relation (3.23) is valid in $G\langle Y\rangle \otimes G\langle Y\rangle$ as well. Thus, (3.22) implies that S_{it} is a skew-primitive polynomial of $G\langle Y\rangle$. □

In a perfect analogy, we have

Lemma 3.4 *If* $H^- = G\langle y_1^-, y_2^-, \ldots, y_n^- \parallel S_{it}^-(y_i^-, y_t^-) = 0, 1 \leq i \neq t \leq n\rangle$ *maintains the Hopf algebra structure, then* $S_{it}^- \sim [\ldots[[y_t^-, y_i^-], y_i^-], \ldots, y_i^-]$ *and* S_{it} *are skew-primitive polynomials in* $G\langle Y^-\rangle$.

Thus, if H, H^+, and H^- maintain the Hopf algebra structure then the defining relations are simplified. Moreover, if the coefficient α_i is nonzero in the relation R_{ii}, see (3.16), we may simplify that relation further by the substitution $y_i \leftarrow \alpha_i y_i$, so we may suppose $\alpha_i = 1$. We can now define the main class of Hopf algebras in which we are interested.

Definition 3.1 A Hopf algebra H is a *quantization* of a Kac-Moody algebra $\mathfrak{g}$ if it is generated by skew-primitive semi-invariants y_i, y_i^-, $1 \leq i \leq n$ with $\chi_-^i = (\chi^i)^{-1}$ and is defined over the group G by the relations (3.6)–(3.9) that satisfy conclusions of Lemmas 3.1–3.4:

$$[y_i, y_t^-] = 0 \text{ if } i \neq t, \quad [y_i, y_i^-] = h_i h_i^- - f_i f_i^-; \tag{3.24}$$

$$[[\ldots[[y_t, \underbrace{y_i], y_i], \ldots], y_i}_{1-a_{it}}] = 0, \quad [[\ldots[[y_t^-, \underbrace{y_i^-], y_i^-], \ldots], y_i^-}_{1-a_{it}}] = 0 \text{ if } i \neq t. \tag{3.25}$$

We stress that the conclusions of Lemmas 3.1–3.4 imply that these relations are skew-primitive as elements of the free character Hopf algebra $G\langle Y, Y^-\rangle$. Therefore, the algebras H, H^+, H^- defined by the above relations maintain the Hopf algebra structure.

In variables $x_i = h_i^{-1} y_i$, $1 \leq i \leq 2n$, the defining relations have the following form in terms of the left adjoint action:

$$(\operatorname{ad} x_i) x_t^- = 0 \text{ if } i \neq t, \quad (\operatorname{ad} x_i) x_i^- = 1 - g_i g_i^-; \tag{3.26}$$

$$(\operatorname{ad} x_i)^{1-a_{it}} x_t = 0, \quad (\operatorname{ad} x_i^-)^{1-a_{it}} x_t^- = 0 \text{ if } i \neq t. \tag{3.27}$$

Indeed, by definition, the left adjoint action has the form $(\operatorname{ad} v)u = \sum_{(v)} v^{(1)} u (\sigma v^{(2)})$, where in Sweedler notations $\Delta(v) = \sum_{(v)} v^{(1)} \otimes v^{(2)}$ and σ is the antipode. Considering that $\Delta(x_i) = x_i \otimes 1 + g_i \otimes x_i$, $g_i = h_i^{-1} f_i$, $1 \leq i \leq 2n$, we have

$$(\operatorname{ad} x_i)u = x_i \cdot u \cdot 1 - g_i \cdot u \cdot g_i^{-1} x_i = x_i u - \chi^u(g_i^{-1}) u x_i \sim u x_i - \chi^u(g_i) x_i u = [u, x_i].$$

Theorem 3.1 *The Hopf algebra H is completely defined by* n^2 *parameters* p_{is}, $1 \leq i, s \leq n$, *whereas the algebra structure of H is completely defined by* $n(n+1)/2$ *parameters* p_{ii}, $1 \leq i \leq n$, $p_{is}p_{si}$, $1 \leq i < s \leq n$, *in which case the algebra structures of* H^+ *and* H^- *are completely defined by only n parameters* p_{ii}, $1 \leq i \leq n$. *The latter parameters are not independent yet:*

$$(p_{ii}^{a_{is}} - p_{ii})(p_{ss}^{a_{si}} - p_{ss})(p_{ii}^{a_{is}} - p_{ss}^{a_{si}}) = 0, \quad 1 \leq i, s \leq n. \tag{3.28}$$

Additionally, if $p_{is}p_{si} \neq p_{ii}^{a_{is}}$, *then*

$$(p_{is}p_{si})^{1-a_{is}} = p_{ii}^{1-a_{is}} = 1. \tag{3.29}$$

Proof We start with the following auxiliary statement.

Lemma 3.5 *We have* $\chi^i(g_s^-) = \chi^i(g_{n+s}) = p_{si}$, $1 \leq i \leq n$. *In particular, the group-like elements* g_i, $g_i^- = g_{n+i}$, $1 \leq i \leq n$ *have the following representation (3.5):*

$$g_i = \mathrm{diag}(p_{1i}, p_{2i}, \ldots, p_{ni}, p_{1i}^{-1}, p_{2i}^{-1}, \ldots, p_{ni}^{-1}); \tag{3.30}$$

$$g_i^- = \mathrm{diag}(p_{i1}, p_{i2}, \ldots, p_{in}, p_{i1}^{-1}, p_{i2}^{-1}, \ldots, p_{in}^{-1}). \tag{3.31}$$

Proof In the proof of Lemma 3.1, we found the equality $q_{ik}'q_{ki}' = q_{ik}q_{ki}$, $k = n + s$, $i \neq s \leq n$, see (3.15). Because $p_{ik} = q_{ik}'q_{ik}^{-1}$ by definition, $p_{ik}p_{ki} = 1$, or, in terms of the characters, $\chi^i(g_s^-) \cdot \chi_-^s(g_i) = 1$, $1 \leq i \neq s \leq n$. Equation (3.20) demonstrates that if $i = s$, the latter equality is valid as well. Considering that $\chi_-^i = (\chi^i)^{-1}$, we obtain $\chi^i(g_s^-) = \chi^s(g_i) = p_{si}$. To see representations (3.30) and (3.31), we recall that by definition, if $g = \mathrm{diag}(\lambda_1, \lambda_2, \ldots, \lambda_{2n})$, then $\chi^i(g) = \lambda_i$, $1 \leq i \leq 2n$. □

Representations (3.30) and (3.31) demonstrate that g_i, $1 \leq i \leq 2n$ are completely defined by the parameters p_{is}, $1 \leq i, s \leq n$. Due to equalities $\chi_-^i = (\chi^i)^{-1}$ and $\chi^i(g_s^-) = p_{si}$, all of the commutation rules are:

$$x_i g_s = p_{is} g_s x_i, \; x_i g_s^- = p_{si} g_s^- x_i, \; x_i^- g_s = p_{is}^{-1} g_s x_i^-, \; x_i^- g_s^- = p_{si}^{-1} g_s^- x_i^-. \tag{3.32}$$

All coefficients of the relations (3.26) and (3.27) are polynomials in p_{is}, p_{is}^{-1}. Thus the Hopf algebra structure of H is completely defined by values of the parameters p_{is}, $1 \leq i, s \leq n$.

Let us check the algebra structure. To this end, we consider a new set of generators $\tilde{y}_i = t_i x_i$, $\tilde{y}_i^- = t_i^{-1} x_i^-$, where

$$t_i = \mathrm{diag}(\tau_{1i}, \tau_{2i}, \ldots, \tau_{ni}, \tau_{1i}^{-1}, \tau_{2i}^{-1}, \ldots, \tau_{ni}^{-1}),$$

while the τ's are defined as follows:

$$\tau_{is} = \begin{cases} 1 & \text{if } n \geq i \geq t, \\ p_{si} & \text{if } i < s \leq n. \end{cases}$$

In other words, the t_i's are defined as group-like elements from G with $\chi^i(t_s) = \tau_{is}$, $1 \leq i, s \leq n$. If $i < s$, we have

$$[\tilde{y}_s, \tilde{y}_i] = \chi^i(t_s)\tilde{y}_s\tilde{y}_i - \chi^s(t_i g_i)\tilde{y}_i\tilde{y}_s = \tau_{is}\tilde{y}_s\tilde{y}_i - \tau_{si}p_{si}\tilde{y}_i\tilde{y}_s \sim \tilde{y}_s\tilde{y}_i - \tilde{y}_i\tilde{y}_s;$$

That is, the bracket $[\tilde{y}_s, \tilde{y}_i]$ is proportional to the ordinary commutator which is completely independent of the parameters.

More generally, if u is a homogeneous polynomial in $\tilde{y}_i, \tilde{y}_s$ linear in $\tilde{y}_s$ and of degree k in $\tilde{y}_i$, then $[u, \tilde{y}_i] \sim u\tilde{y}_i - p_{ii}^k \tilde{y}_i u$. Indeed, $h_u = t_s t_i^k$ and $\chi^u = \chi^s(\chi^i)^k$. Additionally, by definition $\chi^i(t_s) = \tau_{is} = p_{si}$, $\chi^s(t_i) = \tau_{si} = 1$. Therefore,

$$[u, \tilde{y}_i] = \chi^i(h_u)u\tilde{y}_i - \chi^u(t_i f_i)\tilde{y}_i u = \tau_{is}\tau_{ii}^k u\tilde{y}_i - \tau_{ii}^k \tau_{si}(p_{ii})^k p_{si}\tilde{y}_i u \sim u\tilde{y}_i - p_{ii}^k \tilde{y}_i u.$$

Similarly, if u^- is linear in $\tilde{y}_s^-$ and of degree k in $\tilde{y}_i^-$, then

$$[\tilde{y}_i, u] = \chi_-^u(t_i^{-1})\,\tilde{y}_i^- u^- - \chi_-^i(t_i^{-k} t_s^{-1}(g_i^-)^k g_s^-)\,u^- \tilde{y}_i^-$$
$$= \tau_{si}\tau_{ii}^k \tilde{y}_i^- u^- - \tau_{ii}^{-k}\tau_{is}p_{ii}^{-k}p_{si}^{-1}\,u^- \tilde{y}_i^- = \tilde{y}_i^- u^- - p_{ii}^{-k}\,u^- \tilde{y}_i^-,$$

where we have used $\tau_{ii}^k = 1$, and $\chi_-^i(g_s^-) = p_{si}^{-1}$.

By (2.63) in the free character Hopf algebra the following proportion holds

$$[[\ldots [[\underbrace{\tilde{y}_s, \tilde{y}_i], \tilde{y}_i], \ldots], \tilde{y}_i]}_{1-a_{is}} \sim \underbrace{[\tilde{y}_i, [\tilde{y}_i, \ldots, [\tilde{y}_i, \tilde{y}_s]}_{1-a_{is}} \ldots]]. \tag{3.33}$$

This implies that all coefficients of defining relations (3.25) with $i < s$ are polynomials in p_{ii} or p_{ii}^{-1}.

In a perfect analogy, we demonstrate that if $i > s$ then $[\tilde{y}_i, u] \sim \tilde{y}_i u - p_{ii}^k u\tilde{y}_i$ and $[u^-, \tilde{y}_i^-] \sim u^- \tilde{y}_i^- - p_{ii}^{k-} \tilde{y}_i^- u^-$. In this case, $\tau_{si} = p_{is}$, whereas $\tau_{is} = \tau_{ii} = 1$. Therefore,

$$[\tilde{y}_i, u] = \chi^u(t_i)\,\tilde{y}_i u - \chi^i(t_i^k t_s f_i^k f_s)\,u\tilde{y}_i$$
$$= \tau_{si}\tau_{ii}^k \tilde{y}_i u - \tau_{ii}^k \tau_{is}(p_{ii})^k p_{is}\,u\tilde{y}_i \sim \tilde{y}_i u - p_{ii}^k u\tilde{y}_i.$$

Similarly,

$$[u^-, \tilde{y}_i^-] = \chi_-^i(t_i^{-k} t_s^{-1})\,u^- \tilde{y}_i^- - \chi_-^u(t_i^{-1} g_i^-)\,\tilde{y}_i^- u^-$$
$$= \tau_{ii}^k \tau_{is}\,u^- \tilde{y}_i^- - \tau_{ii}^k \tau_{si}(p_{ii})^{-k} p_{is}^{-1}\tilde{y}_i^- u^- = u^- \tilde{y}_i^- - p_{ii}^{-k}\tilde{y}_i^- u^-.$$

Again, due to proportion (3.33), we see that all coefficients of (3.25) with $i > s$ are polynomials in p_{ii} or p_{ii}^{-1} (up to a common scalar factor). Thus, the algebraic structures of H^+ and H^- depend only on p_{ii}, $1 \le i \le n$.

To check the algebraic structure of H, we must analyze relations (3.24), where

$$[\tilde{y}_i, \tilde{y}_s^-] = \chi_-^s(t_i)\,\tilde{y}_i \tilde{y}_s^- - \chi_i(t_s^{-1} g_s^-)\,\tilde{y}_s^- \tilde{y}_i = \tau_{si}^{-1}\,\tilde{y}_i \tilde{y}_s^- - \tau_{is}^{-1} p_{si}\tilde{y}_s^- \tilde{y}_i.$$

If $i < s$ then $\tau_{is} = p_{si}$, $\tau_{si} = 1$. Thus, the relation takes the form

$$\tilde{y}_i \tilde{y}_s^- = \tilde{y}_s^- \tilde{y}_i. \tag{3.34}$$

If $i > s$ then $\tau_{is} = 1,\ \tau_{si} = p_{is}$, whereas the relation reduces to

$$\tilde{y}_i \tilde{y}_s^- = (p_{is} p_{si}) \tilde{y}_s^- \tilde{y}_i. \tag{3.35}$$

Finally, if $i = s$ then the relation transforms to

$$\tilde{y}_i \tilde{y}_i^- - p_{ii} \tilde{y}_i^- \tilde{y}_i = 1 - g_i g_i^-, \tag{3.36}$$

whereas

$$g_i g_i^- = \text{diag}\left(p_{1i}p_{i1}, p_{2i}p_{i2}, \ldots, p_{ni}p_{in}, (p_{1i}p_{i1})^{-1}, (p_{2i}p_{i2})^{-1} \ldots, (p_{ni}p_{in})^{-1}\right).$$

Thus, the algebraic structure of H depends on additional $n(n-1)/2$ parameters $p_{is}p_{si},\ 1 \le i < s \le n$.

Relations (3.28) and (3.29) follow from the existence conditions for skew-primitive polynomials given in Theorem 2.5. If $p_{ii}^{a_{is}} \neq p_{ii}$ then p_{ii} is not a primitive mth root of 1 with $m|(1 - a_{is})$. Thus, $p_{ii}^{a_{is}} = p_{is}p_{si}$. Similarly, if $p_{ss}^{a_{si}} \neq p_{ss}$ then p_{ss} is not a primitive mth root of 1 with $m|(1 - a_{ii})$, and $p_{ss}^{a_{si}} = p_{is}p_{si} = p_{ii}^{a_{is}}$, which proves (3.28). If $p_{is}p_{si} \neq p_{ii}^{a_{is}}$, then p_{ii} is a primitive mth root of 1 with $m|(1 - a_{is})$ and $(p_{is}p_{si})^m = 1$. This result implies (3.29). □

3.4 Triangular Decomposition

Relations (3.26) allows one to transform each word in $X \cup X^-$ to linear combination of G-words where all "negative" variables precede to "positive" variables. In particular, the linear map $H^- \otimes_{\mathbf{k}[G]} H^+ \to H,\ w^- \otimes u \mapsto w^- u$ is an epimorphism. In this section, we shall prove that this map is an isomorphism of coalgebras and $\mathbf{k}[G]$-bimodules. We consider a more general setting when instead of relations (3.27) appear arbitrary polynomial relations of upper degree grater than 1. Recall that by definition the *upper degree* of a polynomial is the minimal length of its monomials. The set of all polynomials in $X = \{x_i \mid i \in I\}$ of upper degree grater than 1 coincides with the ideal Λ^2 generated by the monomials $x_i x_s,\ i, s \in I$.

Let H_1^+ be an algebra defined as a G-algebra by the generators $x_1, \ldots, x_n$ and polynomial relations $\varphi_s = 0,\ \varphi_s \in \mathbf{k}\langle X \rangle,\ s \in S$,

$$H_1^+ = G\langle x_1, \ldots, x_n \,||\, \varphi_s, s \in S \rangle. \tag{3.37}$$

Respectively, H_1^- is an algebra defined as a G-algebra by the generators $x_1^-, \ldots, x_n^-$ and polynomial relations $\psi_t^- = 0,\ \psi_t^- \in \mathbf{k}\langle X^- \rangle,\ t \in T$.

$$H_1^- = G\langle x_1, \ldots, x_n \,||\, \psi_t^-, t \in T \rangle. \tag{3.38}$$

Consider the algebra

$$H_1 = G\langle x_1, \ldots, x_n, x_1^-, \ldots, x_n^- \,||\, F_1, F_2, F_3\rangle, \tag{3.39}$$

where $F_1 = \{\varphi_s, s \in S\}$, $F_2 = \{\psi_t^-, t \in T\}$, and F_3 are relations (3.26): $[x_i, x_s^-] = \delta_i^s(1 - g_i g_i^-)$. We still assume that $\chi_-^i = (\chi^i)^{-1}$ and $\chi^i(g_s^-) = p_{si}$ (see Lemma 3.5). By evident induction on length of words, the latter equality is equivalent to $\chi^u(g_v) = \chi^v(g_u^-)$, where u, v are words in X and u^- appears from u under replacements $x_i \leftarrow x_i^-$. Obviously, those two conditions are equivalent to commutation rules (3.32).

Lemma 3.6 *If H_1^+ and H_1^- maintain the Hopf algebra structure, then so does H_1.*

Proof By Theorem 2.5 with $n \leftarrow 1$, the polynomials $[x_i, x_s^-]$ are skew-primitive because $\chi^i(g_s^-)\chi_-^s(g_i) = p_{si}p_{si}^{-1} = 1$. Consider the ideals of relations $I_1 = \text{id}(F_1)$ and $I_2 = \text{id}(F_2)$ of H^+ and H^- respectively. In the present context, they are Hopf ideals of $G\langle X\rangle$ and $G\langle X^-\rangle$, respectively. Hence, $V = I_1 + I_2 + \sum_{i,s} \mathbf{k}([x_i, x_s^-] - \delta_i^s(1-g_i g_i^-))$ is an antipode stable coideal of $G\langle X\rangle$. Consequently the ideal generated by V is a Hopf ideal, see Proposition 1.6. □

Lemma 3.7 *If H_1^+ and H_1^- maintain the Hopf algebra structure, then every hard in H_1 super-letter belongs to either H_1^+ or H_1^-, and it is hard in the related algebra.*

Proof If a standard word u in $X \cup X^-$contains at least one "positive" letter, then it has to start with one of them. If u contains a "negative" letter, then it has a sub word of the form $x_i x_s^-$. The substitution $x_i x_s^- \leftarrow p_{si} x_s^- x_i + \delta_i^s(1 - g_i g_s^-)$ shows that value of u in H is a linear combination of lesser words of the same degree and G-words of lesser degrees. By Corollary 2.1, the super-letter $[u]$ is not hard. □

The converse statement is much more complicated. We shall proceed with a number of lemmas. Firstly we consider the case $F_1 = F_2 = \emptyset$. Let

$$\tilde{H} = G\langle x_1, \ldots, x_n, x_1^-, \ldots, x_n^- \,||\, F_3\rangle. \tag{3.40}$$

Lemma 3.8 *There is a natural isomorphism of linear spaces*

$$\tilde{H} \cong \mathbf{k}\langle x_1^-, \ldots, x_n^-\rangle \otimes \mathbf{k}[G] \otimes \mathbf{k}\langle x_1, \ldots, x_n\rangle.$$

Proof Since $w^- gu \sim gw^- u$ for all $v \in (X)^*$, $w^- \in (X^-)^*$, and $g \in G$, it follows that we have to demonstrate that the set of elements gw^-u, where $g \in G$ and w^-, u run through all words in X^- and X respectively form a basis of $\tilde{H}$. The latter set is precisely the set of all words in $G \cup X \cup X^-$ that have no subwords $x_i^\pm g$, $x_i x_s^-$. By Theorem 1.2 (Composition Lemma), it suffices to check that the table of multiplication of G, commutations rules $x_i^\pm g = \chi_\pm^i(g) g x_i^\pm$, and relations F_3 are closed with respect to the compositions provided that

$$x_1 > \ldots > x_n > x_1^- > \ldots > x_n^- > g_1 > g_2 > \ldots, \quad g_i \in G.$$

The leading words are: gh for the table of multiplication; $x_i^{\pm}g$, for the commutation rules; and $x_ix_s^-$, $1 \le i, s \le n$ for F_3. Compositions between the former two type of relations were resolved in Sect. 1.3.2. There are no compositions between different relations from F_3. It remains only one type of the compositions when the word $w_i'vw_s'$ appeared in (1.17) equals $x_ix_s^- \cdot g = x_i \cdot x_s^- g$:

$$\begin{aligned}
&\{p_{si}x_s^-x_i + \delta_i^s(1 - g_ig_i^-)\} \cdot g - x_i \cdot \chi_-^s(g)gx_s^- \\
&= p_{si}\underline{x_s^-x_ig} + \delta_i^s g(1 - g_ig_i^-) - \chi_-^s(g)\underline{x_igx_s^-} \\
&\to p_{si}\chi^i(g)\chi_-^s(g)gx_s^-x_i + \delta_i^s g(1 - g_ig_i^-) - \chi_-^s(g)\chi^i(g)g\underline{x_ix_s^-} \\
&\equiv p_{si}\chi^i(g)\chi_-^s(g)gx_s^-x_i + \delta_i^s g(1 - g_ig_i^-) \\
&\quad - \chi_-^s(g)\chi^i(g)g\{p_{si}x_s^-x_i + \delta_i^s(1 - g_ig_i^-)\} = 0
\end{aligned}$$

by virtue of the fact that $\delta_i^s = 0$ if $i \ne s$, and $\chi_-^i(g)\chi^i(g) = 1$. □

Lemma 3.9 *In the Hopf algebra $\tilde{H}$, the following relations hold*

$$[u, x_s^-] = \partial_s^*(u) - g_s\partial_s(u)g_s^-, \quad u \in \mathbf{k}\langle X\rangle, \quad 1 \le s \le n; \tag{3.41}$$

$$[x_i, u^-] = \partial_{-i}^*(u^-)p(x_i, u^-)p_{ii}^{-1} - g_ig_i^-\partial_{-i}(u^-), \quad u^- \in \mathbf{k}\langle X^-\rangle, \quad 1 \le i \le n. \tag{3.42}$$

Here ∂_s, ∂_s^* are the partial derivatives on $\mathbf{k}\langle X\rangle$, whereas ∂_{-i}, ∂_{-i}^* are the partial derivatives with respect to x_i^- defined on $\mathbf{k}\langle X^-\rangle$, see Sect. 1.5.7.

Proof Consider the linear maps

$$D_s : u \mapsto \partial_s^*(u) - g_s\partial_s(u)g_s^-, \quad u \in \mathbf{k}\langle X\rangle, \quad 1 \le i \le s. \tag{3.43}$$

Relations (1.81) and (1.86) imply

$$\begin{aligned}
D_s(uv) &= \partial_s^*(uv) - g_s\partial_s(uv)g_s^- \\
&= \chi^s(g_v)\partial_s^*(u)v + u\partial_s^*(v) - g_s\{\partial_s(u)v + \chi^u(g_s)u\partial_s(v)\}g_s^-.
\end{aligned}$$

Let us replace: first, $\chi^s(g_v) = \chi^v(g_s^-)$; then, $vg_s^- = \chi^v(g_s^-)g_s^-v$; and next, $g_s\chi^u(g_s)u = ug_s$. In this way we obtain the following "Leibniz rule" :

$$D_s(u \cdot v) = \chi^v(g_s^-)D_s(u) \cdot v + u \cdot D_s(v).$$

Under substitution $D_s \leftarrow [\text{ - }, x_s^-]$, this rule coincides with (2.9) for $w = x_s^-$:

$$[u \cdot v, x_s^-] = p(v, x_s^-)[u, x_s^-] \cdot v + u \cdot [v, x_s^-].$$

Since $D_s(x_i) = \delta_i^s(1 - g_ig_i^-) = [x_i, x_s^-]$, it follows that $D_s(u) = [u, x_s^-]$.

Similarly, consider the linear maps $T_i : \mathbf{k}\langle X^-\rangle \to G\langle X^-\rangle$, $1 \le i \le n$,

$$T_i(u^-) = \partial^*_{-i}(u^-)p(x_i, u^-)p_{ii}^{-1} - g_i g_i^- \partial_{-i}(u^-). \tag{3.44}$$

Using relations (1.81) and (1.86), we have

$$T_i(v^- \cdot w^-) = p(x_i, v^- \cdot w^-)p_{ii}^{-1}\{\chi^i_-(g_{w^-})\partial^*_{-i}(v^-) \cdot w^- + v^- \cdot \partial^*_{-i}(w^-)\}$$
$$- g_i g_i^- \{\partial_{-i}(v^-) \cdot w^- + \chi^{v^-}(g_i^-)v^- \partial_{-i}(w^-)\}.$$

Let us replace: first, $p(x_i, v^- \cdot w^-) = p(x_i, v^-)p(x_i, w^-)$; then, $\chi^i_-(g_{w^-}) = p(x_i, w^-)^{-1}$; then $g_i g_i^- \chi^{v^-}(g_i^-)v^- = (\chi^{v^-}(g_i))^{-1}v^- g_i g_i^-$; and next,

$$(\chi^{v^-}(g_i))^{-1} = \chi^v(g_i) = \chi^i(g_v^-) = p(x_i, v^-).$$

In this way we obtain a "Leibniz rule" :

$$T_i(v^- \cdot w^-) = T_i(v^-) \cdot w^- + p(x_i, v^-)v^- \cdot T_i(w^-),$$

which coincides with (2.8) under substitution $u \leftarrow x_i$:

$$[x_i, v^- \cdot w^-] = [x_i, v^-] \cdot w^- + p(x_i, v^-)v^- \cdot [x_i, w^-]. \tag{3.45}$$

Since $T_i(x_s^-) = \delta_i^s(1 - g_i g_i^-) = [x_i, x_s^-]$, it follows that $T_i(u^-) = [x_i, u^-]$. □

Now we turn to the case $F_1 = \emptyset$, $F_2 \neq \emptyset$. Consider the algebra

$$\hat{H} = G\langle x_1, \ldots, x_n, x_1^-, \ldots, x_n^- || F_2, F_3\rangle. \tag{3.46}$$

Lemma 3.10 *If H_1^- maintains the Hopf algebra structure and the upper degrees of ψ_t^- are greater than one, then the ideal generated by F_2 in the algebra $\tilde{H}$ equals $I_2\tilde{H} = I_2 \otimes_{\mathbf{k}[G]} G\langle X\rangle$, where I_2 is the ideal generated by F_2 in $G\langle X^-\rangle$.*

Proof It suffices to demonstrate that $I_2\tilde{H}$ admits left multiplication by x_i, $1 \le i \le n$. If v^- is a word in X^-, $h^- \in F_2$, $r \in \tilde{H}$, then

$$x_i v^- h^- r = [x_i, v^- h^-]r + p(x_i, v^- h^-)v^- h^- x_i r. \tag{3.47}$$

The second term belongs to $I_2\tilde{H}$, whereas the first one can be rewritten by (2.8):

$$[x_i, v^- h^-]r = [x_i, v^-]h^- r + p(x_i, v^-)v^-[x_i, h^-]r. \tag{3.48}$$

The first term of the latter sum belongs to $I_2\tilde{H}$ because $[x_i, v^-] \in G\langle X^-\rangle$ due to (3.42). If h^- is a skew-primitive element of $G\langle X^-\rangle^{(2)}$ (this is a case for the quantizations of Kac-Moody algebras!), then we are done, as (3.42) implies $[x_i, h^-] = 0$ in view of the fact that by Lemma 1.27 all partial derivatives $\partial_{-i}(h^-)$, $\partial^*_{-i}(h^-)$ are zero.

In general case, we may proceed by induction on the combinatorial rank of H_1^-. Recall that by Theorem 1.6 each nonzero bi-ideal has a nonzero skew-primitive element. Consider the ascending chain of ideals

$$\{0\} = I_2^{(0)} \subset I_2^{(1)} \subset I_2^{(2)} \subset \ldots \subset I_2^{(m)} \subset \ldots,$$

where $I_2^{(m)}$ is generated by all elements $u^- \in I_2$ such that $u^- + I_2^{(m-1)}$ are skew-primitive in $I_2^{(m)}/I_2^{(m-1)}$. Assume that the ideal of $\tilde{H}$ generated by $I_2^{(m-1)}$ equals $I_2^{(m-1)}\tilde{H}$. Let $h^- + I_2^{(m-1)}$ be a skew-primitive element of $I_2^{(m)}/I_2^{(m-1)}$. By Lemma 1.27 partial derivatives of h^- are zero in $I_2^{(m)}/I_2^{(m-1)}$. In other words, $\partial_{-i}^*(h^-)$ and $\partial_{-i}(h^-)$ belong to $I_2^{(m-1)}$. Relation (3.42) implies that the second term of (3.48) belongs to the ideal of $\tilde{H}$ generated by $I_2^{(m-1)}$, whereas the latter ideal equals $I_2^{(m-1)}\tilde{H} \subseteq I_2^{(m)}\tilde{H}$. Therefore, (3.47) implies that the ideal of $\tilde{H}$ generated by $I_2^{(m)}$ equals $I_2^{(m)}\tilde{H}$. □

The proven lemma implies a triangular decomposition of $\hat{H}$:

$$\begin{aligned}\hat{H} &= G\langle X^-\rangle \otimes_{\mathbf{k}[G]} G\langle X\rangle/I_2 \otimes_{\mathbf{k}[G]} G\langle X\rangle \\ &\cong (G\langle X^-\rangle/I_2) \otimes_{\mathbf{k}[G]} G\langle X\rangle = H_1^- \otimes_{\mathbf{k}[G]} G\langle X\rangle.\end{aligned} \tag{3.49}$$

Lemma 3.11 *If H_1^+ and H_1^- maintain the Hopf algebra structure and the upper degrees of φ_s, ψ_t^-, are greater than one, then an ideal generated by F_1 in $\hat{H}$ equals $I_1\hat{H} = H_1^- \otimes_{\mathbf{k}[G]} I_1$, where I_1 is the ideal generated by F_1 in $G\langle X\rangle$.*

Proof The proof almost literally coincides with the proof of the above lemma. It suffices to check that $I_1\hat{H}$ admits left multiplication by x_s^-, $1 \le s \le n$. If v is a word in X, $h \in F_1$, $r \in \hat{H}$, then

$$x_s^- vhr = -p(vh, x_s^-)^{-1}[vh, x_s^-]r + p(vh, x_s^-)^{-1}vhx_s^- r. \tag{3.50}$$

The latter term belongs to $I_1\hat{H}$, whereas the former one can be rewritten by (2.9):

$$[vh, x_s^-]r = p(h, x_s^-)[v, x_s^-]h + v[h, x_s^-]. \tag{3.51}$$

The first term of the latter sum belongs to $I_1\hat{H}$ because $[v, x_s^-] \in G\langle X\rangle$ due to (3.41). If h is skew-primitive, then (3.42) implies $[h, x_s^-] = 0$ in view of Lemma 1.27.

In general case, we may proceed by induction on the combinatorial rank of H_1^+ in the perfect analogy with the proof of the above lemma. □

Theorem 3.2 *Let H_1^+, H_1^-, and H_1 be defined by* (3.37), (3.38), *and* (3.39) *respectively. If H_1^+ and H_1^- maintain the Hopf algebra structure, and the upper degrees of all polynomials ψ_t^- and φ_s are greater than one, then we have an isomorphism of coalgebras and $\mathbf{k}[G]$-bimodules*

$$H_1 \cong H_1^- \otimes_{\mathbf{k}[G]} H_1^+ \tag{3.52}$$

provided that commutation rules (3.32) *hold.*

Proof Using Lemmas 3.8, 3.10, 3.11, and decomposition (3.49), we have isomorphisms of $\mathbf{k}[G]$-bimodules

$$H_1 \cong \hat{H}/I_1\hat{H} \cong H_1^- \otimes_{\mathbf{k}[G]} G\langle X\rangle / H_1^- \otimes_{\mathbf{k}[G]} I_1 \cong H_1^- \otimes_{\mathbf{k}[G]} (G\langle X\rangle / I_1) = H_1^- \otimes_{\mathbf{k}[G]} H_1^+.$$

To check that the resulting isomorphism is a $\mathbf{k}[G]$-coalgebra map, we are reminded that the coproduct on $H_1^- \otimes_{\mathbf{k}[G]} H_1^+$ is defined as follows:

$$\Delta(a\underline{\otimes}b) = \sum_{(a),(b)} (a^{(1)}\underline{\otimes}b^{(1)}) \otimes (a^{(2)}\underline{\otimes}b^{(2)}), \tag{3.53}$$

where $\underline{\otimes}$ denotes the tensor product of $\mathbf{k}[G]$-bimodules, $ag\underline{\otimes}b = a\underline{\otimes}gb$, $g \in G$, while $\otimes$ is the ordinary tensor product of spaces. Clearly, this definition is compatible with the $\mathbf{k}[G]$-bimodule structure, $\Delta(ag\underline{\otimes}b) = \Delta(a\underline{\otimes}gb)$, $g \in G$. The converse to isomorphism (3.53) map ζ acts as follows $\zeta(a\underline{\otimes}b) = a\cdot b$, in which case

$$\Delta(a\cdot b) = \Delta(a)\cdot\Delta(b) = (\sum_{(a)} a^{(1)}\otimes a^{(2)})(\sum_{(b)} b^{(1)}\otimes b^{(2)}) = \sum_{(a),(b)} a^{(1)}b^{(1)}\otimes a^{(2)}b^{(2)}.$$

Therefore, ζ is a coalgebra map. □

Remark 3.1 Informally, the fact that (3.52) is a coalgebra isomorphism means that the basis decomposition of the coproduct of a basis super-word does not use (3.26).

Proposition 3.1 *Under the conditions of the above theorem, the space spanned by the skew-primitive elements of* H_1 *equals the sum of these spaces for* H_1^- *and* H_1^+.

Proof Let π be an algebra homomorphism $\pi : H_1^+ \to \mathbf{k}[G]$ defined by $\pi(g) = g$, $g \in G$, $\pi(x_i) = 0$, $1 \leq i \leq n$. This homomorphism is well-defined as all φ_s have upper degree grater than one. Moreover, π is a homomorphism of Hopf algebras. Similarly we define a Hopf algebra homomorphism $\pi^- : H_1^- \to \mathbf{k}[G]$. Consider the map $\mathrm{id}\underline{\otimes}\pi : H_1^- \otimes_{\mathbf{k}[G]} H_1^+ \to H_1^-$. This is a $\mathbf{k}[G]$-coalgebra map because so is π. Similarly $\pi^-\underline{\otimes}\mathrm{id}$ is also a $\mathbf{k}[G]$-coalgebra map.

Inasmuch as the "negative" variables are less than the "positive" ones, the increasing super-words for H_1 have the form $V^-\cdot W$, where V^-, W are increasing super-words for H_1^-, H_1^+ respectively. The isomorphism ξ defined in (3.52) acts as follows $\xi(gV^-\cdot W) = gV^-\underline{\otimes}W$.

Let $T = \sum \alpha_t g_t V_t^- \cdot W_t$, $g_t \in G$, $\alpha_t \neq 0$ be the basis decomposition of a skew-primitive element. Clearly, $(\mathrm{id}\underline{\otimes}\pi)(\xi(T))$ is the sum of all terms $\alpha_t g_t V_t^- \cdot W_t$ with empty W_t, whereas $(\pi^-\underline{\otimes}\mathrm{id})(\xi(T))$ is the sum of all the terms with empty V_t^-. Both of those elements are skew-primitive because all maps are homomorphisms of coalgebras and $\mathbf{k}[G]$-bimodules. By the same reason, the element

$$T' = T - (\mathrm{id}\underline{\otimes}\pi)(\xi(T)) - (\pi^-\underline{\otimes}\mathrm{id})(\xi(T))$$

is skew-primitive too. The basis decomposition of T' is precisely the sum of all terms $\alpha_t g_t V_t^- \cdot W_t$, where both V_t^- and W_t are nonempty minus all the terms where both V_t^- and W_t are empty.

However, if $T' \notin \mathbf{k}[G]$, then by Theorem 2.3 the leading term of T' is a power of a super-letter, while by Lemma 3.7 all super-letters of H_1 belong to either H^+ or H^-. Hence $T' \in \mathbf{k}[G]$, and T is the sum of three skew-primitive elements: one of them belongs to H^-, another one belongs to H^+, and the third one, T', belongs to both algebras H^+ and H^-. □

3.5 Indecomposable Generalized Cartan Matrices

In this section, we return to the quantization H of a Kac-Moody algebra studied in Sect. 3.3. We consider more thoroughly the case in which the generalized Cartan matrix is indecomposable, i.e., there is no partition of the set $\{1, 2, \ldots, n\}$ into two nonempty subsets such that $a_{is} = 0$ whenever i belongs to the first subset, and s belongs to the second. Every generalized Cartan matrix is a diagonal sum of its indecomposable components, $A = \bigoplus A_\lambda$. The Kac-Moody algebra $\mathfrak{g}$ defined by A is a direct sum of Kac-Moody algebras $\mathfrak{g}_\lambda$ defined by A_λ. Respectively, all quantizations of $\mathfrak{g}$ are smash-products of quantizations of $\mathfrak{g}_\lambda$. In this section we always suppose that $A = ||a_{is}||$ is an indecomposable generalized Cartan matrix.

We maintain the notations used in the above sections. Particularly, p_{is}, $1 \leq i, s \leq n$ are the parameters that define the quantization H according to Theorem 3.1.

3.5.1 Regular and Exceptional Quantizations

Definition 3.2 A quantization H is called *regular* if the relations

$$p_{is} p_{si} = p_{ii}^{a_{is}} \tag{3.54}$$

are valid for all pairs (i, s), $1 \leq i \neq s \leq n$. Otherwise, H is called *exceptional.*

By Theorem 3.1 the algebraic structure of a regular quantization is completely defined by n parameters p_{ii}, $1 \leq i \leq n$.

Theorem 3.3 *Only a finite number of algebraic structures exists for exceptional quantizations of a given Kac-Moody algebra.*

Proof By Theorem 3.1, we must demonstrate that there are only a finite number of admissible values for each parameter p_{ii}, $p_{is}p_{si}$, $1 \leq i, s \leq n$. Let (i, s) be a pair such that $p_{is}p_{si} \neq p_{ii}^{a_{is}}$. Then, (3.29) demonstrates that both p_{ii} and $p_{is}p_{si}$ are $(1 - a_{is})$th roots of 1. In particular, each parameter has no more than $1 - a_{is}$ admissible values.

Let us consider an index i such that p_{ii} is an Nth root of 1 and then choose an arbitrary k, $1 \leq k \leq n$ such that a_{ik} or, equivalently, a_{ki} is a nonzero integer number. Consider the following four options:

1. $p_{kk}^{a_{ki}} = p_{ik}p_{ki} = p_{ii}^{a_{ik}}$. Because p_{ii} is a Nth root of 1, $p_{ik}p_{ki}$ is as well, whereas p_{kk} is a $a_{ki}N$th root of 1. In particular, both p_{kk} and $p_{ik}p_{ik}$ have a finite number of admissible values.
2. $p_{kk}^{a_{ki}} \neq p_{ik}p_{ki} = p_{ii}^{a_{ik}}$. Due to (3.29), the fist inequality implies that both p_{kk} and $p_{ik}p_{ik}$ are $(1 - a_{ki})$th roots of 1.
3. $p_{kk}^{a_{ki}} = p_{ik}p_{ki} \neq p_{ii}^{a_{ik}}$. The second inequality and (3.29) imply that $p_{ik}p_{ki}$ is a $(1 - a_{ki})$th root of 1. The first equality demonstrates that p_{kk} is a $a_{ki}(1 - a_{ki})$th root of 1.
4. $p_{kk}^{a_{ki}} \neq p_{ik}p_{ki} \neq p_{ii}^{a_{ik}}$. In this case (3.29) implies that p_{kk} and $p_{ik}p_{ki}$ are $(1 - a_{ik})$th roots of 1.

Thus, in all cases p_{kk} and $p_{ik}p_{ki}$ are roots of 1. Because the Cartan matrix is indecomposable, all p_{ii}, $p_{is}p_{si}$, $1 \leq i, s \leq n$ are roots of 1 of a bounded degree. □

In the proof of the above theorem, we have observed that all parameters that define the algebraic structure of an exceptional quantization are roots of 1. Therefore, we have the following statement:

Corollary 3.1 *If one of the parameters $p_{is}p_{si}$, $1 \leq i, s \leq n$ is not a root of 1 then the quantization H is regular.*

Theorem 3.1 and identities (3.54) demonstrate that the algebraic structure of a regular quantization is defined by the parameters p_{ii}, $1 \leq i \leq n$. These parameters satisfy the equations

$$p_{ii}^{a_{is}} = p_{ss}^{a_{si}}, \tag{3.55}$$

which define an algebraic variety $\mathfrak{P}(A)$ over the algebraic closure $\bar{\mathbf{k}}$ of the field $\mathbf{k}$. This variety is an algebraic group because it is invariant with respect to the term-by-term product. Thus, one may identify $\mathfrak{P}(A)$ with an algebraic group of diagonal matrices. Let $\mathfrak{P}_0(A)$ be the connected component of the unit. It is well-known that $\mathfrak{P}_0(A)$, as well as an arbitrary connected algebraic group of diagonal matrices (torus), is rationally isomorphic to a direct product $(\bar{\mathbf{k}}^*)^{\times m}$, where m is dimension of the variety and $\mathfrak{P}(A)$ is generated as a group by $\mathfrak{P}_0(A)$ and periodic elements (see, for example, [110]). In other words, the algebraic structure of H is defined by m "continuous" independent parameters, and one discrete parameter that runs through the quotient group $\mathfrak{P}(A)/\mathfrak{P}_0(A)$.

Let us denote by $\Gamma(A)$ a simply-laced version of the Coxeter (or Dynkin) diagram of A, see [116, p. 51]. More precisely, $\Gamma(A)$ is a graph on n points with labels $1, 2, 3, \ldots, n$, where two points i and s are connected by an edge if and only if $a_{is} \neq 0$. It is clear that A is indecomposable if and only if $\Gamma(A)$ is a connected graph.

3.5.2 Non-symmetrizable Generalized Cartan Matrices

Recall that the matrix A is called *symmetrizable* if there exists a diagonal matrix $D = \operatorname{diag}(d_1, d_2, \ldots, d_n)$ with natural $d_1, d_2, \ldots, d_n$ such that DA is a symmetric matrix, $d_i a_{is} = d_s a_{si}$, $1 \le i, s \le n$. If $d_1, d_2, \ldots, d_n$ have a natural common divisor d then $d_1/d, d_2/d, \ldots, d_n/d$ also symmetrize A. Thus, there exists D with cosimple entries. Furthermore, such D is unique. Indeed, if a value of d_i for a given i is fixed, then $d_s = d_i a_{is}/a_{si}$ is unique provided that the edge (i, s) belongs to $\Gamma(A)$. Because $\Gamma(A)$ is connected, the value of d_1 uniquely defines values of all d_i, $1 \le i \le n$.

Lemma 3.12 *The matrix A is symmetrizable if and only if for every cycle of $\Gamma = \Gamma(A)$, say, $i_1, i_2, \ldots, i_k, i_{k+1} = i_1$, we have a relation*

$$a_{i_1 i_2} a_{i_2 i_3} \cdots a_{i_k i_1} = a_{i_2 i_1} a_{i_3 i_2} \cdots a_{i_1 i_k}. \tag{3.56}$$

In particular, if Γ has no cycles at all, the matrix A is symmetrizable.

Proof If A is symmetrizable, then

$$d_{i_{s+1}}/d_{i_s} = a_{i_s i_{s+1}}/a_{i_{s+1} i_s}, \quad 1 \le s \le k.$$

The product of all left-hand sides equals 1. Hence, the product of all right-hand sides is as well. This result implies (3.56).

Suppose that all (3.56) hold. Certainly, it is sufficient to find a matrix D with positive rational d_i's. We put $d_1 = 1$. For each s, $1 < s \le n$, we fix a sequence of edges of Γ that connects 1 and s, say $1 = i_1, i_2, \ldots, i_k, s$, and define

$$d_s = \frac{a_{i_1 i_2} a_{i_2 i_3} \cdots a_{i_k s}}{a_{i_2 i_1} a_{i_3 i_2} \cdots a_{s i_k}}.$$

If $t \ne s$, and $1 = i_m, i_{m-1}, \ldots, i_{k+3}, i_{k+2} = t$ is the fixed sequence of edges that connects 1 and t, then according to the above definition

$$d_t = \frac{a_{i_m i_{m-1}} a_{i_{m-1} i_{m-2}} \cdots a_{i_{k+3} t}}{a_{i_{m-1} i_m} a_{i_{m-2} i_{m-1}} \cdots a_{t i_{k+3}}}.$$

Now, (3.56) applied to a cycle $1 = i_1, i_2, \ldots, i_{k+1} = s, i_{k+2} = t, \ldots, i_m = 1$ implies that

$$\frac{d_s a_{st}}{d_t a_{ts}} = \frac{a_{i_1 i_2} a_{i_2 i_3} \cdots a_{i_k s} \cdot a_{st} \cdot a_{t i_{k+3}} \cdots a_{i_{m-2} i_{m-1}} a_{i_{m-1} i_m}}{a_{i_2 i_1} a_{i_3 i_2} \cdots a_{s i_k} \cdot a_{ts} \cdot a_{i_{k+3} t} \cdots a_{i_{m-1} i_{m-2}} a_{i_m i_{m-1}}} = 1.$$

□

Remark 3.2 Of course, if a cycle $i_1, i_2, \ldots, i_k, i_1$ has an edge that does not belong to Γ, then both sides of (3.56) equal zero. In fact, when checking that A is symmetrizable, it is sufficient to analyze only the cycles that have no self intersections.

The following statement demonstrates that if the matrix A is not symmetrizable, then $\dim \mathfrak{P}(A) = 0$.

Proposition 3.2 *If the matrix A is not symmetrizable, then there exists a natural number N such that the parameters defining the algebraic structure of a regular quantization have the form*

$$p_{ii} = \xi^{m_i},$$

where ξ is a fixed primitive Nth root of 1 and $(m_1, m_2, \dots, m_n)$ is a modular symmetrization:

$$m_i a_{is} \equiv m_s a_{si} (mod\, N), \;\; 0 \le m_i, \; m_s < N, \;\; 1 \le i, \; s \le n.$$

Proof By the above lemma, there exists a cycle $i_1, i_2, \dots, i_k, i_{k+1} = i_1$ of $\Gamma = \Gamma(A)$ such that $L \neq R$, where L and R are the left- and right-hand sides of (3.56), respectively. Using the basic relations $p_{i_s i_s}^{a_{i_s i_{s+1}}} = p_{i_{s+1} i_{s+1}}^{a_{i_{s+1} i_s}}$, we obtain $p_{i_1 i_1}^{L} = p_{k+1\,k+1}^{R}$. Because $p_{i_1 i_1} = p_{k+1\,k+1}$, the parameter $p_{i_1 i_1}$ is a $|L - R|$th root of 1.

If p_{ii} is a Mth root of 1, and the edge (i, s) belongs to Γ, then relation $p_{ss}^{a_{si}} = p_{ii}^{a_{is}}$ demonstrates that p_{ss} is an $a_{si}M$th root of 1. Because Γ is connected, all p_{ii}'s are Nth roots of 1, where $N = |L - R| \prod_{(i,s) \in \Gamma} a_{is}$. In particular, if ξ is a fixed primitive Nth root of 1, then $p_{ii} = \xi^{m_i}$ for a suitable natural $m_i \le N$. The equality $p_{ii}^{a_{is}} = p_{ss}^{a_{si}}$ reduces to $\xi^{m_i a_{is}} = \xi^{m_s a_{si}}$. This equality is equivalent to $m_i a_{is} \equiv m_s a_{si} (\mathrm{mod}\, N)$, $1 \le i, s \le n$ because ξ is a primitive Nth root of 1. □

3.5.3 *Symmetrizable Generalized Cartan Matrices*

Consider a symmetrizable generalized Cartan matrix A. Let $d_1, \dots, d_n$ be cosimple natural numbers that symmetrize A. Let Γ' be a maximal subgraph of $\Gamma = \Gamma(A)$ that has no cycles. Because Γ is connected, the subgraph Γ' contains all vertices $1, 2, \dots, n$. The graph Γ' corresponds to a generalized Cartan matrix A' that results from A if one replaces all a_{is}, $(i, s) \notin \Gamma'$ with zero, $\Gamma' = \Gamma(A')$. Denote by Γ_i', $1 \le i \le n$ an oriented graph that appears from Γ' as follows:

Because the graph Γ' is connected and has no cycles, for every point $s \neq i$, there exists precisely one sequence

$$i = i_1, i_2, i_3, \dots, i_k = s \tag{3.57}$$

such that the edges E_t, $1 \le t < k$ connecting i_t with i_{t+1} belong to Γ'. We replace each edge E_t with an arrow $i_t \to i_{t+1}$.

Let $N_i(\Gamma')$ be the product of all $|a_{sr}|$ with $r \to s \in \Gamma_i'$:

$$N_i(\Gamma') = \prod_{r \to s \in \Gamma_i'} |a_{sr}|. \tag{3.58}$$

If i and s are connected by an edge of Γ' then the products in (3.58) that define $N_i(\Gamma')$ and $N_s(\Gamma')$ have the same factors with only one exception: the arrow $i \to s$ provides a factor a_{si} in $N_i(\Gamma')$, whereas the arrow $s \to i$ provides a factor a_{is} in $N_s(\Gamma')$. This fact implies that $N_i(\Gamma')/a_{si} = N_s(\Gamma')/a_{is}$, or $N_i(\Gamma')a_{is} = N_s(\Gamma')a_{si}$. Thus, $N_i(\Gamma')$, $1 \leq i \leq n$ symmetrize the matrix A'. At the same time, the coprime numbers d_i, $1 \leq i \leq n$ symmetrize A' as well. This implies that $N_i(\Gamma') = d_i N(\Gamma')$, $1 \leq i \leq n$ for a suitable natural $N(\Gamma')$. Of course, $N(\Gamma')$ is the maximal common divisor of the numbers $N_i(\Gamma')$, $1 \leq i \leq n$:

$$N_i(\Gamma') = d_i N(\Gamma'), \ \ 1 \leq i \leq n; \ \ N(\Gamma') = \text{mcd}\,\{N_i(\Gamma') \mid 1 \leq i \leq n\}. \tag{3.59}$$

Denote the maximal common divisor of all $N(\Gamma')$ by N when Γ' runs through the set of all maximal subgraphs without cycles:

$$N = \text{mcd}\,\{N(\Gamma') \mid \Gamma' \text{ has no cycles and connects all vertices}\}. \tag{3.60}$$

Consider symmetrzations of A modulo N. The set of all modular symmetrizations $\mathfrak{M}'$ is an additive group with respect to a term-by-term summation. There is a subgroup $\mathfrak{M}_0$ of *trivial* modular symmetrizations that are induced by the non modular symmetrization, $m_i \equiv l d_i (\text{mod} N)$, $0 \leq l < N$. Denote a fixed transversal of $\mathfrak{M}_0$ in $\mathfrak{M}'$ by $\mathfrak{M}$.

Theorem 3.4 *If A is symmetrizable, then* $\dim \mathfrak{P}(A) = 1$, *whereas* $\mathfrak{P}(A)/\mathfrak{P}_0(A)$ *is isomorphic to* $\mathfrak{M}'/\mathfrak{M}_0$. *The algebraic structure of a regular quantization is defined by two independent parameters* $q \in \bar{\mathbf{k}}^*$ *and* $\mathfrak{m} \in \mathfrak{M}$ *such that*

$$p_{ii} = q^{d_i} \xi^{m_i}, \quad \mathfrak{m} = (m_1, m_2, \ldots, m_n),$$

where ξ is a fixed Nth primitive root of 1. *The number* $|\mathfrak{M}|$ *of values of the second parameter is a divisor of* N.

Proof Let $P_0 = \{D(q) \stackrel{\text{df}}{=} \text{diag}(q^{d_1}, q^{d_2}, \ldots, q^{d_n}) \mid q \in \bar{\mathbf{k}}^*\}$, whereas

$$P_i(A) = \{\mathfrak{a} = \text{diag}(\alpha_1, \alpha_2, \ldots, \alpha_n) \in \mathfrak{P}(A) \mid \alpha_i = 1\}. \tag{3.61}$$

The sets $P_i(A)$, $0 \leq i \leq n$ are subgroups of $\mathfrak{P}(A)$. For each i, $1 \leq i \leq n$, we have a decomposition $\mathfrak{P}(A) = P_0 P_i(A)$. Indeed, if $B = \text{diag}(\beta_1, \beta_2, \ldots, \beta_n) \in \mathfrak{P}(A)$ then we can find $q \in \bar{\mathbf{k}}$ such that $q^{d_i} = \beta_i$. In this case, $BD(q)^{-1} \in P_i(A)$, which implies that $B \in P_0 P_i(A)$.

All groups $P_i(A)$, $1 \leq i \leq n$ are finite because if $\alpha_i = 1$ and i is connected with s by a path (3.57) with edges form Γ, then

$$\alpha_s^r = \alpha_i^t = 1, \text{ where } r = \prod_{v=1}^{k} a_{i_{v+1}i_v}, \quad t = \prod_{v=1}^{k} a_{i_v i_{v+1}}. \tag{3.62}$$

This result implies that the quotient group

$$\mathfrak{P}(A)/P_0 = P_0P_i(A)/P_0 \cong P_i(A)/P_i(A) \cap P_0 \tag{3.63}$$

is also finite. Because P_0 is a connected subgroup of $\mathfrak{P}(A)$, we have $P_0 = \mathfrak{P}_0(A)$. Furthermore, by definition P_0 is isomorphic to $\bar{\mathbf{k}}^*$ via $\text{diag}(q^{d_1}, q^{d_2}, \ldots, q^{d_n}) \mapsto q$, hence $\dim \mathfrak{P}(A) = 1$.

Let Γ' be an arbitrary maximal subgraph of $\Gamma = \Gamma(A)$ that has no cycles, and let A' be the matrix that appears from A upon replacement of all a_{is}, $(i,s) \notin \Gamma'$ with zero. Because the d_i's still centralize A', we have $\dim \mathfrak{P}(A') = 1$, $\mathfrak{P}_0(A') = P_0$, whereas $P_i(A) \subseteq P_i(A')$.

The order of the group $P_i(A')$ equals $N_i(\Gamma')$ [see (3.58)]. Indeed, let $\mathfrak{a}$ be defined by (3.61) with A' in place of A. If a value of α_k is fixed and k and s are connected by an arrow of Γ_i' then the equation $\alpha_k^{a_{ks}} = x^{a_{sk}}$ has precisely a_{sk} solutions in $\bar{\mathbf{k}}$. Because the value $\alpha_i = 1$ is fixed and there exists precisely one path (3.57) for each s that connects i and s in Γ_i', we see that the order of $P_i(A')$ is a product of all $|a_{i_{v+1}i_v}|$ that appears on the paths (3.57) for different final vertices s of Γ'. By definition this product equals $N_i(\Gamma') = N(\Gamma')d_i$, see (3.59).

We have $P_i(A) \subseteq \bigcap_{\Gamma_i'} P_i(A')$. In particular, the order of $P_i(A)$ is a divisor of all $N_i(\Gamma') = N(\Gamma')d_i$ when Γ' runs through all maximal subgraphs without cycles. The left-hand side of (3.63) is independent of i, whereas the right-hand side is a homomorphic image of $P_i(A)$. This property implies that the order of $\mathfrak{P}(A)/\mathfrak{P}_0(A)$ is a divisor of all orders of $P_i(A)$, $1 \leq i \leq s$. The maximal common divisor of all $N(\Gamma')d_i$ equals N defined in (3.60). Thus, the order of $\mathfrak{P}(A)/\mathfrak{P}_0(A)$ is a divisor of N.

In particular, $\mathfrak{p}^N \in \mathfrak{P}_0(A)$ for each $\mathfrak{p} \in \mathfrak{P}(A)$. In greater detail, for each $\mathfrak{p} = \text{diag}(p_{11}, \ldots, p_{nn}) \in \mathfrak{P}$ there exists $q_1 \in \bar{\mathbf{k}}$ such that $\mathfrak{p}^N = D(q_1)$. We stress that the element q_1 is uniquely defined by $\mathfrak{p}$ because if $D(q_1) = D(q_1')$ then $(q_1/q_1')^{d_i} = 1$ for the coprime numbers d_i, $1 \leq i \leq n$, which implies that $(q_1/q_1') = 1$. Let us choose $q \in \bar{\mathbf{k}}$ such that $q^N = q_1$. The element q is defined by $\mathfrak{p}$ up to a factor which is an Nth root of 1. Now, $p_{ii}q^{-d_i}$, $1 \leq i \leq n$ are Nth root s of 1. If ξ is any primitive Nth root of 1 then there exist integer numbers m_i, $0 \leq m_i < N$, $1 \leq i \leq n$, uniquely defined by ξ and q, such that $p_{ii}q^{-d_i} = \xi^{m_i}$, or equivalently,

$$p_{ii} = q^{d_i}\xi^{m_i}.$$

Because $\text{diag}(\xi^{m_1}, \ldots, \xi^{m_n}) \in \mathfrak{P}$, we have $\xi^{m_i a_{is}} = \xi^{m_s a_{si}}$, $1 \leq i, s \leq n$. This implies $m_i a_{is} \equiv m_s a_{si} (\text{mod}\, N)$; that is, $\mathfrak{m} = \text{diag}(m_1, \ldots, m_n)$ is a modular symmetrization of A, $\mathfrak{m} \in \mathfrak{M}'$. If instead of q we choose another element $q' = q\xi^l$, $1 \leq l \leq N$, then related symmetrization will be changed via

$$\text{diag}(m_i') = \text{diag}(m_i + ld_i) \equiv \text{diag}(m_i) \,(\text{mod}\, \mathfrak{M}_0).$$

Thus, when $\mathfrak{p}$ runs through $\mathfrak{P}$, the map $\varphi : \mathfrak{p} \mapsto \mathfrak{m}$, is a homomorphism from $\mathfrak{P}$ to $\mathfrak{M}'/\mathfrak{M}_0$. Moreover, the kernel of φ equals $\mathfrak{P}_0(A)$. Indeed, if $\mathfrak{m} \in \mathfrak{M}_0$ then $m_i = ld_i$, $1 \leq i \leq n$. This result implies $p_{ii} = q^{d_i}\xi^{ld_i} = (q\xi^l)^{d_i}$, and therefore $\mathfrak{p} = D(q\xi^l) \in \mathfrak{P}_0(A)$. This proves the required isomorphism $\mathfrak{P}(A)/\mathfrak{P}_0(A) \cong \mathfrak{M}'/\mathfrak{M}_0$. □

3.5.4 Cartan Matrices of Finite Type

We may apply Theorem 3.4 to describe regular quantizations of Kac-Moody algebras of finite type.

Lemma 3.13 *If a symmetrizable generalized Cartan matrix A has not more than one a_{ik}, $i \neq k$ different from -1 and 0, then $|\mathfrak{M}| = 1$, i. e., the discrete parameter in Theorem 3.4 does not appear at all.*

Proof If $a_{is} \neq -1$ then $N_i(\Gamma') = 1$ because a_{is} does not appear among the factors a_{sr} with $r \to s \in \Gamma_i'$. Thus, in Theorem 3.4, we have $N = 1$, and $|\mathfrak{M}| = |\mathfrak{P}(A)/\mathfrak{P}_0(A)| = 1$; therefore, $\mathfrak{P}(A) = \mathfrak{P}_0(A)$. □

Since all Cartan matrices of finite type satisfy the conditions of the lemma, it follows that the discrete parameter in Theorem 3.4 for regular quantizations does not appear at all. With regard to the values of the main parameter q, we stress that $q = p_{11} \in \mathbf{k}$ for all Cartan matrices of finite type, except two cases: B_n and G_2 when $q = p_{nn}$ but still $q \in \mathbf{k}$.

Further we consider in more details exceptional quantizations for Cartan matrix of type A_n, where as usual $a_{ii} = 2$, whereas $a_{is} = a_{si} = -1$ if $s = i + 1$, $1 \leq i < n$ and $a_{is} = 0$ otherwise.

Lemma 3.14 *If the Dynkin diagram is simply-laced and connected then all parameters p_{ii}, $p_{is}p_{si}$, $1 \leq i < s \leq n$ defining an exceptional quantization according to Theorem 3.1 have values ± 1. If $a_{is} = 0$ then $p_{is}p_{si} = 1$. If $a_{is} = -1$ then for the triple $(p_{ii}, p_{is}p_{si}, p_{ss})$ there are just the following options $(\pm 1, 1, \pm 1)$ or $(-1, -1, -1)$. In other words, if $p_{is}p_{si} = -1$ then $p_{ii} = p_{ss} = -1$.*

Proof The Dynkin diagram is simply-laced if and only if all non-diagonal a_{is}'s are -1 or 0. If $a_{is} = 0$ then condition (2.51) as well as (2.52) imply that $p_{is}p_{si} = 1$. If $a_{is} = -1$ then either condition (2.51), $p_{is}p_{si} = p_{ii}^{-1}$, or condition (2.52), $p_{ii} = -1$, $p_{is}p_{si} = \pm 1$, holds. In the latter case, if $p_{is}p_{si} = -1$, then the first condition still holds, and thus, the second option reduces to $p_{ii} = -1$, $p_{is}p_{si} = 1$.

If i is connected with s by an edge of Γ then $a_{is} = a_{si} = -1$. Therefore, condition (2.52) or (2.51) under substitution $i \leftrightarrow s$ holds as well: either $p_{is}p_{si} = p_{ss}^{-1}$

or $p_{ss} = -1$, $p_{is}p_{si} = 1$. Thus, we have either $p_{ii} = p_{ss} = (p_{is}p_{si})^{-1}$ or $p_{is}p_{si} = 1$, $p_{ii} = \pm 1$, $p_{ss} = \pm 1$.

This result implies that $p_{ii} = \pm 1$ is valid for all i, $1 \leq i \leq n$. Indeed, consider the set B of all i with $p_{ii} = \pm 1$. No single vertex $s \notin B$, $p_{ss} \neq \pm 1$, is connected by an edge with any $i \in B$ because $p_{ss} = p_{ii}$ otherwise. We obtain either $B = \emptyset$ or $B = \{1, 2, \ldots, n\}$. In the former case conditions (2.51) are valid for all (i, s); that is, the quantization is regular. □

If char $\mathbf{k} = 2$, then the above lemma demonstrates that there are no exceptions in the simply-laced cases. Therefore, we suppose that char $\mathbf{k} \neq 2$.

Let us label each vertex i of the graph Γ by p_{ii} and each edge $(i, s) \in \Gamma$ by $p_{is}p_{si}$.

To find the total number of different parameters values that define the algebraic structure for exceptional quantizations, we must find the number of all possible options to put labels ± 1 on the graph Γ such that, if (i, s) is labeled by -1 then i and s are labeled by -1 as well.

Proposition 3.3 *If A is a Cartan matrix of type A_n, then there exists $\varphi_{2n} - 2$ options for values of parameters that define the algebraic structure for exceptional quantizations. Here φ_i, $0 \leq i$ is the Fibonacci sequence* $1, 1, 2, 3, 5, 8, 13, \ldots,$ $\varphi_{n+1} = \varphi_n + \varphi_{n-1}$.

Proof Denote by M_n the number of all options for labeling Γ of type A_n so that if an edge is labeled by -1, then both its vertices are labeled by -1.

$$\overset{p_{11}}{\circ} \quad \underline{p_{12}p_{21}} \quad \overset{p_{22}}{\circ} \quad \underline{p_{23}p_{32}} \quad \ldots \quad \overset{p_{n-1\,n-1}}{\circ} \quad \underline{p_{n-1\,n}p_{n\,n-1}} \quad \overset{p_{nn}}{\circ} \tag{3.64}$$

Let M_n^+ be the number of options among them with $p_{nn} = 1$.

$$\overset{p_{11}}{\circ} \quad \underline{p_{12}p_{21}} \quad \overset{p_{22}}{\circ} \quad \underline{p_{23}p_{32}} \quad \ldots \quad \overset{p_{n-1\,n-1}}{\circ} \quad \underline{p_{n-1\,n}p_{n\,n-1}} \quad \overset{1}{\circ} \tag{3.65}$$

Let M_n^- be the number of options with $p_{nn} = -1$.

$$\overset{p_{11}}{\circ} \quad \underline{p_{12}p_{21}} \quad \overset{p_{22}}{\circ} \quad \underline{p_{23}p_{32}} \quad \ldots \quad \overset{p_{n-1\,n-1}}{\circ} \quad \underline{p_{n-1\,n}p_{n\,n-1}} \quad \overset{-1}{\circ} \tag{3.66}$$

We have $M_n = M_n^+ + M_n^-$. One can easily find the recurrence relations for $M_n^{\pm}$ because a subgraph of a correctly labeled graph is correctly labeled. There is only one option for extending the graph (3.65) to the left by one edge and one vertex so that the resulting graph has $p_{n+1\,n+1} = 1$, and one option to extend the graph (3.66). Hence, $M_{n+1}^+ = M_n^+ + M_n^-$. Similarly, there is only one option to extend the graph (3.65) so that the resulting graph has $p_{n+1\,n+1} = -1$, and two options to extend the graph (3.66). Thus, $M_{n+1}^- = M_n^+ + 2M_n^-$. In matrix form, these relations reduce to

$$\begin{pmatrix} M_{n+1}^+ \\ M_{n+1}^- \end{pmatrix} = \begin{pmatrix} 1 & 1 \\ 1 & 2 \end{pmatrix} \begin{pmatrix} M_n^+ \\ M_n^- \end{pmatrix}.$$

Because $M_2^+ = 2$ and $M_2^- = 3$, this recurrence relation implies that

$$\begin{pmatrix} M_{n+1}^+ \\ M_{n+1}^- \end{pmatrix} = \begin{pmatrix} 1 & 1 \\ 1 & 2 \end{pmatrix}^{n-1} \begin{pmatrix} 2 \\ 3 \end{pmatrix} = \begin{pmatrix} 1 & 1 \\ 1 & 2 \end{pmatrix}^{n} \begin{pmatrix} 1 \\ 1 \end{pmatrix}.$$

Now, an ordinary induction demonstrates that

$$\begin{pmatrix} 1 & 1 \\ 1 & 2 \end{pmatrix}^{n} = \begin{pmatrix} \varphi_{2n-2} & \varphi_{2n-1} \\ \varphi_{2n-1} & \varphi_{2n} \end{pmatrix}, \quad n \geq 1,$$

which yields $M_n^+ = \varphi_{2n-2}$, $M_n^- = \varphi_{2n-1}$, and $M_n = \varphi_{2n}$. Thus, there are φ_{2n} options for labeling Γ by ± 1 according to Lemma 3.14, and two of the options ($p_{ii} = p_{is}p_{si} = p_{ss} = 1$ all $i \neq s$, and $p_{ii} = p_{is}p_{si} = p_{ss} = -1$ all $i \neq s$) are regular. □

3.5.5 Isomorphism Problem

We did not discuss whether different admissible collections of the parameters define abstractly different algebraic structures of quantizations. If a generalized Cartan matrix has a symmetry, $a_{is} = a_{\pi(i)\,\pi(s)}$, with respect to some permutation of vertices π, and $p_{ii} = \alpha_i$, $p_{is}p_{si} = \beta_{is} = \beta_{si}$ is an admissible collection of parameters values, then $p_{ii} = \alpha_{\pi(i)}$, $p_{is}p_{si} = \beta_{\pi(i)\,\pi(s)}$ is also an admissible collection. In this manner, the group of symmetries acts on the set of all admissible collections. In a regular case, of course, $d_{\pi(i)} = d_i$; however, this may be $m_{\pi(i)} \neq m_i (\text{mod } N)$ for some index i, so that $q^{d_i}\xi^{m_i} \neq q^{d_{\pi(i)}}\xi^{m_{\pi(i)}}$. In exceptional cases, it may be that $\alpha_i \neq \alpha_{\pi(i)}$ or $\beta_{is} \neq \beta_{\pi(i)\,\pi(s)}$ for some indices as well. Nevertheless, the **k**-algebra defined by values $p_{ii} = \alpha_i$, $p_{is}p_{si} = \beta_{is} = \beta_{si}$ in variables x_i, x_i^-, $1 \leq i \leq n$ is isomorphic to the **k**-algebra defined by the values $p'_{ii} = \alpha_{\pi(i)}$, $p'_{is}p'_{si} = \beta_{\pi(i)\,\pi(s)}$ in variables x'_i, $(x'_i)^-$, $1 \leq i \leq n$. The isomorphism acts on the group G via the permutation π^{-1}:

$$\text{diag}(\lambda_1, \ldots, \lambda_n, \lambda_1^{-1}, \ldots, \lambda_n^{-1}) \xrightarrow{\varphi} \text{diag}(\lambda_{\pi^{-1}(1)}, \ldots, \lambda_{\pi^{-1}(n)}, \lambda_{\pi^{-1}(1)}^{-1}, \ldots, \lambda_{\pi^{-1}(n)}^{-1}).$$

It similarly acts on the generators: $x_i \xrightarrow{\varphi} x'_{\pi^{-1}(i)}$, $x_i^- \xrightarrow{\varphi} (x'_{\pi^{-1}(i)})^-$. Importantly, the definition of the t_i's in Theorem 3.1 is not invariant under the permutations (it depends on the order of vertices), therefore $\varphi(\tilde{y}_i) \neq \tilde{y}'_{\pi^{-1}(i)}$ in general. Hence, different collections from the same orbit define the same abstract algebraic structure. We formulate a precise problem: *do collections from the different orbits define non-isomorphic* **k**-*algebras*? Although it is likely that the isomorphism problem has an affirmative solution, the arguments should include an analysis of possible non-homogeneous isomorphisms. However, the existence of the famous Lusztig's automorphisms demonstrates that such analysis may be difficult.

Among 25 types of generalized Cartan matrices related to affine Lie algebras of Kac classification, [116, pp. 53–55], the matrices of 10 types have nontrivial symmetries, whereas matrices of the other 15 types do not. In line with the isomorphism problem, it is interesting to find the number of orbits for the former cases. For example, in case A_n there is just one nontrivial symmetry $\pi(i) = n - i + 1,\ 1 \leq i \leq n$. Therefore, each orbit has either one or two elements. The number of one element orbits equals the number of symmetric admissible (exceptional) collections. It is then easy to see that this number is $\varphi_{n+1} - 2$. Thus, the total number of orbits is $\frac{1}{2}(\varphi_{2n} + \varphi_{n+1}) - 2$.

3.6 Chapter Notes

The classification of finite-dimensional semisimple Lie algebras, being useful in many areas of mathematics and physics, has an about 80-year history, starting with a paper by Killing [141] and ending with Jean-Pierre Serre's representation by generators and relations [207]. The modern treatment of the theory is given in a book by Humphreys [102]. The Kac-Moody algebras appeared in papers by Moody [168] and Kac [112]. The book by Kac [116] is a canonical text for learning the theory underlying Kac-Moody algebras.

A major event in developing the Hopf algebra theory was the discovery by Drinfeld [65, 66] and Jimbo [106] of a class of Hopf algebras, now called Drinfeld–Jimbo quantum groups, which can be considered as one-parameter deformations of universal enveloping algebras of semisimple complex Lie algebra. Since then, numerous books [46, 54, 71, 105, 107, 120, 142, 153, 160, 176, 213] and articles have been published on the quantizations of Lie algebras.

The multiparameter quantizations appeared in different versions in papers by Reshetikhin [193], Cotta-Ramusino and Rinaldi [56], and Constantini and Varagnolo [55]. In [194], Ringel developed an original approach based on deformations of Hall algebras. Benkart and Witherspoon [22, 23] and then Bergeron et al. [29] introduced a special two-parameter version from another perspective. Kang in [118] and subsequently Benkart et al. in [24] considered quantum deformations of generalized Kac-Moody algebras.

The analysis of quantizations of Kac-Moody algebras given in the book is based on the author [136]. In the tableaux below, we provide the maximum possible numbers, found in [136], of exceptional parameter values for Kac-Moody algebras of finite or affine types in terms of the Kac classification, [116, pp. 53–55], where $\varphi_i,\ i \geq 0$ is the Fibonacci sequence $1, 1, 2, 3, 5, 8, 13, 21, \ldots$.

A_n	$\varphi_{2n} - 2$	G_2	12	$E_7^{(1)}$	632
B_n	$2\varphi_{2n} - 2$	$A_1^{(1)}$	6	$E_8^{(1)}$	4344
C_n	$2\varphi_{2n-2} + 4$	$A_l^{(1)}$	$\varphi_{2n} + \varphi_{2n-2} - 2$	$A_2^{(2)}$	24
E_6	240	$C_l^{(1)}$	$4\varphi_{2n-4} + 28$	$A_{2l}^{(2)}$	$4\varphi_{2n-2} + 16$
E_7	632	$G_2^{(1)}$	38	$D_{l+1}^{(2)}$	$4\varphi_{2n} + 2$
E_8	1658	$F_4^{(1)}$	91	$E_6^{(2)}$	80
F_4	40	$E_6^{(1)}$	635	$D_4^{(3)}$	19

D_n	$\varphi_{2n} + \varphi_{2n-7} - 2$	$D_l^{(1)}$	$15\,\varphi_{2n-6} + 11\,\varphi_{2n-8} - 2$
$A_{2l-1}^{(2)}$	$2\varphi_{2n-2} + 2\varphi_{2n-9} + 2$	$B_l^{(1)}$	$\varphi_{2n+1} + 5\varphi_{2n-5} - 2$

Chapter 4
Algebra of Skew-Primitive Elements

Abstract In this chapter we consider the skew-primitive polynomials of the free character Hopf algebra to be quantum Lie operations. We discuss linearization and specialization processes and criteria for a polynomial to be classified as a quantum Lie operation. We also classify multilinear quantum Lie operations in two, three, and four variables.

In this chapter we consider the skew-primitive polynomials of the free character Hopf algebra to be quantum Lie operations. We discuss linearization and specialization processes and criteria for a polynomial to be classified as a quantum Lie operation. We also classify multilinear quantum Lie operations in two, three, and four variables.

4.1 Quantum Lie Operations

According to the Friedrichs criteria, Lie polynomials are characterized as primitive elements of free associative algebra with primitive free generators:

$$\Delta(x_i) = x_i \otimes 1 + 1 \otimes x_i.$$

Because every Lie polynomial may be considered as a multivariable operation on Lie algebras, this fact yields an idea to define quantum Lie operations as polynomials of free algebra that are skew-primitive for all "admissible" values of variables. In line with this idea, a quantum analog of a Lie algebra is the subspace of a Hopf algebra span by skew-primitive elements and equipped by quantum Lie operations.

The definition of a skew-primitive element includes two group-like elements,

$$\Delta(u) = u \otimes h_u + f_u \otimes u, \quad h_u, \; f_u \in G.$$

V. Kharchenko, *Quantum Lie Theory*, Lecture Notes in Mathematics 2150,
DOI 10.1007/978-3-319-22704-7_4

As the skew-primitive elements are closed with respect to the multiplication by group-like elements, we may normalize one of these group-likes, say,

$$\Delta(h_u^{-1}u) = h_u^{-1}u \otimes 1 + h_u^{-1}f_u \otimes h_u^{-1}u.$$

This allows us to concentrate our attention mainly only on *skew-primitive elements* when $h_u = 1$. This diminishes twice the number of parameters. Of course, every operation $f(x_1, x_2, \dots, x_n)$ on normalized skew-primitive elements has an extension to arbitrary ones by means of the substitution $x_i \leftarrow h_i^{-1}y_i$ and followed then multiplication by group-likes h_i. Let us proceed with the exact definitions.

4.1.1 Quantum Variables

We call a variable x as a *quantum variable* if an element g_x of a fixed Abelian group G and a character $\chi^x \in G^*$ are associated with it. The parameters g_x and χ^x associated with a quantum variable say that an element a in a Hopf algebra H may be considered as a value of this quantum variable only if a is a skew-primitive semi-invariant with the same parameters, that is

$$\Delta(a) = a \otimes 1 + g_x \otimes a, \quad g^{-1}ag = \chi^x(g)a, \quad g \in G, \tag{4.1}$$

where we suppose that the elements of G have some fixed interpretation in H as group-like elements.

A noncommutative polynomial in quantum variables is called a *quantum Lie operation* if all of its values in all Hopf algebras are skew-primitive for all values of the quantum variables. In particular, a *homogeneous* quantum Lie operation has the form:

$$[x_1, \dots, x_n] = \sum_{\pi \in S_n} \alpha_\pi x_{\pi(1)} \cdots x_{\pi(n)},$$

where $x_1, \dots, x_n$ are not necessarily distinct quantum variables. If those variables are mutually distinct (but not necessarily with different χ, g), the operation is called *multilinear*.

Let $x_1, \dots, x_n$ be a set of quantum variables. The skew group algebra $G\langle X\rangle$ becomes a free character Hopf algebra if we define the coproduct

$$\Delta(x_i) = x_i \otimes 1 + g_{x_i} \otimes x_i, \quad 1 \le i \le 1, \quad \Delta(g) = g \otimes g, \ g \in G.$$

Hence the x_i's have skew-primitive values in $G\langle X\rangle$. By this means the quantum Lie operations can be identified with skew-primitive polynomials of the free character Hopf algebra $G\langle X\rangle$.

Example 4.1 (Commutator) If G is a trivial group, then the usual commutator $xy - yx$ is a quantum Lie operation. If the ground field has a positive characteristic $l > 0$, there exists a nonlinear quantum Lie operation x^l. The Friedrichs criteria says that all other operations (if, of course, $G = \mathrm{id}$) are superpositions of these two.

Example 4.2 (Skew Commutator) Let x_1 and x_2 be quantum variables. If $p_{12}p_{21} = 1$ then the skew commutator $[x_1, x_2] = x_1x_2 - p_{12}x_2x_1$ is a quantum Lie operation.

Example 4.3 (Unary Restriction Operation) Let x be a quantum variable such that $p_{11} = \chi^x(g_x)$ is a primitive tth root of 1. Then x^t is a quantum Lie operation.

Indeed, we have $(x \otimes 1)(g \otimes x) = p_{11}(g \otimes x)(x \otimes 1)$. Lemma 1.2 implies $\Delta(x^t) = \Delta(x)^t = x^t \otimes 1 + g_x^t \otimes x^t$. Hence x^t is a quantum Lie operation. Similarly, if the characteristic l of $\mathbf{k}$ is positive, then x^{tl^n}, $n > 0$ are quantum Lie operations as well.

Example 4.4 (Pareigis Quantum Operation) Let ζ be a primitive nth root of unity and $p_{ij}p_{ji} = \zeta^2$. Then

$$P_n(x_1, \ldots, x_n) = \sum_{\pi \in S_n} \Big(\prod_{i<j \& \pi(i)>\pi(j)} (\zeta^{-1}\chi^{x_{\pi(j)}}(g_{x_{\pi(i)}})) \, x_{\pi(1)} \cdots x_{\pi(n)}$$

is a quantum Lie operation (see [184, Theorem 3.1, p. 147], and [183, 185]).

Example 4.5 (Serre Quantum Operation) Let x_1, x_2 be quantum variables. Then Theorem 2.5 demonstrates that

$$W = [\ldots[[x_1, \underbrace{x_2], x_2], \ldots, x_2]}_{n}$$

is a quantum Lie operation provided that either $p_{12}\,p_{21} = p_{22}^{1-n}$ or p_{22} is a primitive mth root of unity, $m|n$, and $p_{12}^m p_{21}^m = 1$.

4.1.2 Linearization and Specialization

The quantum Lie operations admit the well-known linearization process. Recall that a *multidegree* $D(u)$ of a word u is a sequence of non-negative integers $(m_1, m_2, \ldots, m_n)$ such that u is of degree m_1 in x_1, $\deg_1(u) = m_1$; of degree m_2 in x_2, $\deg_2(u) = m_2$; and so on. A linear combination of words $\sum \alpha_i u_i$ is *homogeneous* if all words u_i have the same multidegree. Of course, each polynomial is a sum of its homogeneous components.

Lemma 4.1 *All homogeneous components of a quantum Lie operation are again quantum Lie operations.*

Proof A coproduct of a word $u = a_1a_2 \cdots a_m$, $a_i \in X$, $1 \le i \le m$ is a sum of 2^m tensors $u^A \otimes u^{\bar{A}}$, when A runs through all subsets of $\{1, 2, \ldots, m\}$, whereas u^A is a word that appears from u upon replacement of each $a_i,\ i \in A$ with g_i, and $u^{\bar{A}}$ is a word that appears from u by deleting of all $a_i,\ i \notin A$, see Remark 1.6. In particular, the sum of the multidegrees of u^A and $u^{\bar{A}}$ equals the multidegree of u.

If now $F = \sum \alpha_k u_k$ is a skew-primitive polynomial, then all tensors in

$$\sum \alpha_k \Delta(u_k) - F \otimes 1 - g_f \otimes F \tag{4.2}$$

under the above decomposition of $\Delta(u_k)$ must be canceled. However, if either left or right components of a pair of tensors have different multidegrees, then the tensors may not cancel each other. Therefore, for a given $\mathbf{d} = (d_1, d_2, \ldots, d_n)$, the sum of all tensors $A \otimes B$ of (4.2) with $D(A) + D(B) = \mathbf{d}$ is zero. But the latter sum is precisely $\Delta(F_{\mathbf{d}}) - F_{\mathbf{d}} \otimes 1 - g_f \otimes F_{\mathbf{d}}$, where $F_{\mathbf{d}} = \sum_{D(u_k)=\mathbf{d}} \alpha_k u_k$ is the $\mathbf{d}$-homogeneous component of F. This implies that homogeneous components $F_{\mathbf{d}}$ are skew-primitive. □

Due to the proven lemma, we may concentrate our attention mainly on the homogeneous polynomials. Let $f(x_1, \ldots, x_n)$ be a homogeneous quantum Lie operation of degree m_i in x_i, $1 \leq i \leq n$. Instead of x_i we may introduce m_i new quantum variables y_{ij}, $1 \leq j \leq m_i$ that have the same parameters (χ^{x_i}, g_{x_i}). In this way any linear combination $z_i = \sum_k \varepsilon_k y_{ik}$ is a skew-primitive semi-invariant with the parameters (χ^{x_i}, g_{x_i}) in $G\langle y_{ij}\rangle$. In particular, by the definition of the quantum Lie operation

$$\begin{aligned}
&f(y_{11} + y_{12} + \cdots + y_{1m_1}, x_2, \ldots, x_n)\\
&\quad - \sum_{q=1}^{m_1} f(y_{11} + \cdots + \cdots + \hat{y}_{1q} + \cdots y_{1m_1}, x_2, \ldots, x_n)\\
&\quad + \sum_{1 \leq q_1 < q_2 \leq m_1} f(y_{11} + \cdots + \cdots \hat{y}_{1q_1} + \cdots + \hat{y}_{1q_2} + \cdots + y_{1m_1}, x_2, \ldots, x_n)\\
&\quad - + \cdots + (-1)^{m_1 - 1} \sum_{q=1}^{m_1} f(y_{1q}, x_2, \ldots, x_n)
\end{aligned}$$

is a skew-primitive polynomial that defines a quantum Lie operation in y_{1j}, $1 \leq j \leq m_1$, $x_2, \ldots, x_n$. Here as usual the symbol $\hat{}$ over the addend means that this addend is omitted in the sum. The continuation of this process for $x_2, x_3, \ldots$ will lead to a multilinear quantum operation in y_{ij}. We call the obtained operation $L(f)$ as *complete linearization* of the given one.

Conversely, if some of the variables x_i have the same parameters, say, $\chi^{x_1} = \chi^{x_2}, g_{x_1} = g_{x_2}$, then we may substitute $x_1 = x_2$ in the operation. We will obtain a new operation that is called a *specialization* of the given one. In particular we may substitute $y_{ij} = x_i$ to the complete linearization $L(f)$. In this way we get the initial operation multiplied by a natural number.

For instance, the complete linearization of the operation from Example 4.3 is

$$L(x^r) = \sum_{\pi \in S_r} y_{\pi(1)} y_{\pi(2)} \dots y_{\pi(r)}, \quad r = tl^n,$$

whereas its complete specialization equals $r!\,x^r$. Of course, if the characteristic l is positive, then the specialization is zero. So that, in the case of positive characteristic, there exist operations that are not specializations of the multilinear ones.

However, over a field of zero characteristic, every constitution homogeneous quantum Lie operation is a specialization of a multilinear one. By this reason the investigation of the multilinear operations is of the primary importance.

4.2 Criteria for Quantum Lie Operations

Additionally to the Freiderichs criterion that characterizes the Lie polynomials as primitive elements of the free associative algebra, there exist two more criteria for the Lie polynomials: the Finkelstein and Specht–Wever criteria. We are going to demonstrate that the Finkelstein criterion remains valid for the quantum Lie operations, whereas the Specht–Wever condition is valid for the quantum Lie operations, but this is not a criterion any more. Also we prove a new criterion for quantum Lie operations linear at least in one variable, which reduces the identification to the problem of linear dependence of some special polynomials.

4.2.1 *Left and Right Primitive Polynomials*

Definition 4.1 A polynomial W is called *left primitive in* $x_1 \in X$ if $\Delta(W) - W \otimes 1$ is a linear combination of tensors $A \otimes B$ with $\deg_1(A) = 0$; that is, all left-hand sides of the tensors are independent of x_1. Similarly, W is called *right primitive in* x_1 if $\Delta(W) - g_w \otimes W$ is a linear combination of tensors $A \otimes B$ with $\deg_1(B) = 0$.

Proposition 4.1 *A linear in* x_1 *polynomial* F *is left primitive in* x_1 *if and only if it is a linear combination of long skew commutators:*

$$F = \sum_{\mathbf{i}=(i_1,i_2,\dots,i_k)} \alpha_{\mathbf{i}} [\dots[[x_1, x_{i_1}], x_{i_2}], \dots, x_{i_k}]. \tag{4.3}$$

Proof First we prove by induction on k that all summands on the right of (4.3) are left primitive in x_1. If $k = 0$, then $\Delta(x_1) - x_1 \otimes 1 = g_1 \otimes x_1$, whereas $\deg_1(g_1) = 0$.

Let $i_{k+1} = j \neq 1$, and let $\Delta(U) = U \otimes 1 + \sum u_i \otimes d_i$, where the G-words u_i are independent of x_1. Then, if we take into account that $Ug_j = p(U, x_j)g_j U$ and neglect tensors whose left-hand sides are independent of x_1, we obtain

$$\begin{aligned}\Delta([U,x_j]) &= \Delta(U)\Delta(x_j) - p(U,x_j)\Delta(x_j)\Delta(U)\\ &= (U\otimes 1+\sum u_i\otimes d_i)(x_j\otimes 1+g_j\otimes x_j)\\ &\quad -p(U,x_j)(x_j\otimes 1+g_j\otimes x_j)(U\otimes 1+\sum u_i\otimes d_i)\\ &\equiv Ux_j\otimes 1+Ug_j\otimes x_j - p(U,x_j)(x_jU\otimes 1+g_jU\otimes x_j)\equiv [U,x_j]\otimes 1.\end{aligned}$$

To prove the converse statement, we return to Remark 1.6. The tensor $w^A\otimes w^{\bar{A}}$ is proportional to a tensor of the form $gu\otimes v$, $g\in G$, where both words u, v appear from w by deleting of some letters. The word w may be reconstructed from the words u, v by means of the so called shuffle construction: a word is a *shuffle of* u, v if it appears from the word uv by means of the moving of some letters of v to the left so that the order of letters from v is not changed. For example, the shuffles of $u = ab$, $v = xy$ are $abxy$, $axby$, $axyb$, $xaby$, $xayb$, $xyab$. We see that a tensor $gu\otimes v$ appears in the decomposition of $\Delta(w)$ if and only if w is a shuffle of u, v.

Assume that W is left primitive in x_1. Let w be the leading word of the decomposition of W in a linear combination of words: $W = \alpha w+\sum \alpha_i w_i,\ w_i < w,\ \alpha\neq 0.$ If $w = x_1x_{i_1}x_{i_2}\dots x_{i_k}$ begins with x_1, then the leading word of

$$W' = W - \alpha[\dots[[x_1,x_{i_1}],x_{i_2}],\dots,x_{i_k}]$$

is less than w because $[\dots[[x_1,x_{i_1}],x_{i_2}],\dots,x_{i_k}]$ has only one word, w, that begins with x_1. As W' is still left primitive in x_1, one may apply induction on the leading word.

If $w = ux_1v$ and u is a nonempty word, then the decomposition of $\Delta(w)$ has a tensor $gx_1v\otimes u$ with a nonzero coefficient because $w = ux_1v$ is a shuffle of x_1v, u. At the same time, other shuffles of x_1v, u are greater than ux_1v as x_1 is grater than each letter of u. In particular, no one of the words $w_i < w$ is a shuffle of x_1v, u. This implies that the tensor $gx_1v\otimes u$ remains uncanceled in $\Delta(W)$, so that W is not left primitive in x_1. □

In a similar way, we may describe all right primitive in x_1 polynomials. To this end, we have to use the dual brackets (1.67):

$$[u,v]^* = uv - p_{v,u}^{-1}vu = -p_{v,u}^{-1}[v,u]. \tag{4.4}$$

Proposition 4.2 *A linear in x_1 polynomial F is right primitive in x_1 if and only if it is a linear combination of dual long skew commutators:*

$$F = \sum_{\mathbf{i}=(i_1,i_2,\dots,i_k)} \alpha_{\mathbf{i}}[\dots[[x_1,x_{i_1}]^*,x_{i_2}]^*,\dots,x_{i_k}]^*$$

Proof The proof is quite similar to that of the above proposition. □

4.2.2 Polynomial Criterion

For each word $u = x_{i_1}x_{i_2}\dots x_{i_k}$ in $x_2, x_3, \dots, x_n$ consider the following polynomial

$$D_u = [\dots[[x_1, x_{i_1}], x_{i_2}], \dots, x_{i_k}] - [\dots[[x_1, x_{i_1}]^*, x_{i_2}]^*, \dots, x_{i_k}]^*, \tag{4.5}$$

where the dual brackets are defined in (4.4). The following theorem implies that there exists a quantum Lie operation of multidegree $(1, m_2, m_3, \dots, m_n)$ if and only if the polynomials D_u are linearly dependent.

Theorem 4.1 *Each linear dependence* $\sum \beta_u D_u = 0$ *where* u *runs through all words in* $x_2, \dots, x_n$ *of multidegree* $(m_2, m_3, \dots, m_n)$ *defines a quantum Lie operation:*

$$W = \sum_u \beta_u[\dots[[x_1, x_{i_1}], x_{i_2}], \dots, x_{i_k}] = \sum_u \beta_u[\dots[[x_1, x_{i_1}]^*, x_{i_2}]^*, \dots, x_{i_k}]^*. \tag{4.6}$$

Conversely, every quantum Lie operation W *of multidegree* $(1, m_2, m_3, \dots, m_n)$ *has a representation by* (4.6).

Proof If $\sum_u \beta_u D_u = 0$ then the second equality of (4.6) fulfills and it defines the element W. By Proposition 4.1 the element W is left primitive in x_1, whereas by Proposition 4.2 it is right primitive in x_1. Since $\deg_1(W) = 1$, it follows that W is a skew-primitive polynomial.

Conversely, if f is a skew-primitive polynomial of multidegree $(1, m_2, \dots, m_n)$, then it is left primitive in x_1. By Proposition 4.1, we have

$$f = \sum_u \beta_u[\dots[[x_1, x_{i_1}], x_{i_2}], \dots, x_{i_k}]. \tag{4.7}$$

Applying Proposition 4.2, we obtain

$$f = \sum_u \beta'_u[\dots[[x_1, x_{i_1}]^*, x_{i_2}]^*, \dots, x_{i_k}]^*. \tag{4.8}$$

Thus, we have the equality

$$\sum_u \beta_u[\dots[[x_1, x_{i_2}], x_{i_3}], \dots, x_{i_k}] = \sum_u \beta'_u[\dots[[x_1, x_{i_2}]^*, x_{i_3}]^*, \dots, x_{i_k}]^*. \tag{4.9}$$

If we compare coefficients at $x_1x_{i_1}\cdots x_{i_m} = x_1u$ in both sides of this equality, we get $\beta_u = \beta'_u$, and hence $\sum_u \beta_u D_u = 0$. □

Corollary 4.1 *If there exists a multilinear quantum Lie operation, then*

$$\prod_{1\le i\neq s\le n} p_{is} = 1. \tag{4.10}$$

Proof By definition a multilinear polynomial satisfies $m_2 = m_3 = \ldots = m_n = 1$. In this case, $u = x_{\pi(2)}x_{\pi(3)}\cdots x_{\pi(n)}$, where π is a permutation of symbols $2, 3, \ldots, n$. Let us compare coefficients at $x_{\pi(n)}\cdots x_{\pi(3)}x_{\pi(2)}x_1$ in both sides of (4.9). We arrive at a system of $(n-1)!$ equalities

$$\beta_u(-1)^{n-1}\prod_{k=2}^{n}(\prod_{i=1}^{k-1} p_{\pi(i)\pi(k)}) = \beta_u(-1)^{n-1}\prod_{k=2}^{n}(\prod_{i=1}^{k-1} p^{-1}_{\pi(k)\pi(i)}).$$

Clearly, each of these equalities is equivalent to (4.10) if $\beta_u \neq 0$. □

Using one additional variable, we may apply the criterion of Theorem 4.1 to arbitrary quantum Lie operations.

Proposition 4.3 *A homogeneous polynomial f in quantum variables $x_2, x_3, \ldots, x_n$ defines a quantum Lie operation if and only if so does $[x_1, f] = x_1 f - f x_1$, where x_1 is a quantum variable with $\chi^{x_1} = \mathrm{id}$, $g_{x_1} = \mathrm{id}$.*

Proof The coproduct has a decomposition

$$\Delta(f) = 1 \otimes f + g_f \otimes f + \sum_{u,v} \alpha_{u,v} g_v u \otimes v, \tag{4.11}$$

where u, v run though nonempty words in $x_2, x_3, \ldots, x_n$. We have,

$$\Delta([x_1,f]) = 1\otimes[x_1,f]+g_f\otimes[x_1,f]+\sum_{u,v}\alpha_{u,v}g_v([x_1,u]\otimes v+u\otimes[x_1,v]). \tag{4.12}$$

If f is skew-primitive, then $\alpha_{u,v} = 0$, and $[x_1,f]$ is skew-primitive as well. If $[x_1,f]$ is skew-primitive, then

$$\sum_{u,v}\alpha_{u,v}g_v([x_1,u]\otimes v + u\otimes[x_1,v]) = 0, \tag{4.13}$$

which implies $\alpha_{u,v} = 0$. □

4.2.3 Finkelstein Criterion and Specht–Wever Condition

Theorem 4.2 *A homogeneous polynomial $f = \sum_{\mathbf{i}} \alpha_{\mathbf{i}} x_{i_1} x_{i_2} \ldots x_{i_k}$ in quantum variables $x_2, x_3, \ldots, x_n$ is a quantum Lie operation if and only if*

$$[x_1,f] = \sum_{\mathbf{i}=(i_1,i_2,\ldots,i_k)} \alpha_{\mathbf{i}}[\ldots[[x_1,x_{i_1}],x_{i_2}]\ldots,x_{i_k}]. \tag{4.14}$$

Proof If f is skew-primitive, then Proposition 4.1 under substitution $x_2 \leftarrow f$ implies that $[x_1,f]$ is left primitive in x_1. The same proposition applied to $\mathbf{k}\langle X\rangle$ provides a representation

$$[x_1,f] = \sum_{\mathbf{i}=(i_1,i_2,\dots,i_k)} \beta_{\mathbf{i}}[\dots[[x_1,x_{i_1}],x_{i_2}]\dots,x_{i_k}].$$

The polynomial $[\dots[[x_1,x_{i_1}],x_{i_2}]\dots,x_{i_k}]$ has only one word with the first letter x_1. Comparing the words starting with x_1 in both sides of the latter equality, we obtain $\beta_{\mathbf{i}} = \alpha_{\mathbf{i}}$.

Conversely. If identity (4.14) is valid, then $[x_1,f]$ is left primitive in x_1. Let

$$\Delta(f) = f \otimes 1 + g_f \otimes f + \sum_i f_i^{(1)} \otimes f_i^{(2)}, \tag{4.15}$$

where $f_i^{(2)}$ are linearly independent homogeneous polynomials and $f_i^{(1)} \notin \mathbf{k}[G]$. We have,

$$\Delta([x_1,f]) - [x_1,f] \otimes 1 = \sum_{i\geq 1}(x_1 f_i^{(1)} - p(x_1,g_f)f_i^{(1)}x_1) \otimes f_i^{(2)} + \cdots, \tag{4.16}$$

where by the dots we denote a sum of tensors whose left-hand side is independent of x_1. Since $[x_1,f]$ is left primitive in x_1, it follows that $x_1 f_i^{(1)} = p(x_1,g_f)f_i^{(1)}x_1$, $i \geq 1$. These equalities are possible in $G\langle X\rangle$ only if $f_i^{(1)} \in \mathbf{k}[G]$. Hence, the sum in (4.15) is empty. □

In perfect analogy, we have a dual criterion.

Theorem 4.3 *A homogeneous polynomial* $f = \sum_{\mathbf{i}} \alpha_{\mathbf{i}} x_{i_1}x_{i_2}\dots x_{i_k}$ *in quantum variables* $x_2, x_3, \dots, x_n$ *is a quantum Lie operation if and only if*

$$[x_1,f]^* = \sum_{\mathbf{i}=(i_1,i_2,\dots,i_k)} \alpha_{\mathbf{i}}[\dots[[x_1,x_{i_1}]^*,x_{i_2}]^*\dots,x_{i_k}]^*. \tag{4.17}$$

In the equalities (4.14), (4.17) the variables x_i, $1 \leq i \leq n$ are algebraically independent. Therefore this identity is valid for arbitrary, not necessarily skew-primitive, values $x_i = u_i$ provided that the skew commutators are defined by the same coefficients. This proves the following statement.

Corollary 4.2 *If* $f(x_i) = \sum_{\mathbf{i}} \alpha_{\mathbf{i}} x_{i_1}x_{i_2}\dots x_{i_k}$ *is a quantum Lie operation in quantum variables* $x_2, \dots, x_n$ *then the following* ad-*identities hold*

$$[z,f(u_i)] = \sum_{\mathbf{i}=(i_1,i_2,\dots,i_k)} \alpha_{\mathbf{i}}[\dots[[z,u_{i_1}],u_{i_2}]\dots u_{i_k}], \tag{4.18}$$

$$[z,f(u_i)]^* = \sum_{\mathbf{i}=(i_1,i_2,\dots,i_k)} \alpha_{\mathbf{i}}[\dots[[z,u_{i_1}]^*,u_{i_2}]^*\dots u_{i_k}]^* \tag{4.19}$$

provided that z, u_i, $1 < i \leq n$ *are arbitrary homogeneous polynomials such that* $p(u_i, u_j) = p_{ij}$, $1 < i, j \leq n$.

In order to understand more closely the sense of the ad-identities, let us consider a number of simple examples.

Example 4.6 If $G = \{id\}$, then the commutator $[x_1, x_2] = x_1x_2 - x_2x_1$ is a (quantum) Lie operation. In this case the *ad*-identity (4.18) turns into the Jacobi identity in the following form:

$$[z, [u_1, u_2]] = [[z, u_1], u_2] - [[z, u_2], u_1]. \tag{4.20}$$

Example 4.7 More generally, if $p_{ij}p_{ji} = 1$ then the skew commutator $[x_1, x_2] = x_1x_2 - p_{12}x_2x_1$ is a quantum Lie operation. The ad-identity takes up a conditional identity

$$[z, [u, v]] = [[z, u], v] - p_{u,v}[[z, v], u], \tag{4.21}$$

under the condition $p_{u,v}p_{v,u} = 1$. This condition is universally true if the bicharacter p is symmetric, $p_{ij}p_{ji} = 1$; in this case (4.21) is the Jacobi identity for color Lie super-algebras.

Example 4.8 If $\chi^x(g_x) = \zeta$ is a tth primitive root of 1 and either $n = t$, or $n = tl^r$ where l is the characteristic of the ground field, then x^n is a quantum Lie operation. Thus, we obtain the identities:

$$[z, u^n] = [\dots[[[z, \underbrace{u], u], u], \dots, u]}_{n}, \quad [z, u^n]^* = [\dots[[[z, \underbrace{u]^*, u]^*, u]^*, \dots, u]^*}_{n} \tag{4.22}$$

provided that $p_{u,u} = \zeta$.

For a polynomial $f(x_i) = \sum_{\mathbf{i}} \alpha_{\mathbf{i}} x_{i_1} x_{i_2} \dots x_{i_k}$, we define the operator σ by

$$\sigma(f(x_i)) = \sum_{\mathbf{i}=(i_1,i_2,\dots,i_k)} \alpha_{\mathbf{i}}[\dots[x_{i_1}, x_{i_2}] \dots, x_{i_k}]. \tag{4.23}$$

Theorem 4.4 *If* $f(x_i)$ *is a quantum Lie operation, then* $\sigma(f(x_i)) = kf(x_i)$.

Proof Without loss of generality we may suppose that $f(x_i)$ is a polynomial in $x_2, x_3, \dots, x_n$. Let us introduce a variable x_1 with parameters $\chi^1 = \text{id}$, $g_1 = 1$. Consider a $\mathbf{k}[G]$-algebra H defined by relations $x_1x_i - x_ix_1 - x_i = 0$, $1 < i \leq n$. As left-hand sides of these relations are skew-primitive elements, algebra H maintain the character Hopf algebra structure, and the natural homomorphism $\varphi : G\langle X\rangle \to H$ is a homomorphism of character Hopf algebras. In algebra H, we have $[x_1, x_i] = x_i$. By evident induction this implies $[x_1, x_{i_1}x_{i_2} \dots x_{i_k}] = kx_{i_1}x_{i_2} \dots x_{i_k}$.

Therefore $[x_1, f(x_i)] = kf(x_i)$ in H. At the same time, the right-hand side of (4.14) equals $\sigma(f(x_i))$ in H. Applying φ to the identity (4.14), we obtain $\sigma(f(x_i)) = kf(x_i)$ in H. It remains to note that the restriction of φ on $G\langle x_2, x_3, \dots, x_n\rangle$ is an injection.

Indeed, there are no compositions between the defining relations, whereas the compositions $x_1x_i \cdot g = x_1 \cdot x_ig$ between the defining relations and the commutation rules $x_ig = \chi^i(g)gx_i$ are resolvable:

$$(x_ix_1 + x_i)g - x_1(\chi^i(g)gx_i) = \chi^i(g)g(\chi^1(g)x_ix_1 + x_i) - \chi^i(g)\chi^1(g)g(x_ix_1 + x_i) = 0.$$

By Theorem 1.2 (Diamond Lemma), words that have no subwords x_1x_i are linearly independent in H. In particular, so are all words in $x_2, x_3, \dots, x_n$. □

The following example shows that the Specht-Wever condition is not a criterion for the quantum Lie operations.

Example 4.9 The Specht-Wever condition for the polynomial x^n is

$$[\dots[[x,x],x],\dots,x] = nx^n.$$

If $p = \chi^x(g_x)$, then the above equality reduces to the following one

$$(1-p)(1-p^2)\cdots(1-p^{n-1}) = n. \tag{4.24}$$

For $p = 2, n = 3$ the equality (4.24) is valid, whereas x^3 is not a quantum operation provided that a characteristic of the ground field is zero.

4.3 Bilinear and Trilinear Operations

Theorem 2.5 with $n = 1$ states that there exists only one bilinear operation $[x_1, x_2]$ up to a scalar factor, in which case the existence condition is $p_{12}p_{21} = 1$. This allows us to define the *principle bilinear operation* on skew-primitive elements as follows:

$$[\![a,b]\!] = \begin{cases} [a,b], & \text{if } \chi^a(g_b)\chi^b(g_a) = 1; \\ \text{undefined}, & \text{otherwise}. \end{cases} \tag{4.25}$$

We stress that this operation applies to arbitrary, not necessary normalized, skew-primitive elements. Recall that if $\Delta(a) = a\otimes h_a + f_a\otimes a$ and $\Delta(b) = b\otimes h_b + f_b\otimes b$, then the brackets are defined thus: $[a,b] = \chi^b(h_a)ab - \chi^a(f_b)ba$, see (1.67). In particular, $[a,b] = h_ah_b\chi^a(h_b)\chi^b(h_a)[h_a^{-1}a, h_b^{-1}b]$ is still skew-primitive. By definition $g_a = h_a^{-1}f_a$, $g_b = h_b^{-1}f_b$. Hence, the existence condition remains

$\chi^a(g_b)\chi^b(g_a) = 1$, or it may be written in more symmetric form $\chi^a(f_b)\chi^b(f_a) = \chi^a(h_b)\chi^b(h_a)$.

It is easy to see that the principle bilinear operation satisfies the identities

$$[\![a, b]\!] = -\chi^a(g_b)[\![b, a]\!], \tag{4.26}$$

$$\chi^a(g_c)[\![a, [\![b, c]\!]]\!] + \chi^c(g_b)[\![c, [\![a, b]\!]]\!] + \chi^b(g_a)[\![b, [\![c, a]\!]]\!] = 0, \tag{4.27}$$

subject to the condition that all values $[\![\ ]\!]$ involved are determined.

Theorem 4.5 *For quantum variables x_1, x_2, and x_3, a nonzero trilinear quantum operation exists if and only if*

$$p_{12}p_{21}p_{13}p_{31}p_{23}p_{32} = 1. \tag{4.28}$$

If one of the inequalities

$$p_{12}p_{21} \neq 1,\ p_{13}p_{31} \neq 1,\ p_{23}p_{32} \neq 1 \tag{4.29}$$

holds, then there exists exactly one (up to multiplication by a scalar) such operation. If no one of them holds, then all trilinear operations are linearly expressed in terms of $[\![x_1, [\![x_2, x_3]\!]]\!]$ and $[\![x_2, [\![x_3, x_1]\!]]\!]$ via (4.26) *and* (4.27).

Proof Corollary 4.1 implies that (4.28) is a necessary condition for existence of trilinear operation. To prove that this condition is sufficient, we apply the polynomial criterion given in Theorem 4.1.

For $n = 3$, $m_2 = 1$, $m_3 = 1$, there exists two polynomials D_u corresponding to $u = x_2x_3$ and $u = x_3x_2$, see (4.5). We note that $D_{x_3x_2}$ appears from $D_{x_2x_3}$ by application of the permutation (23) to all indices. Therefore, let D_{id} and $D_{(23)}$ denote the polynomials D_u related to $u = x_2x_3$ and $u = x_3x_2$ respectively. If (4.28) is valid, then

$$D_{id} = (p_{21}^{-1} - p_{12})x_2x_1x_3 + (p_{31}^{-1}p_{32}^{-1} - p_{13}p_{23})x_3x_1x_2,$$
$$D_{(23)} = (p_{21}^{-1}p_{23}^{-1} - p_{12}p_{32})x_2x_1x_3 + (p_{31}^{-1} - p_{13})x_3x_1x_2.$$

If one of the inequalities (4.29) holds (let it be $p_{13}p_{31} \neq 1$ for definiteness), then

$$(p_{21}^{-1} - p_{12})(p_{31}^{-1} - p_{13}) = (p_{31}^{-1}p_{32}^{-1} - p_{13}p_{23})(p_{21}^{-1}p_{23}^{-1} - p_{12}p_{32}),$$

and hence

$$D_{id} - \frac{p_{31}^{-1}p_{32}^{-1} - p_{13}p_{23}}{p_{31}^{-1} - p_{13}} D_{(23)} = 0.$$

Here, $D_{(23)} \neq 0$; that is, the space generated by D_{id} and $D_{(23)}$ is one-dimensional. By Theorem 4.1, there exists the unique trilinear operation up to a scalar factor

$$[[x_1, x_2], x_3] - \frac{p_{31}^{-1} p_{32}^{-1} - p_{13} p_{23}}{p_{31}^{-1} - p_{13}} [[x_1, x_3], x_2]. \tag{4.30}$$

If all products $p_{ij} p_{ji}$, $i \neq j$ are equal to 1, then $D_{id} = D_{(23)} = 0$; that is, there exist exactly two linear dependences between D_{id} and $D_{(23)}$. Hence, there are exactly two linearly independent operations. In this case, on the other hand, all the three values, $[\![x_1, x_2]\!]$, $[\![x_1, x_3]\!]$, and $[\![x_2, x_3]\!]$, of the main bilinear operation are defined. Moreover, since $g_{[\![x_i,x_j]\!]} = g_i g_j$ and $\chi^{[\![x_i,x_j]\!]} = \chi^i \chi^j$, we see that $\chi^{[\![x_i,x_j]\!]}(g_k)\chi^k(g_{[\![x_i,x_j]\!]}) = 1$ holds for $k \neq i, j$; that is, all possible superpositions, too, are defined. Among them, by the above argument, only two may be linearly independent (for example, those specified in the theorem), and the rest are expressed via them using (4.26) and (4.27). □

Note that if exactly one of the inequalities (4.29) fails, say, $p_{12} p_{21} = 1$, then the superposition $[\![x_3, [\![x_1, x_2]\!]]\!]$ will be defined; hence, the unique (by Theorem 4.5) quantum operation will equal that superposition. This circumstance allows us to define the principle trilinear operation thus:

$$[\![a, b, c]\!] = \begin{cases} [[a, b], c] - \frac{p_{31}^{-1} p_{32}^{-1} - p_{13} p_{23}}{p_{31}^{-1} - p_{13}} [[a, c], b], \text{ if } & \prod_{1 \leq i \neq s \leq 3} p_{is} = 1, \text{ and } p_{is} p_{si} \neq 1; \\ \text{undefined}, & \text{otherwise}, \end{cases}$$

where $p_{12} = \chi^a(g_b)$, $p_{13} = \chi^a(g_c)$, etc.

The principle trilinear operation applies to arbitrary, not necessary normalized, skew-primitive elements as well. The existence conditions remain the same, or they may be written in more symmetric form

$$\prod_{i \neq s} q_{is} = \prod_{i \neq s} q'_{is}, \quad \text{and} \quad q_{is} q_{si} \neq q'_{is} q'_{si} \text{ for } i \neq s,$$

where, as usual, $q_{12} = \chi^a(h_b)$, $q'_{12} = \chi^a(f_b)$, $q_{13} = \chi^a(h_c)$, $q'_{13} = \chi^a(f_c)$, etc.

The operation being unique has an implication that if we rename the variables $x_i \to x_{\pi(i)}$, then the value of the main operation on $x_{\pi(1)}, x_{\pi(2)}, x_{\pi(3)}$ (of course, it is defined on that sequence since (4.28) is invariant under such substitutions) should be linearly expressed via its value on x_1, x_2, x_3, that is,

$$[\![x_{\pi(1)}, x_{\pi(2)}, x_{\pi(3)}]\!] = \alpha_\pi [\![x_1, x_2, x_3]\!]. \tag{4.31}$$

If we compare the coefficients at $x_{\pi(1)} x_{\pi(2)} x_{\pi(3)}$ on the right- and left-hand sides of (4.31), we see that $\alpha_\pi = \gamma_{\pi^{-1}}^{\pi} = \gamma_\pi^{-1}$ where γ_π are precisely coefficients in the expansion

$$[\![x_1, x_2, x_3]\!] = \sum \gamma_\pi x_{\pi(1)} x_{\pi(2)} x_{\pi(3)};$$

or by routine computations,

$$\alpha_{id}=1,\ \alpha_{(123)}=\frac{p_{31}-p_{13}^{-1}}{p_{12}-p_{21}^{-1}},\ \alpha_{(132)}=\frac{p_{31}-p_{13}^{-1}}{p_{23}-p_{32}^{-1}},\ \alpha_{(13)}=p_{21}p_{32}p_{31},$$

$$\alpha_{(12)}=p_{21}p_{23}p_{13}\frac{p_{31}-p_{13}^{-1}}{p_{23}-p_{32}^{-1}},\quad \alpha_{(23)}=p_{12}p_{32}p_{13}\frac{p_{31}-p_{13}^{-1}}{p_{12}-p_{21}^{-1}}.$$

4.4 Quadrilinear Operations

We pass to the case $n=4$. Denote by S_4 the permutation group on the set $\{1,2,3,4\}$, and by S_4^1 its subgroup consisting of all permutations leaving 1 fixed. For our goals, both a functional and an exponential notation for the action of S_4 on the index set might seem convenient, while we assume that $i^{(\pi\nu)}=(i^{\pi})^{\nu}=\nu(\pi(i))$. Write $\boldsymbol{\tau}$ to denote the permutation

$$\boldsymbol{\tau}=\begin{pmatrix}1\,2\,3\,4\\4\,3\,2\,1\end{pmatrix}=(14)(23). \tag{4.32}$$

For brevity, we make the convention to write $\pi(A)$ or A^{π} for the permutation π and for an arbitrary expression A, meaning that $\pi(A)$ is obtained from A by applying π to each index occurring in A at letters p_{ij} or at variables x_i. In so doing, we do not require that S_4 acts on the ground field. For instance, it might be the case that, in $\mathbf{k}$, the equality $p_{12}=p_{23}$ is satisfied but $p_{12}^{(123)}=p_{23}^{(123)}$ is not, that is, this is merely a notational convention, which is used only unless it leads to confusion. An arbitrary quadrilinear polynomial, in accordance with the above conventions, can be written in the form

$$W(x_1,x_2,x_3,x_4)=\sum_{\pi\in S_4}\alpha_{\pi}\pi(x_1x_2x_3x_4),\ \ \alpha_{\pi}\in\mathbf{k}. \tag{4.33}$$

To that conventions we add the following:

$$\{p_{ij}p_{kl}\dots p_{rs}\}\stackrel{\text{df}}{=}p_{ij}p_{kl}\cdots p_{rs}-p_{ji}^{-1}p_{lk}^{-1}\dots p_{sr}^{-1}, \tag{4.34}$$

and for the word A depending on p_{ij}, denote by $\overline{A}$ a word obtained from A by replacing all letters p_{ij} with p_{ji}, so that $\{A\}=A-\overline{A}^{-1}$. These are again merely notational conventions: the equality $p_{12}p_{34}=p_{13}p_{32}$ in $\mathbf{k}$ not necessarily implies $p_{21}^{-1}p_{43}^{-1}=p_{31}^{-1}p_{23}^{-1}$ in $\mathbf{k}$.

Lemma 4.2 *Let C, D, and E be some words in p_{ij}. Then*

$$\{CE\}\{D\overline{E}\}-\{C\}\{D\}=\{CD\overline{E}\}\{E\}. \tag{4.35}$$

Proof Using (4.34), we rewrite the left- and right-hand sides of (4.35) in this way:

$$\begin{aligned}\{CE\}\{D\overline{E}\} - \{C\}\{D\} &= (CE - \overline{CE}^{-1})(D\overline{E} - (\overline{D}E)^{-1} - (C - \overline{C}^{-1})(D - \overline{D}^{-1})\\ &= CED\overline{E} - \overline{C}^{-1}\overline{E}^{-1}D\overline{E} - CE\overline{D}^{-1}E^{-1}\\ &\quad + \overline{C}^{-1}\overline{E}^{-1}\overline{D}^{-1}E^{-1} - CD + \overline{C}^{-1}D + C\overline{D}^{-1} - \overline{C}^{-1}\overline{D}^{-1}\\ &= CED\overline{E} + \overline{CED}^{-1}E^{-1} - CD - \overline{CD}^{-1};\\ \{CD\overline{E}\}\{E\} &= (CD\overline{E} - \overline{CD}^{-1}E^{-1})(E - \overline{E}^{-1})\\ &= CD\overline{E}E - \overline{CD}^{-1} - CD + \overline{CDE}^{-1}E^{-1}.\end{aligned}$$

□

Note that the existence condition for the trilinear operation (4.28) can be written via braces thus: $\{p_{12}p_{13}p_{23}\} = 0$, whereas the existence condition for the bilinear operation takes the form $\{p_{12}\} = 0$. It might be useful to point out the following properties of the braces:

$$\{C\} = 0 \rightarrow \{CD\} = C\{D\}, \tag{4.36}$$

$$\{C\} = 0 \;\&\; \{CD\} = 0 \;\rightarrow \{D\} = 0. \tag{4.37}$$

Theorem 4.6 *For quantum variables x_1, x_2, x_3, x_4, a nonzero quadrilinear quantum operation exists if and only if*

$$p_{12}p_{21}p_{13}p_{31}p_{14}p_{41}p_{23}p_{32}p_{24}p_{42}p_{34}p_{43} = 1. \tag{4.38}$$

If this equality holds, and there is a pair of indices i, j such that

$$\{p_{ij}\} \neq 0 \;\&\; \{p_{ij}p_{ik}p_{kj}\} \neq 0 \;\&\; \{p_{ij}p_{is}p_{sj}\} \neq 0, \tag{4.39}$$

where i, j, k, s are distinct indices, then there exist exactly two linearly independent quadrilinear operations. If condition (4.39) fails for all $i \neq j$, then all quadrilinear operations are expressed via bilinear and trilinear principle operations.

Proof Corollary 4.1 implies that (4.38) is a necessary condition for existence of quadrilinear operations. We stress that (4.38) is equivalent to each one of the following 24 equalities

$$\{x_{12}x_{13}x_{14}x_{23}x_{24}x_{34}\}^{\pi} = 0, \quad \pi \in S_4. \tag{4.40}$$

To prove that this condition is sufficient, we apply the polynomial criterion given in Theorem 4.1. For $n = 4$, $m_2 = 1$, $m_3 = 1$, $m_4 = 1$, there exists six polynomials D_u corresponding to different permutations of the word $x_2x_3x_4$, see (4.5). Let D_π

corresponds to the word $u = x_{\pi(2)}x_{\pi(3)}x_{\pi(4)}$, where $\pi \in S_4^1$. Due to our conventions in notations, we have $D_\pi = \pi(D_{id})$. We seek an element D_{id} in an explicit form. Expanding the skew commutators in (4.5) yields

$$\begin{aligned}D_{id} = &-\{p_{12}\}x_2x_1x_3x_4 - \{p_{13}p_{23}\}x_3x_1x_2x_4 - \{p_{14}p_{24}p_{34}\}x_4x_1x_2x_3 \\ &+\{p_{12}p_{13}p_{23}\}x_3x_2x_1x_4 + \{p_{12}p_{14}p_{24}p_{34}\}x_4x_2x_1x_3 \\ &+\{p_{13}p_{23}p_{14}p_{24}p_{34}\}x_4x_3x_1x_2. \end{aligned} \tag{4.41}$$

Now assume that β_π are unknown parameters. Consider the linear combination $\sum \beta_\pi \pi(D_{id})$ and the coefficients at its distinct words. Setting that combination equal to zero, we obtain a homogeneous system of 12 equations (equal to the number of distinct words not beginning with and not ending in x_1) with six unknowns. We have to demonstrate that, under conditions (4.38) and (4.39) for $i = 1, j = 4$, that system has two linearly independent solutions.

Consider the coefficient at $x_2x_1x_3x_4$. If we apply $\pi \in S_4^1$ to (4.41), the element x_1 will be left fixed; therefore, the word $x_2x_1x_3x_4$ arises in $\pi(D_{id})$ only from the first three summands of (4.41). If it arises from the second, then $\pi(3) = 2$, $\pi(2) = 3$, and $\pi(4) = 4$; that is, $\pi = (23)$. If it arises from the third, then $\pi(4) = 2$, $\pi(2) = 3$, and $\pi(3) = 4$; that is, $\pi = (234)$. Therefore, the whole coefficient at $x_2x_1x_3x_4$ is equal to

$$-\{p_{12}\}\beta_{id} - \{p_{12}p_{32}\}\beta_{(23)} - \{p_{12}p_{32}p_{42}\}\beta_{(234)}. \tag{4.42}$$

In a similar way, if we compute coefficients at other six words $\pi(x_2x_1x_3x_4)$, $\pi \in S_4^1$, with x_1 holding second place, we obtain the first group of six equations

$$[-\{p_{12}\}\beta_{id} - \{p_{12}p_{32}\}\beta_{(23)} - \{p_{12}p_{32}p_{42}\}\beta_{(234)}]^\mu = 0, \ \mu \in S_4^1. \tag{4.43}$$

At this point we use the conventions made at the beginning of Sect. 4.4, assuming in addition that permutations μ act on the indices at β by right multiplications: $[\ldots \beta_\pi \ldots]^\mu = \ldots \beta_{\pi\mu} \ldots$.

In perfect analogy, we consider the coefficient at $x_4x_3x_1x_2$. This word arises in $\pi(D_{id})$ from the last three summands of (4.41) only. If it arises from the last but one summand, then $\pi(4) = 4$, $\pi(2) = 3$, and $\pi(3) = 2$; that is, $\pi = (23)$. If it arises from the fourth summand, then $\pi(3) = 4$, $\pi(2) = 3$, and $\pi(4) = 2$, that is, $\pi = (234)$. Therefore, the coefficient is equal to

$$\{p_{13}p_{23}p_{14}p_{24}p_{34}\}\beta_{id} + \{p_{13}p_{14}p_{34}p_{24}\}\beta_{(23)} + \{p_{13}p_{14}p_{34}\}, \beta_{(234)}, \tag{4.44}$$

and we obtain yet other six equations

$$[\{p_{13}p_{23}p_{14}p_{24}p_{34}\}\beta_{id} + \{p_{13}p_{14}p_{34}p_{24}\}\beta_{(23)} + \{p_{13}p_{14}p_{34}\}\beta_{(234)}]^\mu = 0. \tag{4.45}$$

Consider two Eqs. (4.42) and (4.44). Let us check that under condition (4.38), all three minors of that system of two equations equal zero. First, using Lemma 4.2 with $C \leftarrow p_{12}$; $D \leftarrow p_{13}p_{14}p_{34}p_{24}$; and $E \leftarrow p_{32}$, we compute the minor corresponding to the variables β_{id} and $\beta_{(23)}$:

$$-\{p_{12}\}\{p_{13}p_{14}p_{34}p_{24}\} + \{p_{12}p_{32}\}\{p_{13}p_{23}p_{14}p_{24}p_{34}\} = \{p_{12}p_{13}p_{14}p_{34}p_{24}p_{23}\}\{p_{32}\}. \tag{4.46}$$

The first factor of the latter product is zero due to (4.40) with $\pi = id$. Then, by Lemma 4.2 with $C \leftarrow p_{12}$; $D \leftarrow p_{13}p_{14}p_{34}$; and $E \leftarrow p_{32}p_{42}$, we compute the minor corresponding to the variables β_{id} and $\beta_{(234)}$:

$$\begin{aligned} &-\{p_{12}\}\{p_{13}p_{14}p_{34}\} + \{p_{12}p_{32}p_{42}\}\{p_{13}p_{23}p_{14}p_{24}p_{34}\} \\ &= \{p_{12}p_{13}p_{14}p_{34}p_{23}p_{24}\}\{p_{32}p_{42}\} = 0. \end{aligned} \tag{4.47}$$

Next, for the minor corresponding to the variables $\beta_{(23)}$ and $\beta_{(234)}$, we again apply Lemma 4.2 with $C \leftarrow p_{12}p_{32}$; $D \leftarrow p_{13}p_{14}p_{34}$; $E \leftarrow p_{42}$:

$$-\{p_{12}p_{32}\}\{p_{13}p_{14}p_{34}\} + \{p_{12}p_{32}p_{42}\}\{p_{13}p_{14}p_{34}p_{24}\} = \{p_{12}p_{32}p_{13}p_{14}p_{34}p_{24}\}\{p_{42}\}. \tag{4.48}$$

The first factor of the latter product is zero due to (4.40) with $\pi = (23)$.

Besides, if some coefficient in (4.42) equals zero, then by (4.37), the corresponding coefficient in (4.44) equals zero and vise versa. Therefore, Eq. (4.42) is equivalent to (4.44).

Applying $\mu \in S_4^1$ to all of the expressions (4.42), (4.44), and (4.46)–(4.48), we obtain that the whole system of 12 equations is equivalent to the six in (4.43).

We order elements of the group S_4^1 in this way: *id*, (23), (234), (34), (24), (243). The matrix of the system then has the form

$$\begin{pmatrix} \{p_{12}\} & \{p_{12}p_{32}\} & \{p_{12}p_{32}p_{42}\} & 0 & 0 & 0 \\ \{p_{13}p_{23}\} & \{p_{13}\} & 0 & \{p_{13}p_{23}p_{43}\} & 0 & 0 \\ 0 & 0 & \{p_{13}\} & 0 & \{p_{13}p_{43}\} & \{p_{13}p_{43}p_{23}\} \\ 0 & 0 & 0 & \{p_{12}\} & \{p_{12}p_{42}p_{32}\} & \{p_{12}p_{42}\} \\ 0 & \{p_{14}p_{34}p_{24}\} & \{p_{14}p_{34}\} & 0 & \{p_{14}\} & 0 \\ \{p_{14}p_{24}p_{34}\} & 0 & 0 & \{p_{14}p_{24}\} & 0 & \{p_{14}\} \end{pmatrix}$$

If, in this matrix, we delete the first two columns and the third and fourth rows, we obtain a triangular submatrix with the diagonal $\{p_{12}p_{32}p_{42}\}$, $\{p_{13}p_{23}p_{43}\}$, $\{p_{14}\}$. Let us check that under condition (4.39) for $i=1$, $j=4$, no one of those elements is zero.

Condition (4.39) for $i=1$, $j=4$ includes $\{p_{14}\}\neq 0$. Remark (4.37) with $C\leftarrow p_{12}p_{32}p_{42}$ and $D\leftarrow p_{14}p_{13}p_{34}$ implies $\{p_{12}p_{32}p_{42}\}\neq 0$, for otherwise $\{p_{14}p_{13}p_{34}\}=0$. Similarly, (4.37) with $C\leftarrow p_{13}p_{23}p_{43}$ and $D\leftarrow p_{14}p_{12}p_{24}$ implies $\{p_{13}p_{23}p_{43}\}\neq 0$, for otherwise $\{p_{14}p_{12}p_{24}\}=0$.

Thus, the corresponding minor is nonzero and the whole system has not more than two linearly independent solutions. Put $\beta_{id}=1$ and $\beta_{(23)}=0$, and find one solution for the system of the first two and last two equations:

$$\beta_{id}=1,\ \beta_{(23)}=0,\ \beta_{(234)}=-\frac{\{p_{12}\}}{\{p_{12}p_{32}p_{42}\}},\ \beta_{(34)}=-\frac{\{p_{13}p_{23}\}}{\{p_{13}p_{23}p_{43}\}},$$

$$\beta_{(24)}=\frac{\{p_{12}\}\{p_{14}p_{34}\}}{\{p_{14}\}\{p_{12}p_{32}p_{42}\}},\ \beta_{(243)}=-\frac{\{p_{43}\}\{p_{14}p_{24}p_{34}p_{13}p_{23}\}}{\{p_{14}\}\{p_{13}p_{23}p_{43}\}}. \tag{4.49}$$

Using Lemma 4.2, we verify if these values are solutions for the third equation:

$$\begin{aligned}
&-\{p_{13}\}\frac{\{p_{12}\}}{\{p_{12}p_{32}p_{42}\}}+\{p_{13}p_{43}\}\frac{\{p_{12}\}\{p_{14}p_{34}\}}{\{p_{14}\}\{p_{12}p_{32}p_{42}\}}-\{p_{13}p_{43}p_{23}\}\frac{\{p_{43}\}\{p_{14}p_{24}p_{34}p_{13}p_{23}\}}{\{p_{14}\}\{p_{13}p_{23}p_{43}\}}\\
&=-\frac{\{p_{12}\}}{\{p_{12}p_{32}p_{42}\}\{p_{14}\}}(\{p_{13}\}\{p_{14}\}-\{p_{13}p_{43}\}\{p_{14}p_{34}\})-\frac{\{p_{43}\}\{p_{14}p_{24}p_{34}p_{13}p_{23}\}}{\{p_{14}\}}\\
&=\frac{\{p_{43}\}}{\{p_{12}p_{32}p_{42}\}\{p_{14}\}}(\{p_{12}\}\{p_{13}p_{14}p_{34}\}-\{p_{12}p_{32}p_{42}\}\{p_{14}p_{24}p_{34}p_{13}p_{23}\})\\
&=-\frac{\{p_{43}\}\{p_{12}p_{13}p_{14}p_{34}p_{23}p_{24}\}\{p_{32}p_{42}\}}{\{p_{12}p_{32}p_{42}\}\{p_{14}\}}=0.
\end{aligned}$$

Likewise for the fourth equation (with the "−" sign):

$$\begin{aligned}
&\{p_{12}\}\frac{\{p_{13}p_{23}\}}{\{p_{13}p_{23}p_{43}\}}-\{p_{12}p_{32}p_{42}\}\frac{\{p_{12}\}\{p_{14}p_{34}\}}{\{p_{12}p_{32}p_{42}\}\{p_{14}\}}+\{p_{12}p_{42}\}\frac{\{p_{43}\}\{p_{14}p_{24}p_{34}p_{13}p_{23}\}}{\{p_{14}\}\{p_{13}p_{23}p_{43}\}}\\
&=\frac{\{p_{12}\}}{\{p_{14}\}\{p_{13}p_{23}p_{43}\}}(\{p_{13}p_{23}\}\{p_{14}\}-\{p_{13}p_{23}p_{43}\}\{p_{14}p_{34}\})+\dots\\
&=-\frac{\{p_{12}\}\{p_{13}p_{23}p_{14}p_{34}\}\{p_{43}\}}{\{p_{14}\}\{p_{13}p_{23}p_{43}\}}+\{p_{12}p_{42}\}\frac{\{p_{43}\}\{p_{14}p_{24}p_{34}p_{13}p_{23}\}}{\{p_{14}\}\{p_{13}p_{23}p_{43}\}}\\
&=\frac{\{p_{43}\}}{\{p_{14}\}\{p_{13}p_{23}p_{43}\}}(-\{p_{12}\}\{p_{13}p_{23}p_{14}p_{34}\}+\{p_{12}p_{42}\}\{p_{14}p_{24}p_{34}p_{13}p_{23}\})\\
&=\frac{\{p_{43}\}\{p_{12}p_{13}p_{23}p_{14}p_{34}p_{24}\}\{p_{42}\}}{\{p_{14}\}\{p_{13}p_{23}p_{43}\}}=0.
\end{aligned}$$

Thus, by Theorem 4.1, the computed values of β_π determine a quadrilinear operation

$$[\![x_1, x_2, x_3, x_4]\!] = \sum_{\pi \in S_4^1} \beta_\pi [[[x_1, x_{\pi(2)}], x_{\pi(3)}], x_{\pi(4)}]. \quad (4.50)$$

Since $\beta_{id} = 1$, $\beta_{(23)} = 0$, and the word $\pi(x_1x_2x_3x_4)$ in (4.50) occurs only in the summand corresponding to the permutation π, we see that the coefficient at $x_1x_2x_3x_4$ in the expansion (4.33) of the polynomial $[\![x_1, x_2, x_3, x_4]\!]$ equals 1, and the coefficient at $x_1x_3x_2x_4$ is zero.

Consider a sequence of quantum variables $y_1 = x_1$, $y_2 = x_3$, $y_3 = x_2$, and $y_4 = x_4$. This sequence satisfies both conditions (4.38) and (4.39) for $i = 1$, $j = 4$; hence, by the above, there exists a quantum operation $[\![y_1, y_2, y_3, y_4]\!] = [\![x_1, x_3, x_2, x_4]\!]$ such that the coefficient at $x_1x_3x_2x_4$ equals 1 and the one at $x_1x_2x_3x_4$ equals 0. In this way $[\![x_1, x_3, x_2, x_4]\!]$ supplies the second solution for the system under consideration, which proves the first part of the theorem.

Now assume that no one pair $i \neq j$ satisfies condition (4.39).

We call a pair x_i, x_j (respectively, a triple x_i, x_j, x_k) of variables *conforming* if the existence condition for bilinear (trilinear) operation is satisfied: $\{p_{ij}\} = 0$ (respectively, $\{p_{ij}p_{ik}p_{jk}\} = 0$). The failure of condition (4.39) for i, j will mean, then, that the variables x_i and x_j enter some two- or three-element conforming subset. If the pair x_i, x_j is itself conforming, then the value $[\![x_i, x_j]\!]$ is defined, and the set $[\![x_i, x_j]\!]$, x_k, x_l too is conforming. Therefore, one of the superpositions $[\![[\![x_i, x_j]\!], x_k, x_l]\!]$ or $[\![[\![[\![x_i, x_j]\!], x_k]\!], x_l]\!]$ is determined. Similarly, if the triple x_i, x_k, x_j is conforming, then either $[\![[\![x_i, x_j, x_k]\!], x_l]\!]$ or $[\![[\![[\![x_i, x_j]\!], x_k]\!], x_l]\!]$ is defined.

We turn on to consider the possible cases where the six conditions (4.39) fail.

1. All two-element subsets are conforming. The system (4.43) has only zero coefficients, and by Theorem 4.1, we then find six linearly independent operations:

$$\pi([\![[\![[\![x_1, x_2]\!], x_3]\!], x_4]\!]), \quad \pi \in S_4^1.$$

2. All four three-element subsets are conforming. In view of the above, we can assume that one of the two-element subsets is not conforming. Suppose $\{p_{12}\} \neq 0$. Then the system (4.43) splits into three pairs of equations: *id*, (23); (243), (24); (34), (243). Here, the first and third pairs have rank 1 and the second has rank ≤ 1. Thus, if at least one of the inequalities $\{p_{14}\} \neq 0$, $\{p_{13}\} \neq 0$, or $\{p_{43}\} \neq 0$ holds, then the whole system has exactly three solutions, and these, in accordance with Theorem 4.1, yield the following three operations:

$$[\![[\![x_1, x_2, x_3]\!]', x_4]\!]; \ [\![[\![x_1, x_3, x_4]\!]', x_2]\!]; \ [\![[\![x_1, x_2, x_4]\!]', x_3]\!].$$

Here, $[\![]\!]'$ denotes the ternary operation whose uniqueness is asserted by Theorem 4.5, that is, it is either the principle operation or a superposition of the form $[\![[\![\ ]\!],\ ]\!]$. There then exists one more superposition $[\![[\![x_2, x_3, x_4]\!], x_1]\!]$, which should be linearly expressed in terms of the solutions that we have found. Consequently, using the fact that the bilinear (4.26) and trilinear (4.31) operations are symmetric, we arrive at an analog of the Jacobi identity

$$\sum_{k=0}^{3} \xi_k \sigma^k([\![[\![x_1, x_2, x_3]\!], x_4]\!]) = 0, \tag{4.51}$$

where $\sigma = (1234)$ is a cyclic permutation, the coefficients ξ_k are uniquely determined up to multiplication by a common scalar, and all values of the principle operation are assumed determined.

If $\{p_{14}\} = \{p_{13}\} = \{p_{43}\} = 0$, then the second pair of equations disappears, and instead of $[\![[\![x_1, x_3, x_4]\!]', x_2]\!]$, there appear two operations: $[\![[\![[\![x_1, x_3]\!]x_4,]\!]x_2]\!]$ and $[\![[\![[\![x_1, x_4]\!]x_3,]\!]x_2]\!]$.

3. Three three-element subsets are conforming. Condition (4.38) then implies that the fourth subset is also conforming.

4. Exactly two three-element subsets are conforming. To be specific, let

$$\{p_{12}p_{14}p_{24}\} = \{p_{13}p_{14}p_{34}\} = 0, \ \{p_{12}p_{23}p_{13}\} \neq 0, \ \{p_{23}p_{24}p_{34}\} \neq 0.$$

Because condition (4.39) for $i = 2, j = 3$ fails and the two triples involved are not conforming, we have $\{p_{23}\} = 0$. If we write (4.38) in the form $\{p_{12}p_{14}p_{24}p_{13}p_{23}p_{43}\} = 0$, by formulas (4.36) and (4.37), we obtain $0 = \{p_{13}p_{23}p_{43}\} = p_{23}\{p_{13}p_{43}\}$, and similarly $0 = \{p_{12}p_{32}p_{42}\} = p_{32}\{p_{12}p_{42}\}$. In other words, $\{p_{13}p_{43}\} = \{p_{12}p_{42}\} = \{p_{23}\} = 0$, and again condition (4.38) yields $\{p_{14}\} = 0$. In the matrix of (4.43), in particular, the last two columns will disappear, whereas the minor corresponding to the first four columns and last four rows equals $-p_{14}^2\{p_{24}p_{34}\}^2\{p_{13}\}\{p_{12}\}$. Here $\{p_{24}p_{34}\} \neq 0$ due to $0 \neq \{p_{23}p_{24}p_{34}\} = p_{23}\{p_{24}p_{34}\}$.

Now if $\{p_{13}\}\{p_{12}\} \neq 0$, then the whole system has rank 4 and its solutions are determined by arbitrary values of $\beta_{(24)}$ and $\beta_{(243)}$, that is, we obtain two operations:

$$[\![[\![[\![x_1, x_4]\!]x_3,]\!]x_2]\!]; \ [\![[\![[\![x_1, x_4]\!]x_2,]\!]x_3]\!], \tag{4.52}$$

in terms of which all other operations defined in the present case are expressible:

$$[\![[\![x_1, x_2, x_4]\!]', x_3]\!], \ [\![[\![x_1, x_3, x_4]\!]', x_2]\!], \ [\![x_1, [\![x_2, x_3]\!], x_4]\!], \ [\![[\![x_1, x_4]\!], x_2, x_3]\!]'.$$

If $\{p_{13}\}\{p_{12}\} = 0$, in view of the initial conditions being symmetric under the permutation $2 \leftrightarrow 3$, it suffices to consider the case $\{p_{13}\} = 0$. We have $\{p_{12}p_{24}\} = \{p_{13}\} = \{p_{14}\} = \{p_{23}\} = \{p_{34}\} = 0$, $\{p_{12}\} \neq 0$, $\{p_{24}\} \neq 0$ (if not all pairs are conforming). And we face only one additional solution $[\![[\![[\![x_1, x_3]\!]x_4,]\!]x_2]\!]$ because the minor corresponding to the first, fourth, and sixth rows and to the first, second, and fourth columns is $\{p_{12}\}\{p_{12}p_{32}\}\{p_{14}p_{24}p_{34}\} = \{p_{12}\}^2\{p_{24}\}p_{32}p_{14}p_{34} \neq 0$.

5. Only one three-element subset is conforming. Let it be x_2, x_3, x_4. Then the failure of conditions (4.39) for $i = 1, j = 2$; for $i = 1, j = 3$; and for $i = 1, j = 4$ implies that $\{p_{12}\} = \{p_{13}\} = \{p_{14}\} = 0$. In this case $\{p_{34}\} \neq 0$, otherwise the triple x_1, x_3, x_4 would be conforming. Similarly, $\{p_{24}\} \neq 0$ and $\{p_{23}\} \neq 0$. These imply $\{p_{23}p_{24}\} \neq 0$, $\{p_{23}p_{34}\} \neq 0$, and $\{p_{24}p_{34}\} \neq 0$. Indeed, for instance, equality $\{p_{23}p_{24}\} = 0$, combined with $\{p_{1i}\} = 0, i = 2, 3, 4$, and (4.38), yields $\{p_{34}\} = 0$.

Under these conditions, the system splits into three pairs of rank 1 equations: (23), (243); *id*, (24); (234), (34). The first pair agrees with the operation

$$[[[x_1, x_2]], x_3, x_4]],$$

and the two other operations result from it by permutations of indices (23) and (24). All other superpositions defined in the present case are linearly expressed via these three. Specifically, we have an identity of the form

$$[\![x_1, [\![x_2, x_3, x_4]\!]]\!] = \xi_1[\![[\![x_1, x_2]\!], x_3, x_4]\!] + \xi_2[\![x_2, [\![x_1, x_3]\!], x_4]\!] + \xi_3[\![x_2, x_3, [\![x_1, x_4]\!]]\!].$$

6. No one of the three-element subsets is conforming. Then two-element subsets cannot all be conforming; therefore, one of the six conditions (4.39) is satisfied. □

Under conditions (4.38) and (4.39) for $i = 1, j = 4$, the *principle quadrilinear operation* is defined by

$$[\![a_1, a_2, a_3, a_4]\!] = \sum_{\pi \in S_4^1} \beta_\pi [[[a_1, a_{\pi(2)}], a_{\pi(3)}], a_{\pi(4)}], \tag{4.53}$$

where a_i are skew-primitive semi-invariants, $p_{ij} = \chi^{a_i}(g_{a_j})$, and the coefficients β_π are given as in (4.49).

If no proper subset of the set x_1, x_2, x_3, x_4 is conforming, then six conditions (4.39) are satisfied. Therefore, all possible 24 permutation variants $[\![x_{\pi(1)}, x_{\pi(2)}, x_{\pi(3)}, x_{\pi(4)}]\!]$, $\pi \in S_4$, are determined, and by Theorem 4.6, they all are expressible via any pair of them. In order to find that representation, we write the principle operation in the form

$$[\![x_1, x_2, x_3, x_4]\!] = \sum \alpha_\pi x_{\pi(1)} x_{\pi(2)} x_{\pi(3)} x_{\pi(4)}, \tag{4.54}$$

where α_π are particular rational functions in p_{ij}, obtained by expanding the skew commutators in definition (4.53). We have already mentioned that $\alpha_{id} = 1$ and $\alpha_{(23)} = 0$. Given an arbitrary replacement $x_i \leftarrow x_{\mu(i)}$, $\mu \in S_4$, we obtain

$$[\![x_{\mu(1)}, x_{\mu(2)}, x_{\mu(3)}, x_{\mu(4)}]\!] = \sum (\alpha_\pi)^\mu x_{\mu(\pi(1))} x_{\mu(\pi(2))} x_{\mu(\pi(3))} x_{\mu(\pi(4))}.$$

On the right-hand side of the latter equality, the coefficient at $x_1x_2x_3x_4$ equals $\alpha^{\mu}_{\mu^{-1}}$ and the one at $x_1x_3x_2x_4$ equals $\alpha^{\mu}_{(23)\mu^{-1}}$. Therefore, we have a formula that replaces the twisted symmetry in (4.31):

$$[\![x_{\mu(1)}, x_{\mu(2)}, x_{\mu(3)}, x_{\mu(4)}]\!] = \alpha^{\mu}_{\mu^{-1}}[\![x_1, x_2, x_3, x_4]\!] + \alpha^{\mu}_{(23)\mu^{-1}}[\![x_1, x_3, x_2, x_4]\!]. \tag{4.55}$$

4.5 Chapter Notes

Linearization is a process commonly used in modern algebra, see, for example, [210], [236, I, § 5]. The Friedrichs criterion for Lie algebras was discovered in [80] and then proven in three versions by Cohn [51], Lyndon [155], and Magnus [157]. D. Finkelshtein published his criterion in [75]. The Specht–Wever condition, the form with which we begin, appears in N. Jacobson's book [104, Chap. V, Theorem 8].

A generalization of Lie algebras known as n-Lie algebras, with n-linear operations in place of the Lie brackets first appeared in a paper by Filippov [74], and subsequently appeared under the name Nambu–Lie algebras in theoretical research on generalizations of Nambu mechanics by Takhtadjian [223], Dito et al. [63]. Trilinear operation has been considered by Nambu [180], and in numerous papers on generalization of quantum mechanics, see, for example, research on a trilinear oscillator, or on a multilinear commutator by Yamaleev [231–233].

Another group of problems requiring the generalization of Lie algebras corresponds to research on skew derivations of noncommutative algebras. A noncommutative version of the fundamental Dedekind algebraic independence lemma states that the algebraic structure of a Lie algebra and operators with "inner" action define all algebraic dependencies in ordinary derivations (see [121, 122, Chap. 2]). This result was extended to the field of skew derivations by Chuang [49]. His fundamental theorem may be interpreted in the same manner, i.e., the algebraic structure and operators with "inner" action define all algebraic dependences in skew derivations. Hence, the following question arises: which algebraic structure corresponds to the skew derivation operators? This question requires the consideration of n-ary operations that do note reduce to bilinear operations.

According to the Friedrichs criterion, Lie polynomials are characterized as primitive elements of free associative algebra. In these terms, the logical idea to consider spaces spanned by skew-primitive (or primitive, in the case of braided categories) elements was discussed by Larson and Towber [145], and Majid [158, 160]. However, Pareigis in [183–185] first regarded specific skew-primitive polynomials as operations, now known as Pareigis operations, similar to how we are regarding them in this book. Nonetheless, one should remember that multivariable operations are the subject of investigation in the theory of algebraic systems located at the interface between algebra and mathematical logic, the theory of algorithms and computer calculations. See the books by Maltcev [164, 165]. The results presented in this chapter are based on [123, 127].

Chapter 5
Multilinear Operations

Abstract In this chapter, we consider multilinear quantum Lie operations involving more than four variables. Our main goal is to find a necessary and sufficient existence condition to determine the number of linearly independent operations that may exist and to define the principle n-linear operation. Additionally, we discuss symmetric operations, i.e., operations that do not change their values in the context of permutations of variables (up to a scalar factor). We also demonstrate that there are $(n-2)!$ symmetric generic quantum Lie operations.

In this chapter, we consider multilinear quantum Lie operations involving more than four variables. Our main goal is to find a necessary and sufficient existence condition to determine the number of linearly independent operations that may exist and to define the principle n-linear operation. Additionally, we discuss symmetric operations, i.e., operations that do not change their values in the context of permutations of variables (up to a scalar factor). We also demonstrate that there are $(n-2)!$ symmetric generic quantum Lie operations.

5.1 The Basic System of Equations

We are remanded main concepts and notations. A *quantum variable* is a variable x, with which an element g_x of a fixed Abelian group G and a character $\chi^x : G \to \mathbf{k}^*$ are associated, where $\mathbf{k}$ is a ground field. A *quantum operation* in quantum variables $x_1, \ldots, x_n$ is a non-commutative polynomial in these variables that has skew-primitive values in every Hopf algebra H, provided H contains the group G as a subgroup of the set of all group-like elements and every variable x_i has a skew-primitive value $a_i \in H$ such that

$$\Delta(a_i) = a_i \otimes 1 + g_{x_i} \otimes a_i, \quad g^{-1} a_i g = \chi^{x_i}(g) a_i$$

for all $g \in G$.

V. Kharchenko, *Quantum Lie Theory*, Lecture Notes in Mathematics 2150,
DOI 10.1007/978-3-319-22704-7_5

Definition 5.1 A set of quantum variables $x_1, x_2, \ldots, x_n$ is said to be *conforming* if

$$\prod_{1\leq i\neq s\leq n} \chi^{x_i}(g_{x_s}) = 1. \tag{5.1}$$

We are going to prove that there exists a nonzero multilinear quantum operation in a set of quantum variables $x_1, \ldots, x_n$ if and only if this set is conforming. Also we will show that if the set $x_1, \ldots, x_n$ has not many conforming subsets (the intersection of all conforming subsets is nonempty) then the dimension of the space of all multilinear operations equals $(n-2)!$.

Theorem 4.1 gives a way to construct all multilinear quantum operations by means of an investigation of linear dependencies of the following polynomials

$$D_\pi \stackrel{df}{=} \pi([\ldots[[x_1, x_2], x_3], \ldots, x_n]) - \pi([\ldots[[x_1, x_2]^*, x_3]^* \ldots, x_n]^*)$$

in a free associative algebra. Recall that here π is a permutation of the indices, $\pi(1) = 1$; an application of π to an expression of the above formula means its application to all indices of p_{is} and x_i. For every linear dependence $\sum \beta_\pi D_\pi = 0$ there exists an operation

$$W(x_1, \ldots, x_n) = \sum_{\pi\in S_n,\ \pi(1)=1} \beta_\pi D_\pi^+, \tag{5.2}$$

where

$$D_\pi^+ = [\ldots[[x_1, x_2], x_3], \ldots x_n]^\pi.$$

Conversely, every multilinear quantum operation has a representation (5.2) where the coefficients β_π define a linear dependence of D_π.

We fix a set of different quantum variables $x_1, \ldots, x_n$ and the following notations

$$p_{is} = \chi^{x_i}(g_{x_s});\ q_k = \prod_{i=1}^{k-1} p_{ik}. \tag{5.3}$$

Let S_n denotes the permutation group of the set $\{1, 2, \ldots, n\}$, whereas $S_n^{l,m,\ldots,r}$ is a subgroup $\{\pi \in S_n \mid \pi(l) = l, \pi(m) = m, \ldots, \pi(r) = r\}$. If $m < n$, then we identify the group S_m with $S_n^{m+1,m+2,\ldots,n}$. We use both exponential and functional notations for the action of S_n on the set of indices. We consider exponential notation as the basic one; that is, we assume $i^{(\pi\nu)} = (i^\pi)^\nu = \nu(\pi(i))$. In this case, permutations are multiplied from the left to the right. For two arbitrary indices m, k, the symbol $[m; k]$ denotes a monotonous cycle starting with m up to k

$$[m; k] \stackrel{\text{df}}{=} \begin{cases} (m, m+1, \ldots, k), & \text{if } m \leq k \\ (m, m-1, \ldots, k), & \text{if } m \geq k. \end{cases}$$

Clearly $[m;k]^{-1} = [k;m]$ in these notations.

If A is an arbitrary expression and π is a permutation, then by $\pi(A)$ or A^{π}, we infer an expression which appears from A applying π to all indices of p_{is} and x_i. For example, $p_{is}^{\pi} = p_{\pi(i)\pi(s)}$ or $q_k^{\pi} = \prod_{i=1}^{k-1} p_{\pi(i)\pi(k)}$, but not $q_k^{\pi} = q_{\pi(k)}$. We do not suppose that the group S_n acts on the ground field **k**. For instance, it is possible that $p_{12} = p_{23}$ while $p_{12}^{(123)} \neq p_{23}^{(123)}$ in the ground field **k**. According to this agreements an arbitrary multilinear polynomial can be written in the following form:

$$W(x_1, \ldots, x_n) = \sum_{\pi \in S_n} \alpha_{\pi} \pi(x_1 \cdots x_n).$$

For a given word $A = p_{it} p_{kl} \cdots p_{rs}$, we define

$$\{A\} = p_{it} p_{kl} \cdots p_{rs} - p_{ti}^{-1} p_{lk}^{-1} \cdots p_{sr}^{-1}. \tag{5.4}$$

If A, B are two words in p_{is}, then we define a *star product* of braces

$$\{A\} \star \{B\} = \{AB\}.$$

Let $t_{\pi,s}^{\mu}$ denotes an element of **k** defined by the following formula

$$t_{\pi,s}^{\mu} = \{\pi(q_{\pi^{-1}(2)}) \pi(q_{\pi^{-1}(3)}) \cdots \pi(q_{\pi^{-1}(s)})\}^{\mu}, \quad \mu, \pi \in S_n, \ 1 < s < n. \tag{5.5}$$

In what follows, $N^1(s)$ denotes the set of all inverse s-shuffles from S_n^1. According to Definition 1.7, an element π belongs to $N^1(s)$ if and only if

$$\pi(1) = 1; \ \pi^{-1}(2) < \pi^{-1}(3) < \ldots < \pi^{-1}(s); \ \pi^{-1}(s+1) < \ldots < \pi^{-1}(n), \tag{5.6}$$

whereas Lemma 1.16 implies that

$$N^1(s) = \{[2;k_2][3;k_3] \cdots [s;k_s] \mid 1 < k_2 < k_3 < \ldots < k_s \leq n\}, \tag{5.7}$$

in which case $k_i = \pi^{-1}(i)$, $2 \leq i \leq s$.

Theorem 5.1 *If* $\prod_{i \neq t} p_{it} = 1$, *then* $\sum \beta_{\pi} D_{\pi} = 0$ *holds if and only if*

$$\sum_{\pi \in N^1(s)} \beta_{\pi\mu} t_{\pi,s}^{\mu} = 0 \tag{5.8}$$

for all μ, s; $\mu \in S_n^1$, $1 < s < n$, *where* $t_{\pi,s}^{\mu}$ *are defined by* (5.5).

Proof Let us consider a process of developing of brackets $[u, v] = uv - p(u, v) \cdot vu$ from the left to right in $D_{id}^{+} = [\ldots[[x_1, x_2], x_3], \ldots, x_n]$. We obtain that the element

D^+_{id} is a linear combination of the monomials $M_{\mathbf{k},\mathbf{m}} = x_{k_1}x_{k_2}\cdots x_{k_t}x_1x_{m_1}x_{m_2}\cdots x_{m_{n-t}}$, where

$$k_1 > k_2 > \ldots > k_t \text{ and } m_1 < m_2 < \ldots < m_{n-t}. \tag{5.9}$$

Respectively, $D^+_\pi = \pi(D_{id})$ is a linear combination of the monomials

$$M_{\mathbf{k},\mathbf{m},\pi} = x_{\pi(k_1)}x_{\pi(k_2)}\cdots x_{\pi(k_t)}x_1x_{\pi(m_1)}x_{\pi(m_2)}\cdots x_{\pi(m_{n-t})},$$

where the sequences **k** and **m** satisfy (5.9).

Consider a coefficient at $M_s = x_sx_{s-1}\cdots x_1x_{s+1}x_{s+2}\cdots x_n$ in the sum $\sum \beta_\pi D_\pi$. The monomial M_s equals $M_{\mathbf{k},\mathbf{m},\pi}$ only if $t = s$ and $k_1^\pi = s$, $k_2^\pi = s-1, \ldots, k_s^\pi = 2$, $m_1^\pi = s+1, \ldots, m_{n-s}^\pi = n$. Because the sequences **k**, **m** satisfy (5.9), we have $\pi \in N^1(s)$, see (5.6). In this case M_s appears in the decomposition of the long skew commutator D^+_π in the only case when $x_2, x_3, \ldots, x_s$ are moving to the left with respect to x_1 and $x_{s+1}, x_{s+2}, \ldots, x_n$ are moving to the right with respect to x_1. By the formula $[u, x_m] = ux_m - p(u, x_m)x_mu$ we see that the coefficient at M_s equals

$$(-\pi(q_{\pi^{-1}(2)})) \cdot (-\pi(q_{\pi^{-1}(3)})) \cdot \ldots \cdot (-\pi(q_{\pi^{-1}(s)})). \tag{5.10}$$

Here (5.10) equals the word in braces of $t_{\pi,s}$ up to the sign, see (5.5).

Analogously, consider $M_{\mu,s} = x_{\mu(s)}x_{\mu(s-1)}\cdots x_1x_{\mu(s+1)}x_{\mu(s+2)}\cdots x_{\mu(n)}$, where $\mu \in S_n^1$. This monomial appears in D^+_π only if $t = s$, and $k_1^\pi = s^\mu$, $k_2^\pi = (s-1)^\mu$, $\ldots$, $k_s^\pi = 2^\mu$, $m_1^\pi = (s+1)^\mu, \ldots, m_{n-s}^\pi = n^\mu$. We have that $\pi\mu^{-1}$ belongs $N^1(s)$, and the coefficient at $M_{\mu,s}$ equals the product in braces of $t^\mu_{\pi\mu^{-1},s}$ within the factor $(-1)^{s-1}$.

Thus, in the decomposition of $\sum \beta_\pi D_\pi$ the coefficient at $M_{\mu,s}$ equals

$$(-1)^{s-1} \sum_{\pi\mu^{-1}\in N^1(s)} \beta_\pi t^\mu_{\pi\mu^{-1},s}.$$

If we replace the notation $\pi\mu^{-1} \leftarrow \pi$, we will obtain relations (5.8). □

Consider (5.8) as a system of linear equations in β_π. To find a basis of the linear space of multilinear operations, it suffices to find a fundamental system of solutions for (5.8). As all the coefficients $t^\mu_{\pi,s}$ belong the ring $\mathbf{Z}[p_{ij}]$, there exists a fundamental system of solutions in the ring $\mathbf{Z}[p_{ij}]$. Thus we can confine ourself to an investigation of solutions in the ring $\mathbf{Z}[p_{ij}]$ or in the field $\mathbf{F}(p_{ij})$ if it is necessary to normalize one of the coefficients of a quantum operation. Here **F** is the minimal subfield of **k**.

Definition 5.2 The system (5.8) is said to be the *basic system*. Its subsystem corresponding to a fixed number *s* is called an *s-component*.

5.2 Interpretation of Operations in a Crossed Product

Consider a multiplicative Abelian group $\mathcal{F}_n$ freely generated by the symbols P_{it}, $1 \leq i \neq t \leq n$. Let $\mathbf{F}[\mathcal{F}_n]$ be a group algebra of this group over the minimal subfield $\mathbf{F}$ of the ground field $\mathbf{k}$. In other words, $\mathbf{F}[\mathcal{F}_n]$ is an algebra of commutative polynomials in variables P_{it}, P_{it}^{-1} $1 \leq i \neq t \leq n$ with coefficients from the minimal field $\mathbf{F}$. Clearly, $\mathbf{F}[\mathcal{F}_n]$ has a field of fractions that is isomorphic to the field of rational functions $\mathbf{F}(P_{it})$. The action of the symmetric group S_n is well-defined on the ring $\mathbf{F}[\mathcal{F}_n]$ and on the field $\mathbf{F}(P_{it})$ by $P_{it}^{\pi} = P_{\pi(i)\pi(t)}$. Thus, we can define a skew group ring $\mathbf{F}(P_{it}) * S_n$. By Theorem 1.12 this skew group ring is isomorphic to the algebra of all $n! \times n!$ matrices over the Galois field $\mathbf{F}(P_{it})^{S_n}$, and it contains the skew group ring $\mathbf{F}[\mathcal{F}_n] * S_n$. Recall that in a skew group ring, the permutations commute with coefficients according to the formula $A\pi = \pi A^{\pi}$, see Sect. 1.3.3.

If the parameters p_{it}, $1 \leq i \neq t \leq n$ are defined by the quantum variables $x_1, \ldots, x_n$ according to (5.3) then there exists a homomorphism

$$\varphi : \mathbf{F}[\mathcal{F}_n] \to \mathbf{k}, \quad \varphi(P_{it}) = p_{it}, \quad 1 \leq i \neq t \leq n. \tag{5.11}$$

If $A \in \mathcal{F}_n$, then $\overline{A}$ denotes a word appearing from A by replacing of all letters P_{it} with P_{ti}. We call the words A and $\overline{A}$ *conjugated*. We define

$$\{A\} = A - \overline{A}^{-1}.$$

This definition is compatible with (5.4) in the sense that $\varphi(\{A\}) = \{\varphi(A)\}$ if A is a word of $\mathcal{F}_n$. In the same way, the formula $\varphi(A^{\pi}) = (\varphi(A))^{\pi}$ is valid if we assume that $\varphi(A)$ appears from A by replacing P_{it} with p_{it}.

If $A,\ B$ are words of $\mathcal{F}_n$ (possibly empty) then we set

$$\{A\} \star \{B\} = \{AB\}.$$

Note that $\{\emptyset\} = 0$ if as usual the empty word is identified with 1. At the same time the element $\{\emptyset\} \star \{A\} = \{A\}$ can be nonzero.

If $C, D, E \in \mathcal{F}_n$, then by Lemma 4.2 the following equality is valid

$$\{CE\}\{D\overline{E}\} - \{C\}\{D\} = \{CD\overline{E}\}\{E\}. \tag{5.12}$$

Lemma 5.1 *The relation* (5.1) *is equivalent to each one of the relations* $\{W\} = 0$, *where* W *is an arbitrary word in* p_{it}, $1 \leq i \neq t \leq n$ *of length* $C_n^2 = n(n-1)/2$ *that has neither double nor conjugated letters.*

Proof The equality $\{W\} = 0$ is equivalent to $W\overline{W} = 1$. The word $W\overline{W}$ is of the length $n(n-1)$, and it has no double letters. Thus, it has all of the letters p_{it}, $1 \leq i \neq t \leq n$, whereas $W\overline{W}$ coincides with the left-hand side of (5.1) up to an order of factors. □

We fix the following notations: $Q_k = \prod_{i=1}^{k-1} P_{ik}$ and

$$T_{\pi,s} = \{\pi(Q_{\pi^{-1}(2)})\pi(Q_{\pi^{-1}(3)})\cdots\pi(Q_{\pi^{-1}(s)})\}, \quad \pi \in S_n^1,\ s > 1. \tag{5.13}$$

In this case, we have $q_k^\mu = \varphi(Q_k^\mu)$ and $t_{\pi,s}^\mu = \varphi(T_{\pi,s}^\mu)$, $\mu \in S_n$, see (5.3) and (5.5).

Lemma 5.2 *Let $\pi \in N^1(s)$. If $2 \leq k \leq s$, then*

$$\pi(Q_{\pi^{-1}(k)}) = P_{1k}P_{2k}\dots P_{k-1\,k}\cdot P_{s+1\,k}P_{s+2\,k}\dots P_{s+l\,k}, \tag{5.14}$$

where $l = \pi^{-1}(k) - k \geq 0$. If $l = 0$, the second factor of (5.14) *is absent.*

Proof By the definition $\pi(Q_{\pi^{-1}(k)})$ is equal to a word

$$P_{1k}P_{\pi(2)\,k}\dots P_{\pi(\pi^{-1}(k)-1)\,k}.$$

This word, as well as the right hand side of (5.14), is of a length $\pi^{-1}(k) - 1$. By the first chain of inequalities (5.6), the inequality $\pi^{-1}(i) \leq \pi^{-1}(k) - 1$ is valid for $i < k$. This means that the sequence $\pi(2),\ \pi(3),\ \dots,\ \pi(\pi^{-1}(k) - 1)$ contains all of the indices $2, 3, \dots, k-1$. Hence, $\pi(Q_{\pi^{-1}(k)})$ has the first factor of (5.14). Because $\pi^{-1}(i) > \pi^{-1}(k) - 1$ for $k \leq i \leq s$, we see that among the indices $\pi(2),\ \pi(3),\ \dots,\ \pi(\pi^{-1}(k)-1)$ there is no one of the numbers $k, k+1, \dots, s$. Furthermore, if in the sequence $2, 3, \dots, n$ we cross out $\pi^{-1}(2),\ \pi^{-1}(3),\ \dots,\ \pi^{-1}(s)$, then the elements $\pi^{-1}(s+1),\ \pi^{-1}(s+2),\ \dots,\ \pi^{-1}(n)$ remain in the sequence. By the second chain of (5.6), these elements are arranged in the sequence in this very order. It follows that if in the sequence $2, 3, \dots, \pi^{-1}(k) - 1$ we cross out $\pi^{-1}(2),\ \pi^{-1}(3),\ \dots,\ \pi^{-1}(k-1)$, then the elements $\pi^{-1}(s+1),\ \pi^{-1}(s+2),\ \dots,\ \pi^{-1}(s+l)$ remain in the sequence. Therefore, $\pi(Q_{\pi^{-1}(k)})$ has the second factor of (5.14) as well. □

Let us fix the following notations for particular elements of $\mathbf{F}[\mathscr{F}_n] * S_n^1$:

$$V_s = \sum_{\pi \in N^1(s)} \pi T_{\pi,s}, \tag{5.15}$$

where $T_{\pi,s}$ are defined by (5.13). By Lemma 5.2 and decomposition (5.7), we have

$$V_s = \sum_{1<k_2<k_3<\dots<k_s\leq n} [2;k_2][3;k_3][4;k_4]\cdots[s;k_s]T_{k_2:k_3:\dots:k_s},$$

where

$$T_{\pi,s} = T_{k_2:k_3:\dots:k_s} = \{\prod_{m=2}^{s}(P_{1\,m}\cdots P_{m-1\,m}\cdot P_{s+1\,m}\cdots P_{s-m+k_m\,m})\}. \tag{5.16}$$

In particular, if $s = 2$, $k_2 = l$, then

$$T_{[2;l],2} = T_l = \{P_{12} \cdot P_{32} \ldots P_{l2}\}, \quad T_{[2;2],2} = T_2 = \{P_{12}\}.$$

If a permutation π is written as a product of cycles (5.7), then the parameter s is uniquely defined by both the number of factors and the beginning of the last cycle. This fact allows us to use the notation T_π instead of $T_{\pi,s}$.

For an arbitrary sequence of elements $\beta_\pi \in \mathbf{F}[p_{it}]$, $\pi \in S_n^1$, let B denotes an element of $\mathbf{F}[\mathscr{F}_n] * S_n^1$ defined by the formula

$$B = \sum_{\pi \in S_n^1} B_\pi \pi^{-1} \in \mathbf{F}[\mathscr{F}_n] * S_n^1,$$

where B_π are some preimages of β_π in $\mathbf{F}[\mathscr{F}_n]$ with respect to φ.

In this way every quantum Lie operation $\sum \beta_\pi D_\pi^+$, $\beta_\pi \in \mathbf{F}[p_{it}]$ is related to an element B of the skew group algebra.

Theorem 5.2 *If $\prod_{i \neq t} p_{it} = 1$, then an element $B \in \mathbf{F}[\mathscr{F}_n] * S_n^1$ corresponds to a multilinear quantum Lie operation if and only if*

$$B \cdot V_s \in \ker(\varphi) S_n^1, \quad 2 \leq s < n, \tag{5.17}$$

where V_s are defined by (5.15).

Proof By Theorem 4.1, we have to prove that a sequence $\beta_\pi \in \mathbf{F}[p_{it}]$, $\pi \in S_n^1$ is a solution of the basic system if and only if the element B satisfies (5.17). Let us rewrite the left-hand sides of these relations.

$$B \cdot V_s = \sum_{\pi \in S_n^1} B_\pi \pi^{-1} \cdot \sum_{\nu \in N^1(s)} \nu T_{\nu,s} = \sum_{\pi \in S_n^1, \nu \in N^1(s)} B_\pi \pi^{-1} \nu T_{\nu,s}$$

$$= \sum B_\pi T_{\nu,s}^{\nu^{-1}\pi} \pi^{-1} \nu = \sum_{\mu \in S_n^1} \Big(\sum_{\nu \in N^1(s)} B_{\nu\mu} T_{\nu,s}^{\mu} \Big) \mu^{-1}.$$

The latter sum belongs to $\ker(\varphi) S_n^1$ if and only if all of its coefficients belong to $\ker(\varphi)$. Since $\varphi(T_{\nu,s}^\mu) = t_{\nu,s}^\mu$, it follows that (5.17) is equivalent to (5.8). □

Let Σ be an arbitrary multiplicative subset of $\mathbf{F}[\mathscr{F}_n]$ that does not intersect $\ker(\varphi)$. Consider a localization (a ring of quotients) $\Sigma^{-1}\mathbf{F}[\mathscr{F}_n]$. The homomorphism φ has a unique extension up to a homomorphism of $\Sigma^{-1}\mathbf{F}[\mathscr{F}_n]$ into the field $\mathbf{F}(p_{it})$ via

$$\varphi(\sigma^{-1} B) = \varphi(\sigma)^{-1} \varphi(B), \ \sigma \in \Sigma, \ \ B \in \mathbf{F}[\mathscr{F}_n]. \tag{5.18}$$

This allows one to normalize elements corresponding to the quantum Lie operations. For these reasons the following variant of the above theorem is useful.

Theorem 5.3 *If* $\prod_{i\neq t} p_{it} = 1$*, then an element* $B = \sum_{\pi\in S_n^1} B_\pi \pi^{-1}$, $B_\pi \in \Sigma^{-1}\mathbf{F}[\mathscr{F}_n]$ *defines a quantum Lie operation* $\sum_{\pi\in S_n^1} \varphi(B_\pi)D_\pi^+$ *if and only if*

$$B \cdot V_s \in (\Sigma^{-1}\ker(\varphi))S_n^1, \quad 2 \leq s < n. \tag{5.19}$$

Proof It suffices to multiply B by a common denominator $\sigma \in \Sigma$ of all coefficients and to apply Theorem 5.2. □

The set $\ker(\varphi)S_n^1$ is a right ideal and a left $\mathbf{F}[\mathscr{F}_n]$ submodule of $\mathbf{F}[\mathscr{F}_n] * S_n^1$. Inclusions (5.17) signify that $B \cdot V_s$ equal zero in the quotient $(\mathbf{F}[\mathscr{F}_n], \mathbf{F}[\mathscr{F}_n] * S_n^1)$-bimodule $\mathbf{F}[\mathscr{F}_n] * S_n^1 / \ker(\varphi)S_n^1$. In what follows, the symbol $\equiv$ denotes the equality in quotient $(\mathbf{F}[\mathscr{F}_n], \mathbf{F}[\mathscr{F}_n] * S_n^1)$-bimodules. This equality is stable with respect to the right multiplications by elements from $\mathbf{F}[\mathscr{F}_n] * S_n^1$ and with respect to the left multiplications by elements from $\mathbf{F}[\mathscr{F}_n]$. Of course, if $\ker(\varphi)$ is not invariant with respect to the action of S_n^1, then $\equiv$ is not stable with respect to left multiplications by S_n^1.

Similarly, (5.19) are equalities to zero in the $(\Sigma^{-1}\mathbf{F}[\mathscr{F}_n], \mathbf{F}[\mathscr{F}_n] * S_n^1)$-bimodule

$$(\Sigma^{-1}\mathbf{F}[\mathscr{F}_n])S_n^1/(\Sigma^{-1}\ker(\varphi))S_n^1.$$

This bimodule contains the former one because

$$(\Sigma^{-1}\ker(\varphi))S_n^1 \cap \mathbf{F}[\mathscr{F}_n] * S_n^1 = \ker(\varphi)S_n^1.$$

Hence, it is possible to use the same sign $\equiv$ in both cases.

Definition 5.3 A *conforming ideal* is an ideal I of the algebra $\mathbf{F}[\mathscr{F}_n]$ generated by all elements of the form $\{W\}$, where W is an arbitrary (semigroup) word in P_{it}, $1 \leq i \neq t \leq n$ of length $C_n^2 = n(n-1)/2$ that has neither double nor conjugated letters.

By Lemma 5.1, the variables $x_1, \ldots, x_n$ are conforming if and only if the ideal $\ker(\varphi)$ contains the conforming ideal. It is very important to note that the conforming ideal I is invariant with respect to the action of the symmetric group S_n (unlike the ideal $\ker(\varphi)$ itself). Therefore a two-sided ideal of $\mathbf{F}[\mathscr{F}_n] * S_n$ generated by I coincides the right ideal IS_n.

5.3 Co-set Decomposition

The set of transpositions $\{(2,n), (3,n), \ldots, (n-1,n)\}$ is a right transversal of the subgroup $S_n^{1,n} = S_{n-1}^1$ in S_n^1. Indeed, if $\pi \in S_n^1$, then $\pi \cdot (\pi(n), n) \in S_n^{1,n}$, and whence $\pi \in S_n^{1,n}(\pi(n), n)$. This implies that every element $B \in \mathbf{F}[\mathscr{F}_n] * S_n^1$ has a decomposition

$$B = \mathscr{A}_2 \cdot (2,n) + \mathscr{A}_3 \cdot (3,n) + \ldots + \mathscr{A}_{n-1} \cdot (n-1,n) + \mathscr{A}_n \cdot \mathrm{id}, \tag{5.20}$$

where $\mathscr{A}_i \in \mathscr{P} * S_n^{1,n}$.

According to decomposition (5.7) a permutation π belongs to $N^1(2)$ if and only if $\pi = [2; l],\ 2 \le l \le n$. Using commutation rule

$$(l, m, \ldots, r)\pi = \pi(l^{\pi}, m^{\pi}, \ldots, r^{\pi}), \tag{5.21}$$

we obtain

$$\begin{aligned} B \cdot V_2 &= \sum_{i=2}^{n} \mathscr{A}_i(i,n) \cdot \sum_{\pi \in N^1(2)} \pi T_{\pi} \\ &= \sum_{i=2}^{n} \sum_{l=2}^{n-1} \mathscr{A}_i(i,n)[2;l]T_{[2;l]} + \sum_{i=2}^{n} \mathscr{A}_i(i,n)[2;n]T_{[2;n]} \\ &= \sum_{i=2}^{n} \sum_{l=2}^{n-1} \mathscr{A}_i[2;l](i^{[2;l]}, n)T_{[2;l]} + \sum_{i=2}^{n} \mathscr{A}_i(i,n)[2;n]T_{[2;n]}. \end{aligned}$$

Let $\tau_k,\ 2 \le k \le n$ be the permutation $(k^{[n;2]}, n)[2;n](k,n)$; that is,

$$\tau_k = \begin{cases} [2;k-1][k;n-1], & \text{if } 2 < k < n, \\ [2;n-1], & \text{if } k = n, 2. \end{cases} \tag{5.22}$$

In this case $(k^{[n;2]}, n)[2;n] = \tau_k(k,n)$. Therefore

$$B \cdot V_2 = \sum_{k=2}^{n} \Big(\sum_{l=2}^{n-1} \mathscr{A}_{[l;2](k)}[2;l]T_{[2;l]}^{(k,n)} + \mathscr{A}_{[n;2](k)}\ \tau_k T_{[2;n]}^{(k,n)} \Big)(k,n). \tag{5.23}$$

In particular, the inclusion (5.17) with $s = 2$ is equivalent to the following system of $n-2$ equalities in the quotient bimodule $\mathbf{F}[\mathscr{F}_n] * S_n^{1,n} / \ker(\varphi) S_n^1$:

$$\sum_{l=2}^{n-1} \mathscr{A}_{[l;2](k)}[2;l]T_{[2;l]}^{(k,n)} + \mathscr{A}_{[n;2](k)} \tau_k T_{[2;n]}^{(k,n)} \equiv 0, \quad 2 \le k \le n.$$

If $3 \le k \le n$, then the above equality corresponding to k has the form

$$\mathscr{A}_k \sum_{l=2}^{k-1} [2;l]T_{[2;l]}^{(k,n)} + \mathscr{A}_{k-1} \Big(\tau_k T_{[2;n]}^{(k,n)} + \sum_{l=k}^{n-1} [2;l]T_{[2;l]}^{(k,n)} \Big) \equiv 0. \tag{5.24}$$

For $k = 2$, we have the equality

$$D_2 \stackrel{\text{df}}{=} \mathscr{A}_n[2;n-1]T_{[2;n]}^{(2,n)} + \sum_{l=2}^{n-1} \mathscr{A}_l[2;l]T_{[2;l]}^{(2,n)} \equiv 0. \tag{5.25}$$

Let us introduce the following notations:

$$V_{(k)} = \sum_{l=2}^{k-1}[2;l]T_{[2;l]}^{(k,n)} = \text{id}\,\{P_{12}\} + \sum_{l=3}^{k-1}[2;l]\{P_{12}P_{32}\ldots P_{l2}\},\ 3 \le k \le n, \tag{5.26}$$

$$D_{(k)} = \tau_k T_{[2;n]}^{(k,n)} + \sum_{l=k}^{n-1}[2;l]T_{[2;l]}^{(k,n)},\ 3 \le k \le n. \tag{5.27}$$

In this case, (5.23) take up the form

$$B \cdot V_2 = D_2(2,n) + \sum_{k=3}^{n}(\mathscr{A}_k V_{(k)} + \mathscr{A}_{k-1}D_{(k)})(k,n) \tag{5.28}$$

whereas the relations (5.24) with $3 \le k \le n$ reduce to

$$\mathscr{A}_k V_{(k)} + \mathscr{A}_{k-1}D_{(k)} \equiv 0,\quad 3 \le k \le n. \tag{5.29}$$

In particular, the following statement is proven.

Lemma 5.3 *The element (5.20) corresponds to a solution of the second component of the basic system if and only if the equalities (5.29) and (5.25) are valid.*

Proposition 5.1 *Let Σ be a multiplicative $S_n^{1,n}$-invariant subset of $\mathbf{F}[\mathscr{F}_n]$ that does not intersect* ker (φ). *If $V_{(3)},\ V_{(4)},\ \ldots,\ V_{(n)}$ are invertible in $\Sigma^{-1}\mathbf{F}[\mathscr{F}_n] * S_n^{1,n}$, then the dimension of the space of multilinear operations is less than or equal to $(n-2)!$.*

Proof Let the element $B = \sum B_\pi \pi^{-1}$ corresponds to a quantum Lie operation, $\varphi(B_\pi) = \beta_\pi$. Then $B \cdot V_2 \equiv 0$. Consider the decomposition (5.20) of B. Relation (5.29) imply

$$B \equiv \mathscr{A}_2 \sum_{k=2}^{n}(-1)^k(\prod_{i=3}^{k} D_{(i)}V_{(i)}^{-1})(k,n).$$

Thus, the superposition

$$\mathbf{f} \mapsto B \mapsto \mathscr{A}_2 \xrightarrow{\varphi} \mathbf{F}(p_{it})S_n^{1,n}$$

is a linear transformation with zero kernel of the space of multilinear quantum Lie operations into the space of left linear combinations $\sum_{\mu \in S_n^{1,n}} \alpha_\mu \mu$ over $\mathbf{F}(p_{it})$. In other words, the element B is uniquely defined by $\mathscr{A}_2 \in S_n^{1,n}$ up to the relation $\equiv$. Therefore the dimension is less than or equal to the order of $S_n^{1,n}$. □

5.4 Subordinate Sequences

In this section we are going to prove a number of auxiliary results which allows one to harmonize some special elements of the skew group algebra.

Consider a sequence of integer numbers $\mathbf{L} = (l_i, \mid u \le i \le v)$, where u, v are some integer numbers, $u \le v$.

Definition 5.4 For arbitrary indices $k, r,\ u \le k, r \le v$, let $\sigma_k^{(k)} = l_k,\ \gamma_r^{(r)} = l_r + 1$. Furthermore, define by induction $\sigma_k^{(j)}, j \le k$ and $\gamma_r^{(j)}, j \ge k$:

$$\sigma_k^{(j-1)} = \begin{cases} \sigma_k^{(j)} - 1, & \text{if } \sigma_k^{(j)} \le l_{j-1}, \\ \sigma_k^{(j)}, & \text{if } \sigma_k^{(j)} > l_{j-1}; \end{cases} \tag{5.30}$$

$$\gamma_r^{(j+1)} = \begin{cases} \gamma_r^{(j)} + 1, & \text{if } \gamma_r^{(j)} \le l_{j+1}, \\ \gamma_r^{(j)}, & \text{if } \gamma_r^{(j)} > l_{j+1}. \end{cases} \tag{5.31}$$

The sequence $\mathbf{L}' = (l'_j,\ u-1 \le j \le v-1)$ defined by the following formula

$$l'_{j-1} = \begin{cases} l_{j-1} - 1, & \text{if } \sigma_v^{(j)} \le l_{j-1},\ j > u, \\ l_{j-1}, & \text{if } \sigma_v^{(j)} > l_{j-1},\ j > u, \\ \sigma_v^{(u)} - 1, & \text{if } j = u, \end{cases} \tag{5.32}$$

is called a *subordinate sequence for* $\mathbf{L}$.

In the same way the sequence $\mathbf{L}^* = (l_j^*,\ u+1 \le j \le v+1)$ is called *inceptive sequence for* $\mathbf{L}$ if it is defined as follows:

$$l_{j+1}^* = \begin{cases} l_{j+1} + 1, & \text{if } \gamma_u^{(j)} \le l_{j+1},\ j < v, \\ l_{j+1}, & \text{if } \gamma_u^{(j)} > l_{j+1},\ j < v, \\ \gamma_u^{(v)}, & \text{if } j = v. \end{cases} \tag{5.33}$$

Definition 5.5 We say that at the point $m, m > r$, there is a *jump during a motion of* l_r *to the right* if $\gamma_r^{(m-1)} \le l_m$. Analogously we say that at the point $m, m < k$, there is a *jump during a motion of* l_k *to the left* if $\sigma_k^{(m+1)} \le l_m$.

The elements σ and γ with indices are called *right* and *left heads* respectively; that is, the right head traverses from the right to the left (with l_k) and the left one traverses from the left to the right (with l_r). In parentheses, if necessary, we write the name of the initial sequence: $\sigma_k^{(j)} = \sigma_k^{(j)}(\mathbf{L}),\ \gamma_k^{(j)} = \gamma_k^{(j)}(\mathbf{L})$.

Lemma 5.4 *For each sequence* $\mathbf{L}$, *we have* $\mathbf{L}'^* = \mathbf{L} = \mathbf{L}^{*\prime}$. *For all* $j,\ u \le j \le v$, *the following heads relations are valid:*

$$\sigma_v^{(j)}(\mathbf{L}) = \gamma_{u-1}^{(j-1)}(\mathbf{L}'), \tag{5.34}$$

$$\sigma_{v+1}^{(j+1)}(\mathbf{L}^*) = \gamma_u^{(j)}(\mathbf{L}). \tag{5.35}$$

Proof If $j = u$, then (5.34) has the form $\sigma_v^{(u)}(\mathbf{L}) = l'_{u-1}+1$. By the definition (5.32), this equality is valid. We proceed by induction on j. Let (5.34) be valid for a given j, $u \le j < v$. Consider two cases.

If $\sigma_v^{(j+1)}(\mathbf{L}) > l_j$, then $l'_j = l_j$ and $\sigma_v^{(j)}(\mathbf{L}) = \sigma_v^{(j+1)}(\mathbf{L})$. Hence, $\gamma_{(u-1)}^{(j-1)}(\mathbf{L}') > l'_j$. By (5.33) and (5.31) this implies $(l'_j)^\star = l'_j = l_j$ and $\gamma_{u-1}^{(j)}(\mathbf{L}') = \gamma_{u-1}^{(j-1)}(\mathbf{L}')$; that is, (5.34) remains valid for $j+1$.

If $\sigma_v^{(j+1)}(\mathbf{L}) \le l_j$, then $l'_j = l_j - 1$ and $\sigma_v^{(j)}(\mathbf{L}) = \sigma_v^{(j+1)}(\mathbf{L}) - 1$. Using (5.34) for the given j, we have $\gamma_{(u-1)}^{(j-1)}(\mathbf{L}') = \sigma_v^{(j+1)}(\mathbf{L}) - 1 \le l_j - 1 = l'_j$. Definitions (5.33) and (5.31) again imply $(l'_j)^* = l'_j + 1 = l_j$ and $\gamma_{u-1}^{(j)}(\mathbf{L}') = \gamma_{u-1}^{(j-1)}(\mathbf{L}') + 1$; that is, in this case (5.34) remains valid for $j+1$ as well. This completes the proof of (5.34).

To prove (5.35), we shall use the downward induction on j. If $j = v$, then (5.35) is valid by the definitions. Let (5.35) be valid for a given j. Consider two cases.

If $\gamma_u^{(j-1)}(\mathbf{L}) > l_j$, then $l_j^* = l_j$ and $\gamma_u^{(j)}(\mathbf{L}) = \gamma_u^{(j-1)}(\mathbf{L})$. By (5.35) for the given j, we have $\sigma_{v+1}^{(j+1)}(\mathbf{L}^*) > l_j = l_j^*$; that is, by (5.32) we obtain $(l_j^*)' = l_j$, whereas by (5.30) we have $\sigma_{v+1}^{(j)}(\mathbf{L}^*) = \sigma_{v+1}^{(j+1)}(\mathbf{L}^*)$. Thus (5.35) remains valid for $j-1$.

If $\gamma_u^{(j-1)}(\mathbf{L}) \le l_j$, then $l_j^* = l_j + 1$ and $\gamma_u^{(j)}(\mathbf{L}) = \gamma_u^{(j-1)}(\mathbf{L}) + 1$. By (5.35) for the given j, we have $\sigma_{v+1}^{(j+1)}(\mathbf{L}^*) = \gamma_u^{(j)}(\mathbf{L}) = \gamma_u^{(j-1)}(\mathbf{L}) + 1 \le l_j + 1 = l_j^*$. From here by (5.32) we obtain $(l_j^*)' = l_j^* - 1 = l_j$, whereas by the definition (5.30) we have $\sigma_{v+1}^{(j)}(\mathbf{L}^*) = \sigma_{v+1}^{(j+1)}(\mathbf{L}^*) - 1$; that is, the equality (5.35) remains valid for $j-1$. □

Lemma 5.5 *At the point m there is a jump during a motion of l_r to the right if and only if at the point r there is no jump during a motion of l_m to the left.*

Proof Suppose that at the point m there is a jump during a motion of l_r to the right; that is, $\gamma_r^{(m-1)} \le l_m$ or, equivalently, $\gamma_r^{(m-1)} \le \sigma_m^{(m)}$. Using the latter inequality as a basis of induction on d, let us prove that

$$\gamma_r^{(m-d)} \le \sigma_m^{(m-d+1)}, \quad \text{where } 1 \le d \le m - r. \tag{5.36}$$

If $\gamma_r^{(m-d)} < \sigma_m^{(m-d+1)}$, then $\gamma_r^{(m-d-1)} \le \gamma_r^{(m-d)} \le \sigma_m^{(m-d+1)} - 1 \le \sigma_m^{(m-d)}$. Therefore, suppose that $\gamma_r^{(m-d)} = \sigma_m^{(m-d+1)}$.

If $\sigma_m^{(m-d+1)} > l_{m-d}$, then $\sigma_m^{(m-d)} = \sigma_m^{(m-d+1)} = \gamma_r^{(m-d)} \ge \gamma_r^{(m-d-1)}$.

If $\sigma_m^{(m-d+1)} \le l_{m-d}$, then $\sigma_m^{(m-d)} = \sigma_m^{(m-d+1)} - 1$. In this case, the equality $\gamma_r^{(m-d)} = \gamma_r^{(m-d-1)}$ can not be valid, because this equality requires that $\gamma_r^{(m-d-1)} > l_{m-d}$, implying that $\gamma_r^{(m-d)} > l_{m-d}$, which contradicts $\gamma_r^{(m-d)} = \sigma_m^{(m-d+1)}$.

Thus, $\gamma_r^{(m-d-1)} = \gamma_r^{(m-d)} - 1$, and the equality of the heads σ and γ is still valid. This completes the proof of (5.36).

Now, if $d = m - r$, then (5.36) has the form $\gamma_r^{(r)} \le \sigma_m^{(r+1)}$; that is, $l_r + 1 \le \sigma_m^{(r+1)}$ or, equivalently, $l_r < \sigma_m^{(r+1)}$. The latter inequality means that at the point r there is no jump during a motion of l_m to the left.

Conversely, let $l_r < \sigma_m^{(r+1)}$. Then $\gamma_r^{(r)} \leq \sigma_m^{(r+1)}$. We shall use this relation as a beginning of the downward induction on d, to prove the following inequality

$$\gamma_r^{(m-d)} \leq \sigma_m^{(m-d+1)}, \quad m-r \geq d \geq 1. \tag{5.37}$$

If $\gamma_r^{(m-d)} < \sigma_m^{(m-d+1)}$, then $\gamma_r^{(m-d+1)} \leq \gamma_r^{(m-d)} + 1 < \sigma_m^{(m-d+1)} + 1 \leq \sigma_m^{(m-d+2)}$. Therefore, it suffices to consider the case $\gamma_r^{(m-d)} = \sigma_m^{(m-d+1)}$.

If $\gamma_r^{(m-d)} > l_{m-d+1}$, then $\gamma_r^{(m-d+1)} = \gamma_r^{(m-d)} = \sigma_m^{(m-d+1)} \leq \sigma_m^{(m-d+2)}$.

If $\gamma_r^{(m-d)} \leq l_{m-d+1}$, then $\gamma_r^{(m-d+1)} = \gamma_r^{(m-d)} + 1$. In this case, the equality $\sigma_m^{(m-d+1)} = \sigma_m^{(m-d+2)}$ is not valid, because it requires that $\sigma_m^{(m-d+2)} > l_{m-d+1}$, implying that $\sigma_m^{(m-d+1)} > l_{m-d+1}$, which contradicts $\gamma_r^{(m-d)} = \sigma_m^{(m-d+1)}$. Thus $\sigma_m^{(m-d+2)} = \sigma_m^{(m-d+1)} + 1$, and the equality of heads is still valid. This completes the proof of (5.37).

If $d = 1$, we have $\gamma_r^{(m-1)} \leq \sigma_m^{(m)} = l_m$; that is, at the point m there is a jump during a motion of l_r to the right. □

Definition 5.6 Denote by $\mathscr{U}(s,t)$ with $t \geq s \geq 3$ a set of all sequences of integer numbers $(l_2, \ldots, l_s, \ldots, l_t)$ that satisfy the following conditions:

$$n-2 \geq l_2 \geq \ldots \geq l_s \geq s-2; \quad n-2 \geq l_j \geq j-2 \text{ for } j > s. \tag{5.38}$$

For $t \geq s \geq 3$, let $\mathscr{W}(s,t)$ denotes a set of all sequences of integer numbers $(w_1, w_2, \ldots, w_t)$ that satisfy the following conditions

$$n-3 \geq w_1 \geq -1; \; n-3 \geq w_2 \geq \ldots \geq w_s \geq s-2; \; n-3 \geq w_j \geq j-2, \; j > s. \tag{5.39}$$

Lemma 5.6 *If $t > s$, then the following equality is valid:*

$$(\mathscr{U}(s,t))' = \mathscr{W}(s,t-1).$$

Proof Let $\mathbf{L} = (l_2, \ldots, l_t) \in \mathscr{U}(s,t)$. Let us show that $\mathbf{L}' \in \mathscr{W}(s,t-1)$. Note that for the head σ_t the following inequalities hold:

$$j-2 \leq \sigma_t^{(j)}(\mathbf{L}) \leq n-2. \tag{5.40}$$

Indeed, for $j = t$ this arise from the definitions. When the parameter j come down, by one step, the head σ_t can only come down, but also no more than by one step. Therefore, (5.40) is saved.

In particular, (5.40) with $j = 2$ implies $0 \leq \sigma_t^{(2)} \leq n-2$ and $-1 \leq l_1' \leq n-3$, because $l_1' = \sigma_t^{(2)} - 1$.

If $l_j = n-2$, then $l_j \geq \sigma_t^{(j+1)}$; hence $l_j' = n-3$. This means that in any case, $l_j' \leq n-3$, because $l_j' \leq l_j$.

Let $j > s$. If $l_j = j - 2$, then by (5.40) we have $\sigma_t^{(j+1)} \geq j - 1 > l_j$; that is, by definition (5.32) we obtain $l'_j = l_j$. If $l_j > j - 2$, then $l'_j \geq l_j - 1 \geq j - 2$. Therefore, the conditions (5.39) are valid, provided that $w_j = l'_j$ and $j > s$.

If $l_{k-1} > l_k = l_{k+1} = \ldots = l_s = s - 2$, then by inequality (5.40) we have $\sigma_t^{(s+1)} \geq s - 1 > s - 2$; hence $l'_{k-1} \geq l'_k = \ldots = l'_s = s - 2$.

Furthermore, if $2 \leq j \leq s$ and the strict inequality $l_j > l_{j+1}$ is valid, then evidently $l'_j \geq l'_{j+1}$, since any member of the subordinate sequence can only be less than the corresponding member of the initial sequence but not by more than one. If $l_j = l_{j+1}$ and $l'_{j+1} = l_{j+1}$, then $\sigma_t^{(j+1)} = \sigma_t^{(j+2)}$; that is, $l'_j = l_j = l'_{j+1}$. If $l_j = l_{j+1}$ and $l'_{j+1} = l_{j+1} - 1$ then $\sigma_t^{(j+2)} \leq l_{j+1}$ and $\sigma_t^{(j+1)} = \sigma_t^{(j+2)} - 1$, from which $\sigma_t^{(j+1)} \leq l_{j+1} - 1 = l_j - 1$, and, still, $l'_{j+1} = l'_j$.

Thus the subordinate sequence satisfies all of the conditions (5.39). By Lemma 5.4, different sequences have different subordinate ones. Therefore, it suffices to show that the sets $\mathscr{U}(s,t)$ and $\mathscr{W}(s,t-1)$ have the same number of elements:

$$|\mathscr{U}(s,t)| = C_{n-1}^{s-1} \cdot \prod_{j=s+1}^{t} (n - j + 1) = (n-1) \cdot C_{n-2}^{s-1} \cdot \prod_{j=s+1}^{t-1} (n - j) = |\mathscr{W}(s, t-1)|,$$

which can be easily verified by direct calculations. □

Definition 5.7 Denote by $\mathscr{L}(i,t)$ with $0 \leq t \leq i \leq n-2$ and $0 \leq t \leq i$ a set of all sequences of integer numbers $(l_1, l_2, \ldots, l_t)$, such that

$$1 \leq l_1 \leq i,\ 2 \leq l_2 \leq i,\ \ldots,\ t \leq l_t \leq i. \tag{5.41}$$

Note that the set $\mathscr{L}(i,0)$ contains only one sequence, the empty one. However, $\mathscr{L}(i,0)$ is a nonempty set itself.

Let $\mathscr{S}(i,t)$ denotes a set of all sequences of integer numbers $(s_0, s_1, \ldots, s_t)$, which satisfy the following conditions

$$0 \leq s_0 \leq i,\ 1 \leq s_1 \leq i,\ \ldots,\ t \leq s_t \leq i. \tag{5.42}$$

Lemma 5.7 *For* $1 \leq t \leq i \leq n-2$ *the following equality holds:*

$$(\mathscr{L}(i,t))' = \mathscr{S}(i-1, t-1).$$

Proof The sets $\mathscr{L}(i,t)$ and $\mathscr{S}(i-1,t-1)$ have the same number of elements $N = i(i-1)\cdots(i-t+1)$. Therefore it suffices to prove that if $\mathbf{L} \in \mathscr{L}(i,t)$ then $\mathbf{L}' \in \mathscr{S}(i-1,t-1)$.

The following inequality is valid, provided that $1 \leq j \leq t$:

$$j \leq \sigma_t^{(j)} \leq i. \tag{5.43}$$

For $j = t$ this inequality follows from the definition, $\sigma_t^{(t)} = l_t$. If the parameter j is decremented by one, then the right head can only be diminished, but not by more than one. Therefore, the inequality is saved.

Inequality (5.43) with $j = 1$ is $1 \leq \sigma_t^{(1)} \leq i$. Hence $s_0 = l'_0 = \sigma_t^{(1)} - 1$ satisfies (5.42) with $i - 1$ in place of i.

If $j < l_j < i$, then the inequalities $l_j - 1 \leq l'_j \leq l_j$ imply that $j \leq l'_j \leq i - 1$.

If $l_j = i$, then $\sigma_t^{(j+1)} \leq i = l_j$; that is, $l'_j = l_j - 1 = i - 1$.

If $l_j = j$, then by (5.43) we have $\sigma_t^{(j+1)} \geq j + 1 > j = l_j$ and $l'_j = l_j = j$. □

Lemma 5.8 *Let* $\mathbf{L} = (l_j \mid u \leq j \leq v)$ *and* $\mathbf{S} = (s_j \mid u - 1 \leq j \leq v - 1) = \mathbf{L}'$. *For a given k,* $u < k < v$, *at the point m,* $u \leq m < k$, *there is a jump during a motion of* l_k *to the left if and only if at this point there is a jump during a motion of* s_k *to the left.*

If $s_k = l_k - 1$, *then for all j,* $u \leq j \leq k$, *the following inequality holds:*

$$\sigma_v^{(j)}(\mathbf{L}) \leq \sigma_k^{(j)}(\mathbf{S}). \tag{5.44}$$

Proof Consider two cases when $s_k = l_k$ and when $s_k = l_k - 1$.

(A) $s_k = l_k$. In this case $\sigma_v^{(k+1)}(\mathbf{L}) > l_k$, and we can write a chain $\sigma_k^{(k)}(\mathbf{L}) = \sigma_k^{(k)}(\mathbf{S}) < \sigma_v^{(k+1)}(\mathbf{L}) = \sigma_v^{(k)}(\mathbf{L})$. We shall use this chain as a beginning of the downward induction on j, $1 \leq j \leq k$, to prove that

$$\sigma_k^{(j)}(\mathbf{L}) = \sigma_k^{(j)}(\mathbf{S}) < \sigma_v^{(j)}(\mathbf{L}). \tag{5.45}$$

Let us make the inductive step considering three possible cases.

1. $\sigma_k^{(j)}(\mathbf{L}) \leq l_{j-1}$. In this case $s_{j-1} = l_{j-1} - 1$, and $\sigma_v^{(j-1)}(\mathbf{L}) = \sigma_v^{(j)}(\mathbf{L}) - 1$. Furthermore, (5.45) demonstrates that at the point j there is a jump during a motion of l_k to the left; that is, $\sigma_k^{(j-1)}(\mathbf{L}) = \sigma_k^{(j)}(\mathbf{L}) - 1$. For s_k we have $\sigma_k^{(j)}(\mathbf{S}) = \sigma_k^{(j)}(\mathbf{L}) \leq l_{j-1} - 1 = s_j$. Therefore, a jump also exists and in particular, $\sigma_k^{(j-1)}(\mathbf{S}) = \sigma_k^{(j)}(\mathbf{S}) - 1$. Thus, in the passage of j to $j - 1$, all members of (5.45) decremented by one, whereas this very condition remain saved.
2. $\sigma_k^{(j)}(\mathbf{S}) = \sigma_k^{(j)}(\mathbf{L}) \leq l_{j-1} < \sigma_v^{(j)}(\mathbf{L})$. In this case $s_{j-1} = l_{j-1}$, $\sigma_v^{(j-1)}(\mathbf{L}) = \sigma_v^{(j)}(\mathbf{L})$, $\sigma_k^{(j-1)}(\mathbf{L}) = \sigma_k^{(j)}(\mathbf{L}) - 1$. For the sequence $\mathbf{S}$ we have, $\sigma_k^{(j)}(\mathbf{S}) \leq L_{j-1} = s_{j-1}$; that is, a jump also exists and $\sigma_k^{(j-1)}(\mathbf{S}) = \sigma_k^{(j)}(\mathbf{S}) - 1$. Thus in the passage of j to $j - 1$ the equal members of (5.45) were decremented by one, whereas the biggest member remain unchanged; that is, (5.45) remain saved.
3. $l_{j-1} < \sigma_k^{(j)}(\mathbf{S}) = \sigma_k^{(j)}(\mathbf{L}) < \sigma_v^{(j)}(\mathbf{L})$. In this case there are no jumps and all the parameters are saved.

As in a motion to the left the heads are changing only at the points where exist jumps, the equality of heads in (5.45) (and the certain induction also) shows that in the case (A) the lemma is true.

(B) $s_k = l_k - 1$. In this case $\sigma_v^{(k+1)}(\mathbf{L}) \leq l_k$ and $\sigma_v^{(k)}(\mathbf{L}) = \sigma_v^{(k+1)}(\mathbf{L}) - 1$. Therefore we can write the following chain:

$$\sigma_v^{(k)}(\mathbf{L}) \leq \sigma_k^{(k)}(\mathbf{S}) = \sigma_k^{(k)}(\mathbf{L}) - 1.$$

Let us prove by a downward induction by j, $1 \leq j \leq k$, that

$$\sigma_v^{(j)}(\mathbf{L}) \leq \sigma_k^{(j)}(\mathbf{S}) = \sigma_k^{(j)}(\mathbf{L}) - 1. \tag{5.46}$$

For $j = k$ the latter chain is written above. Consider three cases.

1. $\sigma_k^{(j)}(\mathbf{L}) \leq l_{j-1}$. In this case $\sigma_k^{(j-1)}(\mathbf{L}) = \sigma_k^{(j)}(\mathbf{L}) - 1$. Using the inductive supposition (5.46), one may write $\sigma_v^{(j)}(\mathbf{L}) < l_{j-1}$. Therefore $s_{j-1} = l_{j-1} - 1$, and $\sigma_v^{(j-1)}(\mathbf{L}) = \sigma_v^{(j)}(\mathbf{L}) - 1$. In addition, $\sigma_k^{(j)}(\mathbf{S}) = \sigma_k^{(j)}(\mathbf{L}) - 1 \leq l_{j-1} - 1 = s_{j-1}$, and so $\sigma_k^{(j-1)}(\mathbf{S}) = \sigma_k^{(j)}(\mathbf{S}) - 1$. Thus in the passage of j to $j-1$, all three members of (5.46) decremented by one.
2. $\sigma_v^{(j)}(\mathbf{L}) \leq l_{j-1} \leq \sigma_k^{(j)}(\mathbf{S})$. In this case still, $s_{j-1} = l_{j-1} - 1$, $\sigma_v^{(j-1)}(\mathbf{L}) = \sigma_v^{(j)}(\mathbf{L}) - 1$, and by inductive supposition, $\sigma_k^{(j)}(\mathbf{L}) > l_{j-1}$; that is, $\sigma_v^{(j-1)}(\mathbf{L}) = \sigma_v^{(j)}(\mathbf{L})$. For the sequence $\mathbf{S}$ we have $s_{j-1} = l_{j-1} - 1 < \sigma_k^{(j)}(\mathbf{S})$. Therefore $\sigma_k^{(j-1)}(\mathbf{S}) = \sigma_k^{(j)}(\mathbf{S})$. Thus in the passage of j to $j-1$ the left hand side of (5.46) decremented, whereas the others remain unchanged; hence (5.46) is saved.
3. $l_{j-1} < \sigma_v^{(j)}(\mathbf{L})$. In this case in the passage of j to $j-1$ all members of (5.46) remain unchanged.

Thus by the proved relation (5.46) in the motion to the left the difference in heads $\sigma_k(\mathbf{L})$ and $\sigma_k(\mathbf{S})$ is always equal to one. Again taking into account that the heads are changing only at the points where exist jumps, we obtain that in case (B) the lemma is also true. □

Lemma 5.9 *Let* $\mathbf{L} = (l_j \mid u \leq j \leq v)$ *and* $\mathbf{S} = (s_j \mid u - 1 \leq j \leq v - 1) = \mathbf{L}'$. *Let* $u \leq m \leq v - 1$. *If* $s_m = l_m$, *then for every* j, $v > j \geq m$,

$$\gamma_m^{(j)}(\mathbf{L}) = \gamma_m^{(j)}(\mathbf{S}). \tag{5.47}$$

If $s_m = l_m - 1$ *then for all* j, $v > j \geq m$ *the following equality is valid:*

$$\gamma_m^{(j)}(\mathbf{L}) = \gamma_m^{(j)}(\mathbf{S}) + 1. \tag{5.48}$$

In both cases,

$$\gamma_m^{(v)}(\mathbf{L}) = \gamma_m^{(v-1)}(\mathbf{S}) + 1. \tag{5.49}$$

Proof Let us use the induction on j. For $j = m$ the both Eqs. (5.47) and (5.48) follow from definition (5.31). By Lemma 5.5, at the point j there is a jump during a motion of l_m to the right if and only if at the point m there is no jump in a motion of l_j to the left. By Lemma 5.8, this condition is equivalent to one in which at the point m there is no jump during a motion of s_j to the left. Again by Lemma 5.5 applied to the sequence **S**, we see that at the point j there is a jump during a motion of l_m to the right if and only if at this point there is a jump during a motion of s_m to the right.

Because the heads change values only by one and only at the points where exist jumps, the equalities (5.47) and (5.48) are proved.

For equality (5.49) we have

$$\gamma_m^{(v)}(\mathbf{L}) = \begin{cases} \gamma_m^{(v-1)}(\mathbf{L}) + 1, & \text{if } \gamma_m^{(v-1)}(\mathbf{L}) \leq l_v; \\ \gamma_m^{(v-1)}(\mathbf{L}), & \text{if } \gamma_m^{(v)}(\mathbf{L}) > l_v. \end{cases}$$

By Lemma 5.5, the condition $\gamma_m^{(v-1)}(\mathbf{L}) \leq l_v$, which means that at the point v there is a jump during a motion of l_m to the right, is equivalent to one in which at the point m there is no jump during a motion of l_v to the left; that is, $l_m = s_m$. Now it remains to use the equalities (5.47) and (5.48) with $j = v - 1$. □

5.4.1 Relations in the Symmetric Group

Let us turn to the symmetric group. For every index l, $0 \leq l \leq n-2$, we fix the following notations

$$[l] = \begin{cases} [n-1; n-l], & \text{if } 1 \leq l \leq n-2, \\ \text{id}, & \text{if } l = 0; \end{cases}$$

$$\lfloor l \rfloor = [n; n-l].$$

For $0 \leq i \leq n-3$ we also define

$$[2; m]_i = \begin{cases} [2; m], & \text{if } 2 \leq m < n, \\ [2; n-i-1][i]^{-1}, & \text{if } m = n. \end{cases} \tag{5.50}$$

Clearly $[l]$ and $[2; m]_i$ belong to $S_n^{1,n}$, whereas $\lfloor l \rfloor$ belongs to S_n^1.

Direct calculations show that the following relations hold:

$$\lfloor l \rfloor [2; n-\sigma] = \begin{cases} [2; n-(\sigma-1)]\lfloor l-1 \rfloor, & \text{if } 1 \leq \sigma \leq l; \\ [2; n-\sigma]\lfloor l \rfloor, & \text{if } l < \sigma \leq n-2. \end{cases} \tag{5.51}$$

If $\mathbf{L} = (l_j \mid u \le j \le v)$ is a sequence of integer numbers such that $1 \le l_j \le n-2$ for all j, then (5.30) and (5.32) with (5.51) show that

$$\lfloor l_{j-1} \rfloor [2; n - \sigma_v^{(j)}] = [2; n - \sigma_v^{(j-1)}] \lfloor l'_{j-1} \rfloor.$$

The multiple application of this relation yields the following formula:

$$\lfloor l_v \rfloor \lfloor l_{v+1} \rfloor \cdots \lfloor l_{u-1} \rfloor [2; n - l_v] = [2; n - \sigma_v^{(u)}(\mathbf{L})] \lfloor l'_v \rfloor \lfloor l'_{v+1} \rfloor \cdots \lfloor l'_{u-1} \rfloor. \tag{5.52}$$

By Lemma 5.4 this relation with $\mathbf{L}^*$ in place of $\mathbf{L}$ takes up the form

$$[2; n - (l_v + 1)] \lfloor l_{v+1} \rfloor \cdots \lfloor l_u \rfloor = \lfloor l^*_{v+1} \rfloor \cdots \lfloor l^*_u \rfloor [2; n - l^*_{u+1}]. \tag{5.53}$$

Analogously, for $1 \le l \le n-2$ the following relations are valid:

$$[l][2; n - \sigma] = \begin{cases} [2; n - (\sigma - 1)][l - 1], & \text{if } 2 \le \sigma \le l; \\ [2; n - \sigma][l], & \text{if } l < \sigma \le n - 2; \\ [2; n - 0]_{l-1}[l - 1], & \text{if } \sigma = 1. \end{cases} \tag{5.54}$$

Therefore if a sequence $\mathbf{L}$ belongs $\mathscr{L}(i, t+1)$ with $2 \le i \le n-1$ and $1 \le t+1 \le i$, then the following relation holds, provided $j \ge 2$:

$$[l_{j-1}][2; n - \sigma_{t+1}^{(j)}] = [2; n - \sigma_{t+1}^{(j-1)}][l'_{j-1}]$$

(see inequalities (5.43)). The multiple application of this relation yields

$$[2; n - \sigma_t^{(1)}(\mathbf{L})][s_1] \cdots [s_t] = [l_1] \cdots [l_t][2; n - l_{t+1}],$$

where $\mathbf{S} = (s_j \mid 0 \le j \le t) = \mathbf{L}'$. If we multiply this equality by $[i]$ from the left, then by the third line of (5.54) we can write

$$[2; n - s_0]_{i-1}[i - 1][s_1] \cdots [s_t] = [i][l_1] \cdots [l_t][2; n - l_{t+1}]. \tag{5.55}$$

5.5 Decreasing Modules

With the help of the definitions (5.22) and (5.50) one can note that $[2; n]_i = \tau_{n-i}$ if $0 \le i \le n-3$. Therefore it is possible to rewrite the formula (5.27) with $k = n - i$ in the following way:

$$D_{(n-i)} = \sum_{\gamma=0}^{i} [2; n - \gamma]_i T^{(n-i,n)}_{[2; n-\gamma]}. \tag{5.56}$$

Because for $2 \le l < k$ the permutation (k, n) does not change the element $T_{[2;l]} = \{P_{12}P_{32} \cdots P_{l\,2}\}$, we can also rewrite (5.26):

$$V_{(k)} = \sum_{\sigma=n-k+1}^{n-2} [2; n-\sigma] T_{[2;n-\sigma]}. \tag{5.57}$$

Definition 5.8 A right module over $\mathbf{F}[\mathscr{F}_n] * S_n^{1,n}$ is called a *decreasing module* if it is generated by elements $A_2, A_3, \ldots, A_n$ such that

$$A_k V_{(k)} + A_{k-1} D_{(k)} = 0, \quad 3 \le k \le n,$$

where $V_{(k)}$ and $D_{(k)}$ are defined by (5.57) and (5.56), respectively.

For example, if B satisfies (5.17) with $s = 2$, then $\mathscr{A}_2, \ldots, \mathscr{A}_n$, defined by (5.20), generate a decreasing submodule of $\mathbf{F}[\mathscr{F}_n] * S_n^{1,n} / \ker(\varphi) S_n^{1,n}$, see Sect. 5.3.

Theorem 5.4 *Let $n \ge 3$ and*

$$X = (\prod_{n>i>j>1} P_{ij})(\prod_{i=1}^{n-1} P_{in})(\prod_{j=2}^{n-1} P_{1j}). \tag{5.58}$$

Every decreasing module satisfies the following relation:

$$D_2 \prod_{k=0}^{n-3} [n-1; 2] V_{(n-k)} = A_2 [n-1; 2] \prod_{k=0}^{n-3} [2; 2+k] V_{(n-k)}^{(1,n)[3;n-1]^k} \cdot \{X\}^{[2;n-1]} (-1)^n, \tag{5.59}$$

where D_2 is defined by (5.25) *by replacing $\mathscr{A}$ with A.*

Let us define a sequence of elements

$$W_0 = D_2, \quad W_{t+1} = W_t [n-1; 2] V_{(n-t)}, \quad 0 \le t \le n-3.$$

In this notations the left-hand side of (5.59) equals W_{n-2}.

Lemma 5.10 *The element W_t has a representation*

$$W_t = \sum_{t \le i \le n-2} A_{n-i}[i] \sum_{\mathrm{L} \in \mathscr{L}(i,t)} (\prod_{k=1}^{t} [l_k])[2; n-1] R(i, l_1, \ldots, l_t), \tag{5.60}$$

where $\mathbf{L} = (l_1, \ldots, l_t)$. *The following recurrence relations are valid:*

$$R(i, l_1, \ldots, l_t) = R^{[n-1;2][2;n-l_t]}(i, l_1, \ldots, l_{t-1})T_{[2;n-l_t]}$$
$$- R^{[n-1,2]}(i-1, s_1, \ldots, s_{t-1})T^{(n-i+1,n)[i-1][s_1][s_2]\cdots[s_{t-1}]}_{[2;n-s_0]}. \quad (5.61)$$

In these relations $(s_0, s_1, \ldots, s_{t-1})$ *is the subordinate sequence for* $\mathbf{L}$.

In addition, the elements $R^{[n-1;2]}(i, l_1, \ldots, l_t)$ *are invariant with respect to the action of all cycles* $[k; l]$ *with* $2 \le k \le l < n-i$.

Proof If $t = 0$, then $\mathscr{L}(i, 0)$ contains just the empty sequence. So the product in (5.60) is empty, and (5.60) takes the form

$$W_0 = \sum_{0 \le i \le n-2} A_{n-i}[i][2; n-1]R(i).$$

The right-hand side of (5.25) is reduced to this form if we replace the index of summation l with $n-i$ and use the relation $[i][2; n-1] = [2; n-i]$ for $i > 0$. In this case $R(i) = T^{(2,n)}_{[2;n-i]}$. Lemma 5.2 demonstrates that

$$T^{(2,n)[n-1;2]}_{[2;n-i]} = \{P_{1n}P_{2n}\cdots P_{n-i-1\,n}\}.$$

In particular, this coefficient commutes with all permutations $[k; l]$ if $2 \le k \le l < n-i$. We may proceed the induction on t.

Let the lemma be true for a given $t \ge 0$. By (5.57) we obtain

$$V_{(n-t)} = V_{(n-i)} + \sum_{l_{t+1}=t+1}^{i} [2; n-l_{t+1}]T_{[2;n-l_{t+1}]}, \quad t \le i \le n-2.$$

Therefore

$$W_{t+1} = \sum_{t \le i \le n-2} A_{n-i}[i] \sum_{\mathbf{L} \in \mathscr{L}(i,t)} (\prod_{k=1}^{t}[l_k])R^{[n-1;2]}(i, l_1, \ldots, l_t)$$
$$\times (V_{(n-i)} + \sum_{l_{t+1}=t+1}^{i} [2; n-l_{t+1}]T_{[2;n-l_{t+1}]}). \quad (5.62)$$

The elements $T_{[2,l]} = \{P_{12}P_{32}\cdots P_{l2}\}$, $l < n-i$, are fixed with respect to the action of all cycles $[i]$, $[l_j]$, because $n - l_j \ge n - i > l$. By the same reasoning the cycles $[2; l]$ commute with $[i]$, $[l_j]$. Thus the element $V_{(n-i)}$ commutes with all of the permutations $[i]$, $[l_j]$. By the inductive suppositions all the cycles $[k, l]$, $2 \le k \le l < n-i$, commute with $R^{[n-1;2]}(i, l_1, \ldots, l_t)$. Therefore $V_{(n-i)}$ commutes with this coefficient also. Hence we can continue (5.62) by taking into account that for $i = t$ the last sum in (5.62) equals zero and $V_{(2)} = 0$:

$$= \sum_{t \le i \le n-3} A_{n-i} V_{(n-i)}[i] \sum_{\mathbf{L} \in \mathscr{L}(i,t)} (\prod_{k=1}^{t} [l_k]) R^{[n-1;2]}(i, l_1, \ldots, l_t)$$

$$+ \sum_{t+1 \le i \le n-2} A_{n-i}[i] \sum_{\mathbf{L} \in \mathscr{L}(i,t)} (\prod_{k=1}^{t} [l_k]) R^{[n-1;2]}(i, l_1, \ldots, l_t)$$

$$\times \sum_{l_{t+1}=t+1}^{i} [2; n - l_{t+1}] T_{[2;n-l_{t+1}]}. \tag{5.63}$$

As the factor $A_{n-i}V_{(n-i)}$ is located in (5.63) at the left margin position, we may replace it with $-A_{n-i-1}D_{(n-i)}$ and use the relation (5.56). Every sequence $(\gamma, l_1, l_2, \ldots, l_t)$ taking part in the obtained then expression belongs to $\mathscr{S}(i, t)$, because the index γ in (5.56) is going from 0 to i. Therefore if in the first line of (5.63) we replace the summation index i with $i - 1$, we obtain

$$W_{t+1} = - \sum_{\mathbf{S} \in \mathscr{S}(i-1,t)} A_{n-i} \sum_{s_0=0}^{i-1} [2; n - s_0]_{i-1} T^{(n-i+1,n)}_{[2;n-s_0]}$$

$$\times [i-1][s_1] \cdots [s_t] R^{[n-1;2]}(i-1, s_1, \ldots, s_t)$$

$$+ \sum_{t+1 \le i \le n-2} A_{n-i}[i] \sum_{\mathbf{L} \in \mathscr{L}(i,t+1)} [l_1] \cdots [l_t][2; n - l_{t+1}]$$

$$\times R^{[n-1;2][2;n-l_{t+1}]}(i, l_1, \ldots, l_t) T_{[2;n-l_{t+1}]}. \tag{5.64}$$

The formula $T\pi = \pi T^{\pi}$ allows us to shift all coefficients to the right margin position. Lemma 5.7 and the relation (5.55) imply

$$W_{t+1} = \sum_{t+1 \le i \le n-2} A_{n-i}[i] \sum_{\mathbf{L} \in \mathscr{L}(i,t+1)} [l_1] \cdots [l_t][2; n - l_{t+1}]$$

$$\times (R^{[n-1;2][2;n-l_{t+1}]}(i, l_1, \ldots, l_t) T_{[2;n-l_{t+1}]}$$

$$- R^{[n-1;2]}(i-1, s_t, \ldots, s_t) T^{(n-i+1,n)[i-1][s_1]\cdots[s_t]}_{[2;n-s_0]}).$$

This proves both (5.60) and (5.61), because $[2; n - l_{t+1}] = [l_{t+1}][2; n-1]$.

Let us check, finally, that the found value of $R^{[n-1;2]}(i, l_1, \ldots, l_{t+1})$ is fixed with respect to the actions of $[k; l]$, provided that $2 \le k \le l < n-i$. By (5.21) the equality $[n-1; 2][k; l] = [k+1; l+1][n-1; 2]$ is valid. The cycle $[k+1; l+1]$ commutes with $[i-1]$, $[s_1]$, …, $[s_t]$, $(n-i+1, n)$, for $s_j \le i-1$ and $n - s_j \ge n - i + 1 > l + 1$. In the same way $[k; l]$ commutes with $[l_{t+1}]$. By this notes and the relation

$[2; n - l_{t+1}][n-1; 2] = [l_{t+1}]$ we can write

$$R^{[n-1;2][k;l]}(i, l_1, \dots, l_{t+1}) = R^{[n-1;2][k;l][l_{t+1}]}(i, l_1, \dots, l_t)T^{[k+1;l+1][n-1;2]}_{[2;n-l_{t+1}]}$$
$$- R^{[n-1;2][k+1;l+1][n-1;2]}(i-1, s_1, \dots, s_t)T^{[k+1;l+1](n-i+1,n)[i-1][s_1]\cdots[s_t][n-1;2]}_{[2;n-s_0]}.$$

By the inductive suppositions in the first factors of the both summands, it is possible to delete $[k;l]$ and $[k+1; l+1]$, respectively. In addition, the condition $l_{t+1} \leq i$ demonstrates that $n - l_{t+1} \geq n - i \geq l + 1 \geq k + 1 \geq 3$. In the same way, $n - s_0 \geq n - (i-1) > l + 1 \geq k + 1 \geq 3$. Therefore, the equalities $T_{[2;n-s]} = \{P_{12}P_{32}P_{42}\cdots P_{n-s\,2}\}$ with $s = l_{t+1}$ and $s = s_0$ show that in the second factors of the both summands it is possible to delete $[k+1; l+1]$. □

Definition 5.9 For a sequence $\mathbf{L} = (l_1, \dots, l_t)$, define

$$E(l_1) = \{P_{n\,n-1}\};$$
$$E(l_1, l_2, \dots, l_k) = \{P_{n\,n-1}, \dots, P_{n-k+m-1\;n-1}, \dots\},\;\; 2 \leq k \leq t, \tag{5.65}$$

where m runs trough a set of all indices such that at the point m there is a jump during a motion of l_k to the left or, equivalently (see Lemma 5.5), at the point k there is no jump during a motion of l_m to the right.

By Lemma 5.8 we may claim that if $\mathbf{S} = \mathbf{L}'$, then for all k, $1 \leq k \leq t-1$,

$$E(s_1, s_2, \dots, s_k) = E(l_1, l_2, \dots, l_k). \tag{5.66}$$

Furthermore, for $0 \leq t \leq i \leq n-2$, define

$$\begin{aligned}
C(i, l_1, \dots, l_t) = \{&P_{1n}P_{2n}\cdots P_{n-i-1\,n} \sqcup P_{n-t\,n}P_{n-t+1\,n}\cdots P_{n-1\,n}\\
&P_{1\,n-1}P_{2\,n-1}\cdots P_{n-\gamma_t^{(t)}(\mathbf{L})\;n-1} \sqcup \qquad \emptyset\\
&P_{1\,n-2}P_{2\,n-2}\cdots P_{n-\gamma_{t-1}^{(t)}(\mathbf{L})\;n-2} \sqcup \qquad P_{n-1\,n-2}\\
&\vdots \qquad\qquad\qquad\qquad \vdots\\
&P_{1\,n-r}P_{2\,n-r}\cdots P_{n-\gamma_{t-r+1}^{(t)}(\mathbf{L})\;n-r} \sqcup P_{n-r+1\,n-r}\cdots P_{n-1\,n-r}\\
&\vdots \qquad\qquad\qquad\qquad \vdots\\
&P_{1\,n-t}P_{2\,n-t}\cdots P_{n-\gamma_1^{(t)}(\mathbf{L})\;n-t} \sqcup P_{n-t+1\,n-t}\cdots P_{n-1\,n-t}\}.
\end{aligned} \tag{5.67}$$

In another words, a letter P_{xy} takes part in $C(i, l_1, \dots, l_t)$ if and only if

either $y = n$ and $(1 \leq x < n - i$ or $n - t \leq x < n)$,
or $n - t \leq y \leq n - 1$ and $(1 \leq x \leq n - \gamma^{(t)}_{y+t+1-n}(\mathbf{L})$ or $y < x < n)$.

In particular, for $t = 0$ this means

$$C(i) = \{P_{1n}P_{2n}\cdots P_{n-i-1\,n}\} = T_{[2;n-i]}^{(2,n)[n-1;2]} = R^{[n-1;2]}(i). \tag{5.68}$$

In the explicit form (5.67) of the element $C(l_1,\dots,l_t)$ the top row is called the *zero row*, the following row is called the *first row* and so on. In this way the rth *row* corresponds to the value $n-r$ of the second index.

Lemma 5.11 *Let* $\mathbf{L} \in \mathscr{L}(i,t)$ *and* $\mathbf{S} = \mathbf{L}'$. *If* $1 \le k \le t-1$, *then*

$$E^{[n-1\,;\,2]^{t-k-1}[l_t]}(l_1,\dots,l_k) = E^{[n-1\,;\,2]^{t-k}}(s_1,\dots,s_k).$$

Proof If a letter $P_{n-k+m-1\,n-1}$ is involved in the writing (5.65) of the element $E(l_1,\dots,l_k)$, then $P_{n-k+m-1\,n-1}^{[n-1\,;\,2]^{t-k-1}} = P_{n-t+m\,n-t+k}$. Therewith $l_t \ge t > t-m$, and so $n-l_t < n-t+m \le n-t+k$. Thus under an additional application of $[l_t]$, as well as under an application of $[n-1;2]$, all of the indices decremented by one. Thus $E^{[n-1;2]^{t-k-1}[l_t]}(l_1,\dots,l_k) = E^{[n-1;2]^{t-k}}(l_1,\dots l_k)$. By (5.66) we are done. □

Lemma 5.12 *The coefficients of* (5.60) *have the following decomposition:*

$$R^{[n-1\,;\,2]}(i,l_1,\dots,l_t) = (-1)^t C(i,l_1,\dots,l_t)\prod_{k=1}^{t} E^{[n-1\,;\,2]^{t-k}}(l_1,\dots,l_k). \tag{5.69}$$

Proof For $t = 0$ formula (5.69) is valid by (5.68). Assume that (5.69) is valid for $t-1$, $t \ge 1$. Then by (5.61) and $[2;n-l_t][n-1;2] = [l_t]$ we obtain

$$\begin{aligned} &R^{[n-1;2]}(i,l_1,\dots,l_t)\\ &= -(-1)^t C^{[l_t]}(i,l_1,\dots,l_{t-1})T_{[2;n-l_t]}^{[n-1;2]}\prod_{k=1}^{t-1}E^{[n-1;2]^{t-k-1}[l_t]}(l_1,\dots,l_k)\\ &\quad+(-1)^t C^{[n-1;2]}(i-1,s_1,\dots,s_{t-1})T_{[2;n-s_0]}^{(n-i+1,n)[i-1][s_1]\cdots[s_{t-1}][n-1;2]}\\ &\quad\times\prod_{k=1}^{t-1}E^{[n-1;2]^{t-k}}(s_1,\dots,s_k). \end{aligned} \tag{5.70}$$

Using Lemma 5.11 and equality (5.66), we may factor out the product $(-1)^t\prod_{k=1}^{t-1}E^{[n-1;2]^{t-k}}(l_1,\dots l_k)$. Therefore it suffices to prove that

$$\begin{aligned} &C^{[n-1;2]}(i-1,s_1,\dots,s_{t-1})T_{[2;n-s_0]}^{(n-i+1,n)[i-1][s_1]\cdots[s_{t-1}][n-1;2]}\\ &\quad= C^{[l_t]}(i,l_1,\dots,l_{t-1})T_{[2;n-l_t]}^{[n-1;2]} + C(i,l_1,\dots,l_t)E(l_1,\dots,l_t). \end{aligned}$$

This follows from (5.12) with $\{C\} \leftarrow T^{[n-1;2]}_{[2;n-l_t]}$, $\{D\} \leftarrow C^{[l_t]}(i, l_1, \ldots, l_{t-1})$, and $\{E\} \leftarrow E(l_1, \ldots, l_t)$, provided that the following three equalities are valid

$$C^{[n-1;2]}(i-1, s_1, \ldots, s_{t-1}) = C^{[l_t]}(i, l_1, \ldots, l_{t-1}) \star \overline{E}(l_1, \ldots, l_t), \tag{5.71}$$

$$T^{(n-i+1,n)[i-1][s_1]\cdots[s_{t-1}][n-1;2]}_{[2;n-s_0]} = T^{[n-1;2]}_{[2;n-l_t]} \star E(l_1, \ldots, l_t), \tag{5.72}$$

$$C(i, l_1, \ldots, l_t) = C^{[n-1;2]}(i-1, s_1, \ldots, s_{t-1}) \star T^{[n-1;2]}_{[2;n-l_t]}. \tag{5.73}$$

Consider the first one. Because $\mathbf{L}$ satisfies (5.41), we have $n - l_t < n - t + 1$. Therefore $[l_t]$ decreases by one all of he second indices of $C(i, l_1, \ldots, l_{t-1})$, but n. This means that in (5.67), with $t-1$ in place of t, all the rows, but zero row, are shifted down by one step, so that the first row becomes empty.

Every first index of a letter located after a gap $\sqcup$ in (5.67) is greater then the second index of this letter. Therefore $[l_t]$ decreases these indices by one as well. This means that all letters located in the rows after the gaps are shifted to the left by one step.

Condition (5.41) implies $n - i - 1 < n - l_t$. Therefore $[l_t]$ does not shift a part of the zero row located before the gap. In particular, the last position of the zero row of $C^{[l_t]}(i, l_1, \ldots, l_{t-1})$ is vacant. For letters located in nonzero rows before the gap consider the following two cases.

If $s_{t-r+1} = l_{t-r+1} - 1$, then at the point $t - r + 1$ there is a jump during a motion of l_t to the left. By Lemma 5.5, at the point t there is no jump in a motion of l_{t-r+1} to the right. Definition 5.5 shows that $n - \gamma^{(t-1)}_{t-r+1}(\mathbf{L}) < n - l_t$. Therefore $[l_t]$ does not change the first indices of letters located in $(r-1)$th row, $r > 1$, before the gap. In this case the last position of the rth row of $C^{[l_t]}(l_1, \ldots, l_{t-1})$ is vacant, and the length of the rth row located before the gap equals $n - \gamma^{(t-1)}_{t-r+1}(\mathbf{L}) = n - \gamma^{(t)}_{t-r+1}(\mathbf{L})$, see (5.31).

If $s_{t-r+1} = l_{t-r+1}$, or, equivalently $n - \gamma^{(t-1)}_{t-r+1}(\mathbf{L}) \geq n - l_t$, then the letter $P_{n-l_t\, n-r+1}$ goes to $P_{n-1\, n-r}$ and occupies a position in the last column. The next letter, $P_{n-l_t+1\, n-r+1}$, goes to $P_{n-l_t\, n-r}$, and so on. Thus in this case the length of the rth row located before the gap is set by the same formula $n - \gamma^{(t-1)}_{t-r+1}(\mathbf{L}) - 1 = n - \gamma^{(t)}_{t-r+1}(\mathbf{L})$, see (5.31).

Furthermore, by the definition $\overline{E}(l_1, \ldots, l_t) = \{P_{n-1\, n}, \ldots, P_{n-1\, n-t+m-1}, \ldots\}$, where at the point m there is a jump during a motion of l_t to the left; that is, $s_m = l_m - 1$. By replacing the index m with $t - r + 1$ in this definition, we get that the letters of $\overline{E}(l_1, \ldots, l_t)$ occupy exactly the positions in the last column that are vacant in $C^{[l_t]}(l_1, \ldots, l_{t-1})$.

Thus the zero row of $C^{[l_t]}(l_1, \ldots, l_{t-1}) \star \overline{E}(l_1, \ldots, l_t)$ has the form

$$P_{1n}P_{2n}\cdots P_{n-i-1\, n} \;\sqcup\; P_{n-t\, n}\cdots P_{n-1\, n}; \tag{5.74}$$

the first row is vacant, and the rth row, $r > 1$, has the form

$$P_{1\,n-r}P_{2\,n-r}\cdots P_{n-\gamma^{(t)}_{t-r+1}(\mathbf{L})\,n-r} \sqcup P_{n-r+1\,n-r}\cdots P_{n-1\,n-r}.$$

Consider the left-hand side of (5.71). The permutation $[n-1;2]$ shifts to the left by one step all gaps of $C(i-1,s_1,\ldots,s_{t-1})$. It also shifts down by one step all rows (because $n-(t-1) > 2$). Therefore the zero row of $C^{[n-1;2]}(i-1,s_1,\ldots,s_{t-1})$ has the form (5.74), the first row is vacant, and the rth row, $r > 1$, equals

$$P_{1\,n-r}P_{2\,n-r}\cdots P_{n-\gamma^{(t-1)}_{t-r+1}(\mathbf{S})-1\,n-r}, \sqcup P_{n-r+1\,n-r}\cdots P_{n-1\,n-r}. \tag{5.75}$$

By (5.49) we have $\gamma^{(t-1)}_{t-r+1}(\mathbf{S})+1 = \gamma^{(t)}_{t-r+1}(\mathbf{L})$. Thus (5.71) is proved.

As (5.75) coincides with the r-th row of (5.67), all rows of $C(i,l_1,\ldots,l_t)$, but the first one coincide with the same rows of $C^{[n-1;2]}(i-1,s_1,\ldots,s_{t-1})$. The first row of $C(i,l_1,\ldots,l_t)$ equals $T^{[n-1;2]}_{[2;n-l_t]} = \{P_{1\,n-1}P_{2\,n-1}\cdots P_{n-l_t-1\,n-1}\}$ since $\gamma^{(t)}_t(\mathbf{L}) = l_t+1$. Thus (5.73) is proved.

Consider (5.72). The right-hand side has only one row (the first one)

$$\{P_{1\,n-1}P_{2\,n-1}\cdots P_{n-l_t-1\,n-1} \sqcup \ldots P_{n-t+m-1\,n-1}\ldots P_{n\,n-1}\}, \tag{5.76}$$

where m runs through the set of all indices with $s_m = l_m - 1$.

By Lemma 5.7 we have $s_0 \leq i-1$; that is, $2 < n-i+1 \leq n-s_0$. Therefore

$$T^{(n-i+1,n)}_{[2;n-s_0]} = \{P_{12} \sqcup P_{32}\cdots P_{n-i\,2} \sqcup P_{n-i+2\,2}\cdots P_{n-s_0\,2} \sqcup P_{n2}\}.$$

From this we obtain

$$T^{(n-i+1)[i-1]}_{[2;n-s_0]} = \{P_{12} \sqcup P_{32}\cdots P_{n-s_0-1\,2} \sqcup P_{n2}\}.$$

Because $\gamma^{(0)}_0(\mathbf{S}) = s_0+1$, we may start an induction on k to prove that

$$T^{(n-i+1)[i-1][s_1]\cdots[s_k]}_{[2;n-s_0]} = \{P_{12} \sqcup P_{32}\cdots P_{n-\gamma^{(k)}_0(\mathbf{S})\,2} \sqcup \cdots P_{n-k+m-1\,2}\cdots \sqcup P_{n2}\}, \tag{5.77}$$

where m runs trough the set of all indices less than k with $l_m = s_m + 1$.

If $l_{k+1} = s_{k+1}$, then by Lemma 5.4 and the definition (5.33) we have $n-\gamma^{(k)}_0(\mathbf{S}) < n-s_{k+1}$. Therefore $[s_{k+1}]$ does not shift the letters located in (5.77) before the second gap. By (5.42) we have $s_{k+1} > k$ and $n-k+m-1 > n-s_{k+1}$ (as $m \geq 1$). This implies that $[s_{k+1}]$ shifts to the left by one step all letters (except P_{n2}) located after the second gap. Since in this case $\gamma^{(k+1)}_0(\mathbf{S}) = \gamma^{(k)}_0(\mathbf{S})$, we may replace k with $k+1$ in (5.77).

If $l_{k+1} = s_{k+1} + 1$, then $n - \gamma_0^{(k)}(\mathbf{S}) \geq n - s_{k+1}$. Therefore $[s_{k+1}]$ shifts the letter $P_{n-s_{k+1}\,2}$ to the place $n - 1 = n - (k+1) - (k+1) - 1$, the next letter $P_{n-s_{k+1}+1\,2}$ to the place $n - s_{k+1}$, and so on. In particular, the segment before the second gap is ended by a letter with the first index $n - \gamma_0^{(k)}(\mathbf{S}) - 1 = n - \gamma_0^{(k+1)}(\mathbf{S})$. As above $[s_{k+1}]$ shifts to the left by one step all letters (except P_{n2}) located after the second gap. Thus in this case in the formula (5.77) we may replace k with $k + 1$ as well. Hence (5.77) is proved.

If we apply $[n - 1; 2]$ to the relation (5.77) with $k = t - 1$ and note that the last head $\gamma_0^{(t-1)}(\mathbf{S})$ equals l_t (see (5.34)), then by (5.76) we obtain (5.72). □

Lemma 5.13 *If a sequence $l_1, \ldots, l_i$ satisfies (5.41) with $t = i$, that is, $\mathbf{L} \in \mathscr{L}(i, i)$, then $\gamma_j^{(i)}(\mathbf{L}) = i + 1$ for all j, $1 \leq j \leq i$.*

Proof Let us use induction by i. If $i = 1$, then $\gamma_1^{(1)} = l_1 + 1 = 2$, because $1 \leq l_1 \leq i = 1$. Suppose that the lemma holds for all sequences of the length $i - 1$. If $\mathbf{S} = \mathbf{L}'$, then by Lemma 5.7 we may apply the inductive supposition to the sequences $s_1, s_2, \ldots, s_{i-1}$. So $\gamma_j^{(i-1)}(\mathbf{S}) = i$ for all j, $1 \leq j \leq i - 1$. The relation (5.49) with $t = i$ shows $\gamma_j^{(i)}(\mathbf{L}) = \gamma_j^{(i-1)}(\mathbf{S}) + 1 = i + 1$, provided that $j < i$. If $j = i$, then $\gamma_i^{(i)}(\mathbf{L}) = l_i + 1$, herewith $i \leq l_i \leq i$. □

Lemma 5.14 *Let $\mathbf{L} = (l_1, \ldots, l_t) \in \mathscr{L}(i, i)$ (and, in particular, $t = l_t = i$). If $\mathbf{S} = (s_0, s_1, \ldots, s_{t-1}) = \mathbf{L}'$, then*

$$[s_1] \cdots [s_{t-1}] E^{[2;n-1]}(l_1, \ldots, l_t) = T^{(1,n)[3;n-1]^{n-2-i}}_{[2;i+2-\sigma_t^{(1)}(\mathbf{L})]} [s_1] \cdots [s_{t-1}]. \tag{5.78}$$

Proof The representations of $E(l_1, \ldots, l_t)$ and $T_{[2;\theta]}$, $\theta = i + 2 - \sigma_t^{(1)}(\mathbf{L})$ by means of braces have the same number of letters. Indeed, by (5.65) the former element contains $1 + \varepsilon$ letters, where ε is the total number of jumps in a motion of l_t to the left. As the heads are changing only at points where there is a jump and only by one to the side of diminution, we may write $\varepsilon = l_t - \sigma_t^{(1)}(\mathbf{L}) = i - \sigma_t^{(1)}(\mathbf{L})$. The element $T_{[2,\theta]} = \{P_{12}P_{32} \cdots P_{\theta\,2}\}$ has $\theta - 1$ letters, and $\theta - 1 = i + 2 - \sigma_t^{(1)}(\mathbf{L}) - 1 = \varepsilon + 1$.

Denote by π the permutation $[s_1] \cdots [s_{t-1}]$. Then we may write $\pi A = A^{\pi^{-1}} \pi$. Because permutations do not change the number of letters of words, it suffices to show that the permutation

$$\nu = [2; n-1]\pi^{-1}[n-1; 3]^{n-2-i}(1, n)$$

shifts each letter of $E(l_1, \ldots, l_t)$ to a letter of $T_{[2;\theta]}$.

We have $P^{\nu}_{n\,n-1} = P_{12}$, as $P^{[2;n-1]}_{n\,n-1} = P_{n\,2}$ and $[s_j]^{-1}$, $[n-1; 3]$ do not move n or 2 (clearly, $n - s_j \geq n - (i - 1) \geq 3$). Let us demonstrate that if $s_m = l_m - 1$, then $P^{\nu}_{n-t+m-1\,n-1} = P_{\omega\,2}$, where $\omega = i + 2 - \sigma_m^{(1)}(\mathbf{S})$.

Evidently, $P^{[2;n-1]}_{n-t+m-1\ n-1} = P_{n-t+m\ 2}$. Using this relation as a beginning of the downward induction on j, $t \geq j > m$, let us show that

$$P^{[s_{t-1}]^{-1}\cdots[s_j]^{-1}}_{n-t+m\ 2} = P_{n-j+m\ 2}, \tag{5.79}$$

provided that for $j = t$ the product is empty. Let $j - 1 > m$, and assume that (5.79) holds for a given j. Then $n-j+m < n-1$ and $n-j+m \geq n-(j-1) \geq n-s_{j-1}$ because $m \geq 1$ and $s_{j-1} \geq j-1$. This implies that $[s_{j-1}]^{-1}$ amplifies by one the first index of the right-hand side of (5.79), and it does not change the second one. Thus in (5.79) it is possible to replace j with $j-1$.

Now by (5.79) with $j = m+1$ we have

$$P^{[s_{t-1}]^{-1}\cdots[s_m]^{-1}}_{n-t+m\ 2} = P^{[s_m]^{-1}}_{n-1\ 2} = P_{n-s_m\ 2}.$$

We may use this equality as the beginning of the downward induction by j to prove the following general formula, provided that $1 \leq j \leq m$:

$$P^{[s_{t-1}]^{-1}\cdots[s_j]^{-1}}_{n-t+m\ 2} = P_{n-\sigma_m^{(j)}(\mathbf{S})\ 2}. \tag{5.80}$$

Suppose that (5.80) holds for a given j. If $n - s_{j-1} \leq n - \sigma_m^{(j)}(\mathbf{S})$, that is, $\sigma_m^{(j)}(\mathbf{S}) \leq s_{j-1}$, then $[s_{j-1}]^{-1}$ amplifies by one the first index of the right-hand side of (5.80). By (5.30) we have $n - (\sigma_m^{(j)}(\mathbf{S}) - 1) = n - \sigma_m^{(j-1)}(\mathbf{S})$. Therefore, we may replace j with $j-1$ in (5.80).

If $\sigma_m^{(j)}(\mathbf{S}) > s_{j-1}$, then $[s_{j-1}]^{-1}$ does not change the first index, and again by (5.30) this index equals $n - \sigma_m^{(j-1)}(\mathbf{S})$. Therefore in this case it is possible to replace j with $j-1$ as well. Thus (5.80) is proved.

The heard $\sigma_m^{(1)}(\mathbf{S})$ is equal to a difference of s_m with the total number of jumps during a motion of s_m to the left; that is, it is not less than $s_m - (m-1) \geq 1$ and not grater than $i - 1 = n - 3$. Thus the transposition $(1, n)$ does not move the index $n - \sigma_m^{(1)}(\mathbf{S})$. If we apply $(1,n)[n-1;3]^{n-2-i} = [n-1;3]^{n-2-i}(1,n)$ to (5.80) with $j = 1$, we obtain $\omega = i + 2 - \sigma_m^{(1)}(\mathbf{S})$. By (5.44) with $j = 1$, $k = m$ we have $\sigma_m^{(1)}(\mathbf{S}) \geq \sigma_t^{(1)}(\mathbf{L})$. Hence $\omega \leq \theta$, and $P^{\pi}_{n-t+m-1} = P_{\omega\ 2}$ occurs in $T_{[2;\theta]}$. □

In Lemmas 5.10–5.14, we accumulated necessary information to start the proof of Theorem 5.4.

Proof Let $\mathscr{L} = \mathscr{L}(n-2, n-2)$. For every $\mathbf{L} \in \mathscr{L}$ define a chain of sequences $\mathbf{L}_1 = \mathbf{L}, \mathbf{L}_2, \ldots, \mathbf{L}_{n-2}$ satisfying (5.41) with $i = t = n-2, n-3, \ldots, 1$, respectively. Let $\mathbf{L}_k = (l_1^{(k)}, \ldots, l_{n-k-1}^{(k)})$ be defined, which satisfies (5.41) with $i = t = n-k-1$. If $\mathbf{S}_k = (s_0^{(k)}, s_1^{(k)} \ldots, s_{n-k-2}^{(k)})$ is a subordinate sequence for $\mathbf{L}_k$, then put $\mathbf{L}_{k+1} = (s_1^{(k)}, \ldots, s_{n-k-2}^{(k)})$.

Denote by $\mathbf{L}^\circ$ the following sequence

$$l_1^\circ = n - \sigma_{n-2}^{(1)}(\mathbf{L}_1) - 1, l_2^\circ = n - \sigma_{n-3}^{(1)}(\mathbf{L}_2) - 1, \dots, l_{n-2}^\circ = n - \sigma_1^{(1)}(\mathbf{L}_{n-2}) - 1.$$

This sequence belongs to $\mathscr{L}$. Indeed, we have $\sigma_{n-k-1}^{(n-k-1)}(\mathbf{L}_k) = l_{n-k-1}^{(k)} = n-k-1$. Because at each step into the left a heard is decremented by not more than one, we obtain

$$n-k-1 \geq \sigma_{n-k-1}^{(1)}(\mathbf{L}_k) \geq (n-k-1) - (n-k-2) = 1.$$

Therefore $k \leq n - \sigma_{n-k-1}^{(1)}(\mathbf{L}_k) - 1 \leq n-2$ and $\mathbf{L}^\circ \in \mathscr{L}$.

The sequence $\mathbf{L}$ can be restored in a unique way from $\mathbf{L}^\circ$ by an inverse process. Indeed, by (5.34) we know all the heads:

$$\gamma_0^{(0)}(\mathbf{S}_k) = \sigma_{n-k-1}^{(1)}(\mathbf{L}_k) = n - l_k^\circ - 1.$$

Thus we know all zero terms of the subordinate sequences $s_0^{(k)} = \gamma_0^{(0)}(\mathbf{S}_k) - 1$. Therefore, starting with the sequence $\mathbf{S}_{n-3} = (s_0^{(n-3)}, s_1^{(n-3)})$, where $s_1^{(n-3)} = l_1^{(n-2)} = 1$, we may restore $\mathbf{L}_{n-2}, \mathbf{L}_{n-3}, \dots, \mathbf{L}_1$ with the help of (5.33) in a unique way, see Lemma 5.4.

Thus $\circ : \mathscr{L} \to \mathscr{L}$ is a one-to-one correspondence. In particular, in (5.60) we may replace the summation index $L \in \mathscr{L}$ with $L^\circ \in \mathscr{L}$, that is,

$$W_{n-2} = A_2[n-1;2] \sum_{\mathbf{L}^\circ \in \mathscr{L}} [l_1] \cdots [l_{n-3}][2; n - l_{n-2}] R(i, l_1, \dots, l_{n-2}). \tag{5.81}$$

By Lemma 5.13 we have $\gamma_j^{(n-2)}(\mathbf{L}) = n-2$ for all j, $1 \leq j \leq n-2$. Now (5.67) demonstrates that $C(n-2, l_1, \dots, l_{n-2}) = \{X\}$ does not depend on $\mathbf{L} \in \mathscr{L}$, herewith the word X is defined by (5.58). Thus, replacing $R(\dots)$ in (5.81) with the help of (5.69), we may factor out $\{X\}^{[2;n-1]}(-1)^n$ to the right-hand side. The rest part of the sum in (5.81) can be rewritten with the help of (5.52):

$$\begin{aligned}\Sigma = &\sum_{\mathbf{L}^\circ \in \mathscr{L}} [2; n - \sigma_{n-2}^{(1)}(\mathbf{L}_1)][s_1] \cdots [s_{n-3}] E^{[2;n-1]}(l_1, \dots, l_{n-2}) \\ &\times \prod_{k=1}^{n-3} E^{[n-1;2]^{n-k-3}}(l_1, \dots, l_k).\end{aligned}$$

Using Lemma 5.14, then (5.66), and the definition of $(l_1^{(2)}, \dots l_{n-3}^{(2)})$, we obtain

$$\begin{aligned}\Sigma = &\sum_{\mathbf{L}^\circ \in \mathscr{L}} [2; l_1^\circ + 1] T_{[2;l_1^\circ+1]}^{(1,n)} [l_2^{(2)}] \cdots [l_{n-3}^{(2)}] E(l_1^{(2)}, \dots, l_{n-3}^{(2)}) \\ &\times \prod_{k=1}^{n-4} E^{[n-1;2]^{n-k-3}}(l_1^{(2)}, \dots, l_k^{(2)}).\end{aligned}$$

Let us replace $[l^{(2)}_{n-3}]$ with an equal permutation $[2; n - l^{(2)}_{n-3}][n - 1; 2]$. Then again by (5.52), (5.66), and the definition of $\mathbf{L}_3$ we obtain

$$\Sigma = \sum_{\mathbf{L}^\circ \in \mathcal{L}} [2; l^\circ_1 + 1]T^{(1,n)}_{[2;l^\circ_1+1]}[2; l^\circ_2 + 1][l^{(3)}_1] \cdots [l^{(3)}_{n-4}]E^{[2;n-1]}(l^{(2)}_1, \dots, l^{(2)}_{n-3})[n-1; 2]$$

$$\times \prod_{k=1}^{n-4} E^{[n-1;2]^{n-k-3}}(l^{(3)}_1, \dots, l^{(3)}_k).$$

By Lemma 5.14 we have

$$\Sigma = \sum_{\mathbf{L}^\circ \in \mathcal{L}} [2; l^\circ_1 + 1]T^{(1,n)}_{[2;l^\circ_1+1]}[2; l^\circ_2 + 1]T^{(1,n)[3;n-1]}_{[2;l^\circ_2]}[l^{(3)}_1] \cdots [l^{(3)}_{n-5}][2; n - l^{(3)}_{n-4}][n-1; 2]^2$$

$$\times \prod_{k=1}^{n-4} E^{[n-1;2]^{n-k-3}}(l^{(3)}_1, \dots, l^{(3)}_k).$$

Let us explore this process further, so that we can see that every new left-hand factor does not depend on the whole sequence $\mathbf{L}^\circ$, but it depends up the only member of $\mathbf{L}^\circ$. Therefore we have

$$\Sigma = \prod_{j=1}^{n-2}\Big(\sum_{l^\circ_j = j}^{n-2} [2; l^\circ_j + 1]T^{(1,n)[3;n-1]^{j-1}}_{[2;l^\circ_j+2-j]}\Big) \cdot [n-1; 2]^{n-2}. \tag{5.82}$$

Let us replace the summation index l°_j with $l = l^\circ_j + 2 - j$ in all of the sums. Then for a given j, the sum of (5.82) take up the form

$$\Sigma_j = \sum_{l=2}^{n-j} [2; l + j - 1]T^{(1,n)[3;n-1]^{j-1}}_{[2;l]}.$$

As $[2; l + j - 1] = [2; j + 1][3; n - 1]^{1-j}[2; l][3; n - 1]^{j-1}$, we may write

$$\Sigma_j = [2; j + 1]\Big(\sum_{l=2}^{n-j} [2; l]T_{[2;l]}\Big)^{(1,n)[3;n-1]^{j-1}} = [2; j + 1]V^{(1,n)[3;n-1]^{j-1}}_{(n-j+1)}.$$

Finally, it is sufficient to insert this value to (5.82) and replace the index j with $k = j - 1$ in the product. □

5.6 Second Components

For a given subset $Y \subseteq \{x_1, x_2, \ldots, x_n\}$, let $W(Y)$ be the set of elements of the form $\{W\}$, where W is a word in P_{ij}, $x_i, x_j \in Y$ of the length $|Y|(|Y|-1)/2$ that does not contain neither double nor conjugated letters. The set Y is conforming if and only if $W(Y) \subseteq \ker(\varphi)$. Let Σ be a multiplicative set generated by $W(Y)$, where Y runs through subsets that do not contain x_n. The set Σ is invariant with respect to the action of the subgroup $S_n^{1,n}$. Therefore, the group $S_n^{1,n}$ acts on the localization $\Sigma^{-1}\mathbf{F}[\mathscr{F}_n]$, and we may define the skew group ring $\Sigma^{-1}\mathbf{F}[\mathscr{F}_n] * S_n^{1,n}$ that are contained in the crossed product.

Lemma 5.15 *The elements* $V_{(m)}$, $3 \le m \le n$ *defined by* (5.26) *are invertible in* $\Sigma^{-1}\mathbf{F}[\mathscr{F}_n] * S_n^{1,n}$.

Proof Let us use induction on n. If $n > 2$, then the set Σ contains $\{P_{12}\}$. Therefore $V_{(3)}^{-1} = \{P_{12}\}^{-1} \cdot \mathrm{id} \in \Sigma^{-1}\mathbf{F}[\mathscr{F}_n] * S_n^{1,n}$.

Assume that $V_{(k)}$, $3 \le k \le m < n$ are invertible in $\Sigma^{-1}\mathbf{F}[\mathscr{F}_n] * S_n^{1,n}$. Consider the set of quantum variables $x_1, \ldots, x_m$. The element V_2 for this set coincides with $V_{(m+1)}$. To apply results of previous sections to $x_1, \ldots, x_m$, we replace notations $n \leftarrow N$ and $m \leftarrow n$. So that now $V_2 = \sum_{l=2}^{n}[2;l]T_{[2;l]}$, whereas $V_{(3)}, \ldots, V_{(n)}$ are invertible in $\Sigma^{-1}\mathscr{P}_N * S_N^{1,N}$. By induction we define a sequence $A_2, \ldots, A_n \in \mathscr{P}_N\Sigma^{-1} * S_N^{1,N}$. Let

$$A_2 = (-1)^n \prod_{k=0}^{n-3}[n-1;2]V_{(n-k)}$$

$$\times \left([n-1;2]\prod_{k=0}^{n-3}[2;2+k]V_{(n-k)}^{(1,n)[3;n-1]^k} \cdot \{X\}^{[2;n-1]}\right)^{-1}, \tag{5.83}$$

where X is given in (5.58). All of the factors in the parentheses are invertible in $\Sigma^{-1}\mathbf{F}[\mathscr{F}_n] * S_n^{1,n}$: the V's are by the inductive hypothesis, and $\{X\}$ is because it belongs to Σ. Furthermore, let

$$A_k = -A_{k-1}D_{(k)}V_{(k)}^{-1}, \quad 3 \le k \le n,$$

where $D_{(k)}$, $3 \le k \le n$ are defined in (5.27). Right multiplication of the above equality by $V_{(k)}$ demonstrates that $A_2, \ldots, A_n$ generate a right decreasing submodule over $\mathbf{F}[\mathscr{F}_n] * S_n^{1,n}$. By Theorem 5.4 we have relation (5.59). This relation with (5.83) imply

$$D_2 \prod_{k=0}^{n-3}[n-1;2]V_{(n-k)} = \prod_{k=0}^{n-3}[n-1;2]V_{(n-k)}.$$

Since by inductive hypothesis all factors $V_{(n-k)}$ are invertible, the element D_2 defined in (5.25) with A in place of $\mathscr{A}$ equals the identity permutation. Consider the element

$$B = A_2(2,n) + A_3(3,n) + \ldots + A_n \,\mathrm{id}.$$

Formula (5.28) and the definition of the A_k's imply

$$B \cdot V_2 = D_2(2,n) + \sum_{k=3}^{n} (A_k V_{(k)} + A_{k-1} D_{(k)})(k,n) = (2,n).$$

Therefore $V_2^{-1} = (2,n)B \in \Sigma^{-1}\mathscr{P}_N * S_N^{1,N}$. □

Corollary 5.1 *If the intersection of all conforming subsets of $x_1, x_2, \ldots, x_n$ is not empty, then there exists not more than $(n-2)!$ linearly independent multilinear quantum Lie operations.*

Proof Without loss of generality we may assume that x_n belongs to all conforming subsets. If a subset Y does not contain x_n, then this set is not conforming; that is, $\varphi(\{W\}) \neq 0$ for each word W in P_{ij}, $x_i, x_j \in Y$ of the length $|Y|(|Y|-1)/2$ that does not contain neither double nor conjugated letters. As φ is a homomorphism into a field, the multiplicative set Σ generated by all $\{W\}$'s has no intersection with ker φ. It remains to apply Proposition 5.1. □

5.7 Components with $s \geq 3$

Theorem 5.5 *If $s \geq 3$ then*

$$V_s \cdot [n;2]^{s-1} \prod_{t=s}^{n-1} (V_{(n-t+2)}[n;2]) \in V_2 \cdot \mathbf{F}[\mathscr{F}_n] * S_n^1 + IS_n^1,$$

where I is the conforming ideal, see Definition 5.3.

Define a sequence H_t, $s \leq t \leq n$, as $H_s = V_s[n;2]^{s-1}$, $H_{t+1} = H_t V_{(n-t+2)}[n;2]$. We have to prove that $H_n \equiv 0$ by modulo a right ideal $\mathscr{V}$ generated by V_2 and I.

Lemma 5.16 *The element H_t has the following representation by modulo $\mathscr{V}$*

$$H_t \equiv \sum_{\mathbf{L} \in \mathscr{U}(s,t)} \lfloor l_2 \rfloor \lfloor l_3 \rfloor \cdots \lfloor l_t \rfloor R(\mathbf{L}). \tag{5.84}$$

Therewith the coefficients $R(\mathbf{L})$ satisfy the following recurrence relations

$$R(l_2, \ldots, l_s) = \{\prod_{r=0}^{s-2} (P_{1\,n-r} P_{2\,n-r} \cdots P_{n-(l_s-r+1)\,n-r} \cdot P_{n-s+2\,n-r} \cdots P_{n-r-1\,n-r})\},$$

where for $r = s-2$ *the second factor in the parentheses is absent;*

$$R(l_2,\ldots,l_t) = R^{\lfloor l_t \rfloor}(l_2,\ldots,l_{t-1})T^{[n;2]}_{[2;n-l_t]} - R^{[n;2]}(w_2,\ldots,w_{t-1})T^{\lfloor w_2 \rfloor \cdots \lfloor w_{t-1} \rfloor [n;2]}_{[2;n-(w_1+1)]}, \tag{5.85}$$

where $(w_1, w_2, \ldots, w_{t-1})$ *is a subordinate sequence for* $(l_2,\ldots,l_t)$.

Proof Consider the case $t = s$. By (5.15) we have $V_s = \sum_{\pi \in N^1(s)} \pi T_{\pi,s}$, and by (5.7), $\pi = [2;k_2][3;k_3]\cdots[s;k_s]$ with $2 \le k_2 < k_3 < \ldots < k_s \le n$. Evidently this chain of inequalities is equivalent to the following:

$$2 \le k_2 \le k_3 - 1 \le k_4 - 2 \le \ldots \le k_s - s + 2 \le n - s + 2.$$

If we define

$$l_i = n - k_i + i - 2, \quad 2 \le i \le s, \tag{5.86}$$

then the above chain is equivalent to $\mathbf{L} = (l_2,\ldots,l_s) \in \mathscr{U}(s,s)$, see (5.38).

Note that $\pi[n;2]^{s-1} = \lfloor l_2 \rfloor \cdots \lfloor l_s \rfloor$ because $\lfloor l_i \rfloor = [2;n-l_i][n;2]$ and

$$[n;2]^{i-2}[2;n-l_i] = [i;n-l_i+i-2][n;2]^{i-2} = [i;k_i][n;2]^{i-2},$$

see (5.21). Thus

$$V_s[n;2]^{s-1} = \sum_{\mathbf{L} \in \mathscr{U}(s,s)} \lfloor l_2 \rfloor \lfloor l_3 \rfloor \cdots \lfloor l_s \rfloor R(\mathbf{L}),$$

where $R(\mathbf{L}) = T^{[n;2]^{s-1}}_{\pi,s}$. Consider representation (5.16) of T_π with $m = s - r$.

$$\{\prod_{r=0}^{s-2}(P_{1\ s-r}\cdots P_{s-r-1\ s-r}\cdot P_{s+1\ s-r}\cdots P_{r+k_{s-r}\ s-r})\}^{[n;2]^{s-1}}$$

$$= \{\prod_{r=0}^{s-2}(P_{1\ n-r}P_{2\ n-r}\cdots P_{r+k_{s-r}-s+1\ n-r}\cdot P_{n-s+2\ n-r}\cdots P_{n-r-1\ n-r})\}.$$

By (5.86) we have $r + k_{s-r} - s + 1 = n - l_{s-r} - 1$ and the required recurrence relations for the coefficients $R(\mathbf{L})$ are proved.

Suppose that (5.84) is valid for a given t. Let $\mathscr{U}(0)$ be a subset of $\mathscr{U}(s,t)$ of all sequences $(w_2, w_3, \ldots, w_t)$ satisfying (5.39) without the condition on w_1. Then by (5.57) and $V_2 = \sum_{\gamma=0}^{n-2}[2; n-\gamma]T_{[2;n-\gamma]}$ we may write

$$H_{t+1} \equiv \sum_{\mathbf{L}\in\mathscr{U}(s,t)} \lfloor l_2 \rfloor \cdots \lfloor l_t \rfloor R(\mathbf{L}) \sum_{\sigma=t-1}^{n-2} [2; n-\sigma]T_{[2;n-\sigma]}[n; 2]$$

$$- \sum_{\gamma=0}^{n-2}[2; n-\gamma]T_{[2;n-\gamma]} \cdot \sum_{\mathbf{W}\in\mathscr{U}(0)} \lfloor w_2 \rfloor \cdots \lfloor w_t \rfloor R(\mathbf{W})[n; 2].$$

Note that the set of all sequences $(\mathbf{L}, \sigma)$ such that $\mathbf{L}$, σ occur in the former line equals $\mathscr{U}(s, t+1)$. In the same way the set of all sequences $(\gamma - 1, \mathbf{W})$ such that γ, $\mathbf{W}$ occur in the latter line equals $\mathscr{W}(s, t)$. Let us move all of the coefficients to the utterly right position and put $\sigma = l_{t+1}, \gamma = w_1 + 1$. Then (5.85) for $t+1$ arises from Lemma 5.6, equality (5.53), and Lemma 5.4. □

Definition 5.10 For a sequence $\mathbf{L} = (l_2, \ldots, l_t)$, define

$$F(l_2, l_3, \ldots, l_k) = \{\ldots, P_{n-k+m\ n}, \ldots\}, \quad s < k \leq t,$$

where m runs through the set of all indices such that at the point m there is a jump during a motion of l_k to the left (in particular, $m \geq 2$). By Lemma 5.8 we may claim that if $\mathbf{W} = (w_1, \ldots, w_{t-1})$ is a subordinate sequence for $\mathbf{L}$, then for all k, $s < k \leq t-1$, the following equalities hold:

$$F(w_2, w_3, \ldots, w_k) = F(l_2, l_3, \ldots, l_k). \tag{5.87}$$

Furthermore, for $s \leq t \leq i \leq n$, define

$$D(l_2, \ldots, l_t) = \{ \prod_{r=0}^{t-s-1} (P_{1\ n-r} \cdots P_{n-\gamma_{t-r}^{(t)}(\mathbf{L})\ n-r} \cdot P_{n-r+1\ n-r} \cdots P_{n\ n-r}) \tag{5.88}$$

$$\times \prod_{r=t-s}^{t-2} (P_{1\ n-r} \cdots P_{n-\gamma_{t-r}^{(t)}(\mathbf{L})\ n-r} \cdot P_{n-t+2\ n-r} \cdots P_{n-r-1\ n-r}$$

$$\times \sqcup P_{s+n-t+1\ n-r} \cdots P_{n\ n-r})\}.$$

In another words, $P_{xy}, x \neq y$ occurs in $D(l_2, \ldots, l_t)$ if and only if

$$\begin{aligned} &\textbf{either } n-t+2 \leq y \leq n-t+s \text{ and} \\ &(1 \leq x \leq n-\gamma_{y+t-n}^{(t)}(\mathbf{L}) \text{ or } n-t+2 \leq x < y \text{ or } n-t+s < x \leq n), \\ &\textbf{or } n-t+s < y \leq n \text{ and } (1 \leq x \leq n-\gamma_{y+t-n}^{(t)}(\mathbf{L}) \text{ or } y < x \leq n). \end{aligned} \tag{5.89}$$

A word corresponding to a value $n-r$ of the second index in (5.88) is called *r-th row* of the element $D(l_2\dots,l_t)$. Respectively, a word in all letters of (5.88) with the second index equal to n is called the *last column*.

Because $\gamma_i^{(i)}=l_i+1>l_j$ provided that $i<j\le s$, we have $\gamma_i^{(s)}=l_i+1$ and

$$D(l_2,\dots l_s)=R(l_2,\dots,l_s). \tag{5.90}$$

Lemma 5.17 *Let* $\mathbf{L}=(l_2,\dots,l_t)\in\mathscr{U}(s,t)$, $t>s>2$, *and* $\mathbf{W}=\mathbf{L}'$. *Then for each* k, $s<k<t$, *the following equality holds*

$$F^{[n;2]^{t-k-1}\lfloor l_t\rfloor}(l_2,\dots,l_k)=F^{[n;2]^{t-k}}(w_2,\dots,w_k).$$

Proof Let $P_{n-k+m\ n}$ occurs in $F(l_2,\dots,l_k)$. Then $P_{n-k+m\ n}^{[n;2]^{t-k-1}}=P_{n-t+m+1\ n-t+k+1}$. Wherewith $l_t\ge t-2>t-m-1$, and so $n-l_t<n-t+m+1\le n-t+k+1$. Thus with the additional application of $\lfloor l_t\rfloor$ as well as the application of $[n;2]$, all the indices are decremented by one, that is,

$$F^{[n;2]^{t-k-1}\lfloor l_t\rfloor}(l_2,\dots,l_k)=F^{[n;2]^{t-k}}(l_2,\dots l_k),$$

and we may use (5.87). □

Lemma 5.18 *The coefficients of* (5.84) *have the following representation*

$$R(l_2,\dots,l_t)=(-1)^{t-s}D(l_2,\dots,l_t)\prod_{k=s+1}^{t}F^{[n;2]^{t-k}}(l_2,\dots,l_k).$$

Proof Let us use induction on t. If $t=s$, then the required equality turns into (5.90). Assume that the required equality is valid for $t-1$. Then by (5.85) we have

$$\begin{aligned}R(l_2,\dots,l_t)&=(-1)^{t-1-s}D^{\lfloor l_t\rfloor}(l_2,\dots,l_{t-1})T_{[2;n-l_t]}^{[n;2]}\prod_{k=s+1}^{t-1}F^{[n;2]^{t-k-1}\lfloor l_t\rfloor}(l_2,\dots,l_k)\\&\quad-(-1)^{t-1-s}D^{[n;2]}(w_2,\dots,w_{t-1})T_{[2;n-(w_1+1)]}^{\lfloor w_2\rfloor\cdots\lfloor w_{t-1}\rfloor[n;2]}\\&\quad\times\prod_{k=s+1}^{t-1}F^{[n;2]^{t-k}}(w_2,\dots,w_k).\end{aligned}$$

Using Lemma 5.17 and (5.87), we may factor out

$$(-1)^{t-s}\prod_{k=s+1}^{t-1}F^{[n;2]^{t-k}}(l_2,\dots l_k).$$

Therefore it suffices to prove that

$$D^{[n;2]}(w_2,\dots,w_{t-1})T^{\lfloor w_2\rfloor\cdots\lfloor w_{t-1}\rfloor[n;2]}_{[2;n-(w_1+1)]}$$
$$-D^{\lfloor l_t\rfloor}(l_2,\dots,l_{t-1})T^{[n;2]}_{[2;n-l_t]}=D(l_2,\dots,l_t)F(l_2,\dots,l_t).$$

This equality will arise from (5.12) with

$$\{C\}=T^{[n;2]}_{[2;n-l_t]},\quad \{D\}=D^{\lfloor l_t\rfloor}(l_2,\dots,l_{t-1}),\quad \{E\}=F(l_2,\dots,l_t)$$

if we prove the following three equalities:

$$D^{[n;2]}(w_2,\dots,w_{t-1})=D^{\lfloor l_t\rfloor}(l_2,\dots,l_{t-1})\star\overline{F}(l_2,\dots,l_t), \tag{5.91}$$

$$T^{\lfloor w_2\rfloor\cdots\lfloor w_{t-1}\rfloor[n;2]}_{[2;n-(w_1+1)]}=T^{[n;2]}_{[2;n-l_t]}\star F(l_2,\dots,l_t), \tag{5.92}$$

$$D(l_2,\dots,l_t)=D^{[n;2]}(w_2,\dots,w_{t-1})\star T^{[n;2]}_{[2;n-l_t]}. \tag{5.93}$$

Let us start with the first one. As $l_t \geq t-2$, we have $n-l_t < n-t+3$. Therefore $\lfloor l_t\rfloor$ decreases by one all of the second indices of $D(l_2,\dots,l_{t-1})$. So $\lfloor l_t\rfloor$ shifts down by one step all of the rows in (5.88) with $t-1$ in place of t.

Every first index of a letter located after a central point in the explicit form (5.88) of $D(l_2,\dots,l_{t-1})$ is greater than or equal to $n-(t-1)+2=n-t+3$. Therefore $\lfloor l_t\rfloor$ also decreases these indices by one. Hence all letters located in the rows after $\cdot$ are shifted to the left by one step.

Consider letters that are located in the rows of the explicit form before the central points. If $w_{t-r-1}=l_{t-r-1}-1$, then by Lemma 5.5 at the point t there is no jump during a motion of l_{t-r-1} to the right, that is, $n-\gamma^{(t-1)}_{t-1-r}(\mathbf{L})<n-l_t$. Thus with application of $\lfloor l_t\rfloor$ the first indices of rth row located before $\cdot$ remain unchanged. In this case the last position of $(r+1)$th row of $D^{\lfloor l_t\rfloor}(l_2,\dots,l_{t-1})$ became vacant, whereas the length of a part located before $\cdot$ equals $n-\gamma^{(t-1)}_{t-r-1}(\mathbf{L})=n-\gamma^{(t)}_{t-r-1}(\mathbf{L})$, see (5.31).

If $w_{t-r-1}=l_{t-r-1}$, then $\gamma^{(t-1)}_{t-r-1}(\mathbf{L})\leq l_t$. Hence the letter $P_{n-l_t\,n-r}$ is shifted by $\lfloor l_t\rfloor$ to $P_{n\,n-r-1}$. It takes the last position of the $(r+1)$th row. The next letter $P_{n-l_t+1\,n-r}$ is shifted to the position of previous one, and so on. Thus the length of a part located before $\cdot$ is decremented by one, and it can be set by the same formula $n-\gamma^{(t-1)}_{t-r-1}(\mathbf{L})-1=n-\gamma^{(t)}_{t-r-1}(\mathbf{L})$, see (5.31).

Furthermore, by Definition 5.10, we have

$$\overline{F}(l_2,\dots,l_t)=\{\dots,P_{n\,n-t+m},\dots\},$$

where $w_m=l_m-1$. Replacing the index m with $t-r-1$ in this definition, we see that the letters of $\overline{F}(l_2,\dots,l_t)$ occupy positions only in the last column and exactly the positions which are vacant in $D^{\lfloor l_t\rfloor}(l_2,\dots,l_{t-1})$.

Thus the zero row of $D^{\lfloor l_t \rfloor}(l_2, \ldots, l_{t-1}) \star \overline{F}(l_2, \ldots, l_t)$ is vacant, whereas the rth, $r > 0$, one is equal to the rth row of (5.88).

Consider the left hand side of (5.91). The permutation $[n; 2]$ shifts to the left by one step all of the gaps with central points of $D(w_2, \ldots, w_{t-1})$ and shifts down by one step all of the rows (because $n - ((t-1) - 2) > 2$). Therefore the zero row of $D^{[n;2]}(w_2, \ldots, w_{t-1})$ is vacant while the rth one, $r > 0$, equals

$$P_{1\, n-r} P_{2\, n-r} \cdots P_{n-\gamma_{t-r}^{(t-1)}(\mathbf{W})-1\, n-r} \cdot \Delta_r,$$

where Δ_r denotes a part of the rth row of (5.88) located after the central point. By (5.49) the equality (5.91) is proved.

By the above consideration all rows (but the zero one) of the explicit representation of $D(l_2, \ldots, l_t)$ coincide with the same rows of $D^{[n;2]}(w_2, \ldots, w_{t-1})$, whereas $T^{[n;2]}_{[2;n-l_t]} = \{P_{1\,n} P_{2\,n} \cdots P_{n-l_t-1\,n}\}$ coincides with the zero row of $D(l_2, \ldots, l_t)$, because $\gamma_t^{(t)}(\mathbf{L}) = l_t + 1$. Therefore (5.93) is valid.

Consider (5.92). Its right-hand side has the only row (a zero one)

$$\{P_{1\,n} P_{2\,n} \cdots P_{n-l_t-1\,n} \quad \cdot \quad \cdots P_{n-t+m\,n} \cdots\},$$

where m runs through all indices such that $l_m = w_m + 1$.

Let us prove the following formula

$$T^{\lfloor w_2 \rfloor \cdots \lfloor w_k \rfloor}_{[2;n-(w_1+1)]} = \{P_{12} P_{32} \cdots P_{n-\gamma_1^{(k)}(\mathbf{W})\,2} \quad \cdot \quad \cdots P_{n-k+m\,2} \cdots\}, \tag{5.94}$$

where m runs through all indices less than or equal to k, such that $l_m = w_m + 1$. For $k = 1$ the formula reduces to

$$T_{[2;n-(w_1+1)]} = \{P_{12} P_{32} \cdots P_{n-\gamma_1^{(1)}(\mathbf{W})\,2}\},$$

which is valid by Definition 5.4. We may start induction on k.

If $l_{k+1} = w_{k+1}$, then because of definition (5.33) and Lemma 5.4 we have $n - \gamma_1^{(k)}(\mathbf{W}) < n - w_{k+1}$. Therefore $\lfloor w_{k+1} \rfloor$ does not shift the letters of (5.94) located before the central points. By (5.39) the inequality $w_{k+1} > k-2$ is valid (for $k < s$ still $w_{k+1} \geq s-2 > k-2$), so $n-k+m > n-w_{k+1}$ (as $m \geq 2$). Hence $\lfloor w_{k+1} \rfloor$ shifts to the left by one step all of the letters located after the central point. Therefore in (5.94), it is possible to replace k with $k+1$, because in this case $\gamma_1^{(k+1)}(\mathbf{W}) = \gamma_1^{(k)}(\mathbf{W})$.

If $l_{k+1} = w_{k+1} + 1$, then by (5.33) we have $n - \gamma_1^{(k)}(\mathbf{W}) \geq n - w_{k+1}$. Therefore $\lfloor w_{k+1} \rfloor$ shifts the letter $P_{n-w_{k+1}\,2}$ to a place $P_{n\,2}$ at the last column, whereas it shifts the next letter $P_{n-w_{k+1}+1\,2}$ to $P_{n-w_{k+1}\,2}$ and so on. Thus the segment located before the central point is decremented by one $n - \gamma_1^{(k)}(\mathbf{W}) - 1 = n - \gamma_1^{(k+1)}(\mathbf{W})$. As above, all letters located after the central point are shifted to the left by one step, wherewith at the end of the row there arise the letter $P_{n\,2}$, the first index of which equals $n - (k + 1) + m$ with $m = k + 1$. Thus in (5.94), it is possible to replace k with $k + 1$ in this case also.

If we apply $[n; 2]$ to (5.94) with $k = t - 1$ and use the fact that the last head $\gamma_1^{(t-1)}(\mathbf{W})$ equals l_t, see (5.34), then we obtain (5.92). □

Lemma 5.19 *If a sequence* $\mathbf{L} = (l_2, \ldots, l_n)$ *satisfies condition* (5.38) *with* $t = n$, *then* $\gamma_j^{(n)}(\mathbf{L}) = n - 1$ *for all* j, $2 \leq j \leq n$.

Proof Consider conditions (5.41) with $n + 1$ in place of n, and with $i = n - 1$. The sequence $\Lambda = (\lambda_1, \ldots, \lambda_{n-1})$ with $\lambda_k = l_{k+1} + 1$, $1 \leq k \leq n - 1$ satisfies these conditions: $k = ((k + 1) - 2) + 1 \leq l_{k+1} + 1 = \lambda_k \leq n - 1$ (for $k + 1 \leq s$ still $l_{k+1} \geq s - 2 \geq (k + 1) - 2$). Evidently, if all members of a sequence increase by one, then all heads will increase by one as well. After this, it remains to use Lemma 5.13. □

Now we accumulate sufficient information to prove Theorem 5.5.

Proof By Lemma 5.19 every word located at a row of (5.88) before the central point has just one letter. The first product of (5.88) has

$$\sum_{r=0}^{n-s-1} (r + 1) = (n - s)(n - s + 1)/2$$

letters, the second one has

$$\left(\sum_{r=n-s}^{n-2} (n - r - 1)\right) + (s - 1)(n - s) = s(s - 1)/2 + (s - 1)(n - s);$$

that is, the total number of letters in $D(l_2, \ldots, l_n)$ equals $n(n - 1)/2 = C_n^2$.

If a letter P_{xy} with $x < y$ occurs in $D(l_2, \ldots, l_n)$, then by the definition (5.89) either $x = 1$, or $2 \leq y \leq s$ and $2 \leq x < y$, that is, the letter P_{xy} appears in (5.88) before the gap $\sqcup$. Definition (5.88) shows that the second index never equals one. This definition also shows that if the first index is greater than the second one, then it is greater than s. Thus the letter P_{yx} with $2 \leq y \leq s$ and $2 \leq x < y$ does not occur in (5.88). Hence $D(l_2, \ldots, l_n)$ has no conjugated letters. By Definition 5.3 we have $D(l_2, \ldots, l_n) \in I$. By Lemmas 5.16 and 5.18 we are done. □

5.8 Existence Condition

Theorem 5.6 *There exists a nonzero multilinear quantum operation in a set of quantum variables* $x_1, \ldots, x_n$ *if and only if this set is conforming.*

Proof Let us use induction on n. For $n = 2, 3, 4$ this statement follows from Theorems 2.5, 4.5, and 4.6. The necessity that $x_1, \ldots, x_n$ conform to have a nonzero multilinear operation is proved in Corollary 4.1.

Let $x_1, \ldots, x_n$ be a conforming set of quantum variables. If this set have a proper conforming subset, say, $x_1, \ldots, x_m$, $2 \leq m < n$, then by inductive supposition there exists a nonzero multilinear quantum operation $W(x_1, \ldots, x_m)$. The operation W as an element of the free character Hopf algebra is a skew-primitive element with a character $\chi = \chi^{x_1}\chi^{x_2}\cdots\chi^{x_m}$ and a group-like element $g = g_{x_1}g_{x_2}\cdots g_{x_m}$; that is,

$$\Delta(W) = W \otimes 1 + g \otimes W, \quad h^{-1}Wh = \chi(h)W, \ h \in G.$$

Consider a new quantum variable z which is related to the character χ and to the group-like element g. For every nonzero multilinear quantum operation

$$W_1(z, x_{m+1}, x_{m+2} \ldots, x_n)$$

we can define a superposition

$$W_1(W(x_1, \ldots, x_m), x_{m+1}, \ldots, x_n)$$

that does not equal to zero as a polynomial. By the inductive hypothesis the operation W_1 exists if the set $z, x_{m+1}, \ldots, x_n$ is conforming; that is,

$$\prod_{i>m} \chi^z(g_{x_i}) \cdot \prod_{i>m} \chi^{x_i}(g_z) \cdot \prod_{m<i\neq j\leq n} \chi^{x_i}(g_{x_j}) = 1.$$

The left-hand side of this formula differs from the left-hand side of (5.1) only by the factor $\prod_{1\leq i\neq j\leq m} \chi^{x_i}(g_{x_j})$. Thus if both sets $x_1, \ldots, x_m$ and $x_1, \ldots, x_n$ are conforming, then $z, x_{m+1}, \ldots, x_n$ does as well.

Therefore it suffices to prove the existence of an operation under additional assumption that the given set of quantum variables has no proper conforming subsets. In the next theorem we will prove a more general statement. □

Theorem 5.7 *If each conforming subset of a conforming set* $x_1, \ldots, x_n$ *contains* x_n, *then the dimension of the space of all multilinear quantum Lie operations equals* $(n-2)!$, *during which there exists an operation* $[\![x_1, \ldots, x_n]\!]$ *such that a basis of the space consists of operations* $[\![x_1, x_2^{\mu}, \ldots, x_{n-1}^{\mu}, x_n]\!]$, *where* μ *runs through the symmetric group* $S_n^{1,n}$.

Proof By Theorem 5.15 the elements $V_{(k)}$, $3 \leq k \leq n$, defined by (5.26) are invertible in $\Sigma^{-1}\mathbf{F}[\mathscr{F}_n] * S_n^{1,n}$. This allows us for each permutation $\mu \in S_n^{1,n}$ to define an element $B^{[\mu]} \in \Sigma^{-1}\mathbf{F}[\mathscr{F}_n] * S_n^1$ by the following formula:

$$B^{[\mu]} = A_2(2, n) + A_3(3, n) + \cdots + A_n id,$$

where A_i are defined by induction

$$A_2 = \mu, \quad A_k = -A_{k-1}D_{(k)}V_{(k)}^{-1}, \ 3 \leq k \leq n,$$

or in the explicit form

$$B^{[\mu]} = \mu\left((2,n) + \sum_{k=3}^{n}(-1)^k\Big(\prod_{i=3}^{k} D_{(i)}V_{(i)}^{(-1)}\Big)(k,n)\right) = \sum_{\pi\in S_n^1} B_\pi \pi^{-1}. \tag{5.95}$$

Denote $\beta_\pi = \varphi(B_\pi)$, where φ is defined by (5.11) and (5.18). Let

$$[\![x_1, x_2, \ldots, x_n]\!]_\mu \stackrel{\text{df}}{=} \sum_{\pi\in S_n^1} \beta_\pi[\ldots[[x_1,x_2],x_3],\ldots x_n]^\pi,$$

Let us show that this is a quantum Lie operation.

By Theorem 5.3 we have to prove that $B^{[\mu]} \cdot V_s \equiv 0$, $2 \le s \le n$. The element $B^{[\mu]}$ has the form (5.20) with $\mathscr{A}_i = A_i$. Therefore formula (5.28) holds. According to the definition of the elements A_k, this formula implies

$$B^{[\mu]} \cdot V_2 = D_2(2,n),$$

where D_2 is defined by (5.25) with A in place of $\mathscr{A}$. The elements $A_2, \ldots, A_n$ generate a right decreasing module over $\mathbf{F}[\mathscr{F}_n] * S_n^{1,n}$. Therefore we may apply Theorem 5.4. Because $\{X\}$ belongs to the conforming ideal, the product on the right-hand side of (5.59) belongs to $\ker(\varphi)S_n^1$, see (5.58) and Definition 5.3. A right multiplication of (5.59) by $(\prod_{k=0}^{n-3}[n-1;2]V_{(n-k)})^{-1}$ shows that $D_2 \equiv 0$, that is, relation (5.19) with $s = 2$ holds.

If $s \ge 3$, then by Theorem 5.5 there exists a representation

$$V_s \cdot [n;2]^{s-1}\prod_{t=s}^{n-1}(V_{(n-t+2)}[n;2]) = V_2 \cdot E + F, \tag{5.96}$$

where $E \in \mathbf{F}[\mathscr{F}_n] * S_n^1$ and $F \in IS_n^1$. Let us multiply (5.96) from the left by $B^{[\mu]}$. Using $B^{[\mu]} \cdot V_2 \equiv 0$ and $B^{[\mu]} \cdot F \in \Sigma^{-1}IS_n^1 \subseteq \Sigma^{-1}\ker(\varphi)S_n^1$, we obtain

$$(B^{[\mu]}V_s) \cdot [n;2]^{s-1}\prod_{t=s}^{n-1}(V_{(n-t+2)}[n;2]) \equiv 0.$$

The second factor of the left-hand side of this equality is invertible. Thus $B^{[\mu]} \cdot V_s \equiv 0$, and by Theorem 5.3 the polynomial $[\![x_1, x_2, \ldots, x_n]\!]_\mu$ is a quantum Lie operation.

Formula (5.95) demonstrates that for $\pi \in S_n^{1,n}(2,n)$, only one of the B_π's does not equal zero, that is, $B_{\mu^{-1}} = 1$. As $x_1x_{\pi(2)}\cdots x_{\pi(n)}$ is the only monomial of $[\ldots[[x_1,x_2],x_3]\ldots,x_n]^\pi$ starting with x_1, we obtain that just one monomial of the type $x_1x_n\cdots$ of $[\![x_1, x_2, \ldots, x_n]\!]_{\mu^{-1}}$ has a nonzero coefficient, that is, $x_1x_nx_3^\mu\cdots x_{n-1}^\mu x_2^\mu$, and this coefficient equals one. In particular, the polynomials $[\![x_1, \ldots, x_n]\!]_\mu$, $\mu \in S_n^{1,n}$ are linearly independent.

Corollary 5.1 states that the dimension of the space of multilinear operations is less than or equal to $(n-2)!$. We have found $(n-2)!$ linearly independent operations. This implies that the operations $[\![x_1, x_2, \ldots, x_n]\!]_\mu$, $\mu \in S_n^{1,n}$ span the space.

Furthermore, for any permutation $\mu \in S_n^{1,n}$, consider a new set of quantum variables $y_1 = x_1$, $y_2 = x_2^\mu$, ..., $y_{n-1} = x_{n-1}^\mu$, $y_n = x_n$. This set is conforming because the left-hand side of (5.1) for the y's differs from the left-hand side for the x's by the order of factors only. Thus the following operation is defined:

$$[\![y_1, y_2, \ldots, y_{n-1}, y_n]\!]_{\text{id}} \stackrel{\text{df}}{=} [\![x_1, x_2^\mu, \ldots, x_{n-1}^\mu, x_n]\!]_{\text{id}}.$$

As the coefficient at $y_1 y_n y_{\nu(3)} \cdots y_{\nu(n-1)} y_{\nu(2)} = x_1 x_n x_3^{\nu\mu} \cdots x_{n-1}^{\nu\mu} x_2^{\nu\mu}$, $\nu \in S_n^{1,n}$ does not equal zero in the only case, $\nu = id$, we have

$$[\![x_1, x_2^\mu, \ldots, x_{n-1}^\mu, x_n]\!]_{\text{id}} = [\![x_1, x_2, \ldots, x_n]\!]_{\mu^{-1}}.$$

Thus the operation $[\![x_1, x_2, \ldots, x_n]\!]_{\text{id}}$ satisfies all properties stated in the theorem.

□

5.9 Interval of Dimensions

Recall that a *conforming ideal* is an ideal I of the algebra $\mathbf{F}[\mathscr{F}_n]$ generated by all elements of the form $\{W\}$, where W is an arbitrary semigroup word in P_{ij} of length $n(n-1)/2$ that has neither double nor conjugated letters. The variables $x_1, \ldots, x_n$ are conforming if and only if the ideal $\ker(\varphi)$ contains I. The conforming ideal is invariant with respect to the action of S_n. In particular, the two-sided ideal of $\mathbf{F}[\mathscr{F}_n] * S_n$ generated by I coincides with the right ideal IS_n.

Consider a field of rational functions $\mathbf{K}$ over $\mathbf{F}$ in $(n(n-1)/2) - 1$ variables t_{is}, $1 \leq i \neq s \leq n$, $(i, s) \neq (1, n)$, and put

$$t_{1n} = (\prod_{(k,s)\neq(1,n)} t_{ks})^{-1}.$$

Then the kernel of the homomorphism $\xi : P_{ij} \to t_{ij}$ coincides with the conforming ideal I, and ξ defines an embedding of $\mathbf{F}[\mathscr{F}_n]/I$ in $\mathbf{K}$.

Consider a new set of quantum variables $X_1, \ldots, X_n$ with which free generators $G_1, \ldots, G_n$ of a free Abelian group are associated, whereas the characters over $\mathbf{K}$ are defined by $\chi^{X_i}(G_j) = t_{ij}$.

Definition 5.11 System (5.8) with t_{ij} in place of p_{ij} is said to be the *generic system*. Its solutions define *generic* quantum operations in $X_1, X_2, \ldots, X_n$ with coefficients from $\mathbf{K}$.

The parameters t_{ij} of the generic system are connected by relations that include all parameters t_{ij}. Therefore the set of generic variables $X_1, \ldots, X_n$ has no proper conforming subsets. By Theorem 5.7 there exists precisely $(n-2)!$ linearly independent generic operations.

All coefficients of the generic system belong to a subalgebra $\mathbf{F}[t_{ij}]$ over the minimal subfield $\mathbf{F}$ generated by t_{ij}, $1 \leq i, j \leq n$. Hence there exists a fundamental system of solutions in the field $\mathbf{K} = \mathbf{F}(t_{ij})$. Moreover, multiplying solutions by suitable elements from $\mathbf{F}[t_{ij}]$, we can find a fundamental system of solutions that belong to $\mathbf{F}[t_{ij}]$. If γ_π, $\pi \in S_n^1$ is a solution with $\gamma_\pi \in \mathbf{F}[t_{ij}]$, then $\beta_\pi = \varphi'(\gamma_\pi)$ are solutions of the basic system (5.8), where $\varphi' = \varphi \circ \xi^{-1}$ is a natural homomorphism from $\mathbf{F}[t_{ij}]$ to $\mathbf{k}$. In this way, the generic operations define operations with arbitrary values of p_{ij}. Nevertheless, the homomorphism φ' not necessary (almost never) has an extension up to a homomorphism of $\mathbf{K} = \mathbf{F}(t_{ij})$. By this reason, there may exist operations that do not appear from the generic ones in the above manner.

Theorem 5.8 *If* $x_1, \ldots, x_n$ *is a conforming set of quantum variables, then the dimension of the space of all multilinear quantum Lie operations in this set is greater then or equal to* $(n-2)!$ *and less than or equal to* $(n-1)!$

Proof To prove the first part of the theorem, it suffices to demonstrate that the rank of the basic system is less then or equal to

$$(n-1)! - (n-2)! = (n-2)!(n-2).$$

This is equivalent to the condition that all minors of the order greater then or equal to $(n-2)!(n-2)$ are zero. Since the minors are integer functions in the matrix coefficients, it suffices to show that this condition is valid for the generic system. By Theorem 5.7 the generic system has precisely $(n-2)!$ solutions. Hence all the minors are zero in $\mathbf{F}[t_{ij}]$. Applying the homomorphism $\varphi' : t_{ij} \mapsto p_{ij}$, we obtain that the minors are zero in $\mathbf{k}$ as well.

The basic system has not more than $(n-1)!$ linearly independent solutions because it has only $(n-1)!$ unknowns. □

Of course, if the dimension is grater than $(n-2)!$, then there must be operations that are not reduced to the generic ones. Nevertheless, it looks likely that if we include superpositions of operations in lesser number of variables, than it would be possible to construct all the operations from the generic ones. Moreover, the detail analysis of the case $n = 4$ given in Theorem 4.6 provides a hypothesis that in this case all the operations are linear combinations of superpositions.

Conjecture 5.1 If the dimension of the space of multilinear operations in conforming set of quantum variables $x_1, \ldots, x_n$ is not $(n-2)!$, then all multilinear operations are linear combinations of superpositions of operations with lesser n.

5.10 Symmetric Operations

A quantum Lie operation $[\![x_1, \ldots, x_n]\!]$ is called *symmetric* (or *skew symmetric*) if for every permutation $\pi \in S_n$ the following equality is valid

$$[\![x_{\pi(1)}, \ldots, x_{\pi(n)}]\!] = \alpha_\pi [\![x_1, \ldots, x_n]\!], \tag{5.97}$$

where $\alpha_\pi \in \mathbf{k}$. In the case of quantum operations, as well as in the case of arbitrary partial operations, we have to explain what does it mean the left hand side of the above equality. Strictly speaking, the left hand side is defined only if $x_{\pi(i)}$ has the same parameters χ, g as x_i does. By definition only in this case the substitution $x_i \leftarrow x_{\pi(i)}$ is admissible. In other word, all parameters p_{it} should be equal each other. This is very rigid condition. It excludes both the color super-brackets and the above defined generic operations.

However, we may suppose that $[\![x_1, \ldots, x_n]\!]$ is a polynomial whose coefficients depend on the quantization parameters, χ^{x_i}, g_{x_i}; that is, there are shown distinguished entries of p_{it} in the coefficients. Then a substitution $x_i \leftarrow y$ means not only the substitution of the variable but also one of the parameters $g_i \leftarrow g_y,\ \chi^{x_i} \leftarrow \chi^y$. In particular, the permutation of variables means the application of this permutation to all indices: $p_{it} \leftarrow p_{\pi(i)\,\pi(t)}$.

This interpretation of symmetry is not contradictory only if the application of the permutation is independent of the way how the coefficients of $[\![x_1, \ldots, x_n]\!]$ are represented as rational functions in p_{it}, p_{it}^{-1}. The action of permutations is independent of the above representation if (and only if) ker (φ) is invariant ideal with respect to the action of S_n.

Definition 5.12 A collection of quantum variables $x_1, \ldots, x_n$ is said to be *symmetric* if ker (φ) is an invariant ideal with respect to S_n, or, equivalently, the action of S_n on the algebra $\mathbf{F}[p_{it}]$ given by $p_{it}^\pi = p_{\pi(i)\pi(t)}$ is well-defined.

The symmetry of a collection has nothing to do with the symmetry of the matrix $||p_{it}||$, while it means the symmetry of relations between the parameters p_{it}.

Thus, to impart a sense to the term "symmetric operation", we should, first, suppose that the coefficients of the operation belong to the field $\mathbf{F}(p_{it})$, which does not affect the generality; and, then, we should consider only symmetric sets of quantum variables. Yet, this is a bounding condition. Nevertheless, this condition excludes no one of the above examples. Moreover, the existence of the symmetric set X_i with generic parameters t_{ij} is a key argument of the proof of both the existence theorem and its corollaries. Therefore, the symmetric collections of variables are of a special interest.

Consider a symmetric polynomial over $\mathbf{F}(p_{it})$:

$$\mathbf{f}(x_1, \ldots, x_n) = \sum \gamma_\mu x_{\mu(1)} \cdots x_{\mu(n)}.$$

Without loss of generality (if necessary by applying a permutation), we may assume that the monomial $x_1x_2\cdots x_n$ has a coefficient 1. Let us compare coefficients at $x_{\pi(1)}x_{\pi(2)}\cdots x_{\pi(n)}$ in the both sides of (5.97). We have $\alpha_\pi = \gamma^\pi_{\pi^{-1}}$. Afterwards the equality (5.97) takes a form

$$\gamma^\pi_{\pi^{-1}}\sum_{\mu\in S_n}\gamma_\mu x_{\mu(1)}\cdots x_{\mu(n)} = (\sum_{\mu\in S_n}\gamma_\mu x_{\mu(1)}\cdots x_{\mu(n)})^\pi$$

$$= \sum_{\mu\in S_n}\gamma^\pi_\mu x_{\pi(\mu(1))}\cdots x_{\pi(\mu(n))} = \sum_{\nu\in S_n}\gamma^\pi_{\nu\pi^{-1}}x_{\nu(1)}\cdots x_{\nu(n)}.$$

This implies $\gamma^\pi_{\mu\pi^{-1}} = \gamma^\pi_{\pi^{-1}}\gamma_\mu$. Let us replace $\nu = \pi^{-1}$ and then apply ν to both sides of the latter equality. We see that the polynomial **f** is symmetric if and only if

$$\gamma_{\mu\nu} = \gamma^\nu_\mu\gamma_\nu,\ \text{ with } \alpha_\pi = \gamma^\pi_{\pi^{-1}} = \gamma_\pi^{-1}. \tag{5.98}$$

In other words, the set of normed symmetric polynomials can be identified with the first cogomology group $H^1(S_n, \mathbf{F}(p_{it})^*)$ with values in the multiplicative group of $\mathbf{F}(p_{it})$.

Now a natural question arises: does there exist a basis of the space of multilinear quantum Lie operations consisting of the symmetric operations, provided that the variables form a symmetric set?

We start with some counterexamples. Firstly we consider the case when the variables are *absolutely symmetric*; that is, $p_{it} = q, 1 \leq i \neq t \leq n$.

Lemma 5.20 *If the set of variables is absolutely symmetric and $n > 3$, then the basis consisting of symmetric operations does not exist. If $n = 3$, then the required basis exists only if $q \neq \pm 1$. The bilinear operation is symmetric.*

Proof If the set of variables is absolutely symmetric, then the group S_n acts identically on the field $\mathbf{F}(p_{it})$. Therefore there exists only two symmetric polynomials up to a scalar factor:

$$S(x_1,\ldots,x_n) = \sum_{\pi\in S_n} x_{\pi(1)}\cdots x_{\pi(n)},$$

$$T(x_1,\ldots,x_n) = \sum_{\pi\in S_n} (-1)^\pi x_{\pi(1)}\cdots x_{\pi(n)}.$$

On the other hand, if the existence condition, $q^{n(n-1)} = 1$, holds, then by Theorem 5.8 the dimension of the space of multilinear operations can not be less than $(n-2)!$.

Thus if $n > 4$, or if $n = 4$ and the characteristic of the ground field equals 2, then wittingly the basis consisting of symmetric operations does not exist.

If $n = 4$, then we may use the analysis from the proof of Theorem 4.6: the dimension is 2 only if $q^{12} = 1$, $q^6 \neq 1$, $q^4 \neq 1$, or, equivalently, $q^6 = -1$,

$q^2 \neq -1$. If under these conditions the polynomials $S,\ T$ are quantum operations, then they should be expressed trough the quadrilinear operation given in (4.50) with β_ν defined in (4.49). The coefficients of that expression are equal to the coefficients at $x_1x_2x_3x_4$ and $x_1x_3x_2x_4$ of S and T respectively:

$$S = [\![x_1, x_2, x_3, x_4]\!] + [\![x_1, x_3, x_2, x_4]\!],$$
$$T = [\![x_1, x_2, x_3, x_4]\!] - [\![x_1, x_3, x_2, x_4]\!].$$

This implies that $2[\![x_1, x_2, x_3, x_4]\!] = S + T$. In particular all coefficients of $[\![x_1, x_2, x_3, x_4]\!]$ at monomials corresponding to odd permutations have to be zero. The explicit formula (4.49) demonstrates that the coefficient at $x_1x_2x_4x_3$ equals

$$-\frac{\{p_{13}p_{23}\}}{\{p_{13}p_{23}p_{43}\}} = -\frac{q^2 - q^{-2}}{q^3 - q^{-3}} \neq 0,$$

for $q^4 \neq 1$. Thus in this case the symmetric basis neither exists.

If $n = 3$, then the existence condition takes the form $q^6 = 1$. If $q \neq \pm 1$, then there exists only one trilinear operation up to a scalar multiplication, and this operation is symmetric, see Theorem 4.5 and formula (4.31). More precisely, if $q^3 = 1$, then (4.30) equals S, whereas if $q^3 = -1$, then it equals T.

If $q = \pm 1$, then the space of operations is generated by two polynomials: $[[x_1, x_2], x_3]$, and $[[x_1, x_3], x_2]$, while $S + T$ is not a linear combination of them. □

Lemma 5.21 *Let the quantization matrix of a symmetric quadruple of quantum variables has the form*

$$||p_{it}|| = \begin{pmatrix} * & p & q & s \\ p & * & s & q \\ q & s & * & p \\ s & q & p & * \end{pmatrix},$$

where p, q, s are pairwise different and $p^2q^2s^2 = 1$.

1. *If the characteristic of the field* **k** *is not equal to 2, then there do not exist nonzero quadrilinear symmetric quantum Lie operations at all.*
2. *If the characteristic is 2, then there exist not more then two linearly independent quadrilinear symmetric operations.*
3. *In both cases the dimension of the whole space of quadrilinear quantum Lie operations equals* 3.

Proof If the parameter matrix has the form given in the lemma, then the action of the group S_4 on the field $\mathbf{F}(p_{it})$ is not faithful. The kernel of this action includes the following four elements

$$\text{id};\ a = (12)(24);\ b = (13)(24);\ c = (14)(23).$$

These elements form a normal subgroup $H \lhd S_4$ isomorphic to $Z_2 \times Z_2$. Let

$$S = \sum_{\pi \in S_4} \gamma_\pi x_{\pi(1)} x_{\pi(2)} x_{\pi(3)} x_{\pi(4)}$$

be some symmetric quantum operation, $\gamma_{\mathrm{id}} = 1$. According to (5.98) with $h = \mu = \nu \in H$ we have $\gamma_h^2 = \gamma_{h^2} = \gamma_{\mathrm{id}} = 1$; that is, $\gamma_h = \pm 1 \in \mathbf{F}$. Moreover, all of the elements γ_h, $h \neq \mathrm{id}$, $h \in H$ may not be equal to -1 because, again by (5.98), the product of every two of them equals the third one. On the other hand, formula (5.98) with $h \in H$, $g \in S_4$ implies that $\gamma^g_{g^{-1}} \gamma_g = 1$ and

$$\gamma_{g^{-1}hg} = \gamma^{hg}_{g^{-1}} \gamma_{hg} = \gamma^g_{g^{-1}} \gamma^g_h \gamma_g = \gamma^g_h = \gamma_h.$$

Therefore all of γ_h, $h \in H$ equal each other and equal to 1.

Furthermore, the polynomial S, as well as any other quantum Lie operation, has a commutator representation:

$$S = \sum_{\nu \in S_4^1} \beta_\nu [[[x_1, x_{\nu(2)}], x_{\nu(3)}], x_{\nu(4)}].$$

If we compare coefficients at monomials $x_1x_2x_3x_4$ and $x_4x_3x_2x_1$, we obtain $1 = \gamma_{\mathrm{id}} = \beta_{\mathrm{id}}$ and $1 = \gamma_{(14)(23)} = \beta_{\mathrm{id}}(-p_{12})(-p_{13}p_{23})(-p_{14}p_{24}p_{34}) = -p^2q^2s^2 = -1$. This completes the first statement.

In both cases, the condition $p^2q^2s^2 = 1$ implies that all three element subsets of the given quadruple are conforming. If some pair of them does as well, say $1 = p_{12}p_{21} = p^2$, then by symmetry all others pairs are conforming too; that is, $q^2 = s^2 = 1$. In this case $p, q, s \in \mathbf{F}$. Thus $p = p^{(23)} = q = q^{(34)} = s$. This contradicts to conditions of the lemma. Therefore by Theorem 4.6, see the second case in the proof of the second part, the operations space is generated by the following three polynomials

$$[W, x_4]; \quad [W^\sigma, x_1]; \quad [W^{\sigma^2}, x_2],$$

where $\sigma = (1234)$ is the cyclic permutation, whereas W is the main trilinear operation in x_1, x_2, x_3. By the definition of this operation, see (4.30), in the case of the characteristic 2, we have

$$W = (x_1x_2x_3 + x_3x_2x_1) + \frac{p + p^{-1}}{q + q^{-1}}(x_2x_3x_1 + x_1x_3x_2)$$
$$+ \frac{s + s^{-1}}{q + q^{-1}}(x_3x_1x_2 + x_2x_1x_3).$$

Let

$$S = \xi[W, x_4] + \xi_1[W^{\sigma}, x_1] + \xi_2[W^{\sigma^2}, x_2].$$

If we compare the coefficients at the monomials $x_1x_2x_3x_4$ and $x_2x_1x_4x_3$, then we obtain $\xi + \xi_1 = \gamma_{\text{id}} = 1$, $\xi_2 = \gamma_{(12)(34)} = 1$. Therefore

$$S = \xi([W, x_4] + [W^{\sigma}, x_1]) + ([W^{\sigma}, x_1] + [W^{\sigma^2}, x_2]).$$

Thus the symmetric operations span not more then two-dimensional subspace. □

Theorem 5.9 *If $x_1, x_2, \ldots, x_n$ is a symmetric but not absolutely symmetric collection of quantum variables, then the space of multilinear quantum Lie operations is spanned by symmetric operations with the only exception given in Lemma 5.21.*

Proof Consider the skew group algebra $M = \mathbf{F}(p_{it}) * S_n$. The permutation group action defines a structure of right M-module on the set of multilinear polynomials:

$$\sum_{\pi} \gamma_{\pi} x_{\pi(1)} \cdots x_{\pi(n)} \cdot \sum_{\nu} \beta_{\nu} \nu = \sum_{\nu,\pi} \beta_{\nu} \gamma_{\pi}^{\nu} x_{\nu(\pi(1))} \cdots x_{\nu(\pi(n))}.$$

A polynomial $\mathbf{f}$ is symmetric if and only if it generates a submodule of dimension one over $\mathbf{F}(p_{it})$. The space of quantum Lie operations is a right M-submodule, provided that the collection of variables is symmetric.

Indeed, let the basic system, see (5.8), is fulfilled for the coefficients β_{π} of a polynomial $\mathbf{f}$ represented by (5.2). The application of a permutation $\nu \in S_n^1$ to the basic system demonstrates that the coefficients of the polynomial $\mathbf{f}^{\nu}$ satisfy the same system up to rename of the variables $x_i \leftarrow x_{\nu(i)}$. Therefore $\mathbf{f}^{\nu}$, $\nu \in S_n^1$ are quantum Lie operations. If we replace the roles of indices 1 with 2, then we obtain that $\mathbf{f}^{\nu}$, $\nu \in S_n^2$ are quantum Lie operations as well. As the subgroups S_n^1 and S_n^2 with $n > 2$ generate S_n, all multilinear quantum Lie operations form an M-submodule.

Assume that S_n acts faithfully on the field $\mathbf{F}(p_{it})$. In this case, M is isomorphic to the trivial crossed product of the field $\mathbf{F}(p_{it})$ with the Galois group S_n. By Theorem 1.12, the skew group algebra M is isomorphic to the algebra of $n!$ by $n!$ matrices over the Galois subfield $\mathbf{F}_1 = \mathbf{F}(p_{it})^{S_n}$. This implies that each right M-module is a direct sum of simple submodules, whereas all simple submodules are isomorphic to the $n!$-rows module over the Galois field $\mathbf{F}_1$, see Corollary 1.4. On the other hand, the dimension of $\mathbf{F}(p_{it})$ over $\mathbf{F}_1$ equals $n!$ too. Because every right M-module is a right space over $\mathbf{F}(p_{it})$, all irreducible right M-modules are of dimension one over $\mathbf{F}(p_{it})$. This proves the theorem in the case of a faithful action.

If $n > 4$ or $n = 3$, while the action is not faithful, then all even permutations act identically. This immediately implies that the collection of variables is absolutely symmetric, $p_{it} = q$.

Let $n = 4$. If the action is not faithful, then all even permutations, id; $a = (12)(24)$; $b = (13)(24)$; $c = (14)(23)$, act identically. This implies that the parameter matrix has the required form. The existence condition for quantum Lie operations is $p^4q^4s^4 = 1$. If $p^2q^2s^2 = 1$, then we obtain the example given in Lemma 5.21. Therefore, assume that $p^2q^2s^2 = -1 \neq 1$.

If p, q, s are pairwise different, then S_4^1 acts faithfully on $\mathbf{F}(p, q, s)$. Therefore $M_1 = \mathbf{F}(p, q, s) * S_4^1$ is the algebra of 6 by 6 matrices over the Galois field $\mathbf{F}_1$. This is a central simple algebra. Thus, by Theorem 1.13 it splits in M as a tensor factor $M = M_1 \otimes Z_1$, where Z_1 is a centralizer of M_1 in M. Let us calculate this centralizer.

First of all, Z_1 is contained in the centralizer of $\mathbf{F}(p, q, s)$, that equals the group algebra $A = \mathbf{F}(p, q, s)[\mathrm{id}, a, b, c]$. This group algebra has a decomposition in a direct sum of idcals

$$A = \mathbf{F}(p, q, s)e_1 \oplus \mathbf{F}(p, q, s)e_2 \oplus \mathbf{F}(p, q, s)e_3 \oplus \mathbf{F}(p, q, s)e_4,$$

where $e_1 = \frac{1}{4}(\mathrm{id} + a + b + c)$, $e_2 = \frac{1}{4}(\mathrm{id} + a - b + c)$, $e_3 = e_2^{(23)}$, $e_4 = e_2^{(34)}$. The stabilizer of e_2 in S_4^1 equals a two-element subgroup $S_4^{1,3}$. Let $\mathbf{F}_2 = \mathbf{F}(p, q, s)^{S_4^{1,3}}$ be a Galois subfield of this subgroup. Then Z_1 equals the centralizer of S_4^1 in A. This consists of the sums

$$\alpha e_1 + \beta e_2 + \beta^{(23)} e_3 + \beta^{(34)} e_4, \quad \alpha \in \mathbf{F}_1, \ \beta \in \mathbf{F}_2.$$

Thus, $Z_1 \simeq \mathbf{F}_1 \oplus \mathbf{F}_2$. Consequently, $M \simeq (\mathbf{F}_1)_{6\times 6} \oplus (\mathbf{F}_2)_{6\times 6}$.

This result means that up to isomorphism there exists just two irreducible right modules over M. One of them equals the 6-rows space over $\mathbf{F}_1$, whereas another one equals the 6-rows space over $\mathbf{F}_2$. The dimensions of these modules over $\mathbf{F}_1$ are equal to respectively 6 and 18. Therefore, the first module is of dimension one over $\mathbf{F}(p, q, s)$, whereas the second one is of dimension three. By Theorem 4.6 the module of quantum Lie operations is of dimension two over $\mathbf{F}(p, q, s)$. Of course, its irreducible submodules may not be of dimension three. Thus all of them are of dimension one. □

Theorem 5.10 *There exists a collection of* $(n-2)!$ *generic symmetric multilinear quantum Lie operations that span the space of generic multilinear quantum Lie operations.*

Proof The set of generic variables is symmetric because the only defining relation, $\prod_{i\neq j} t_{ij} = 1$, is invariant with respect to the action of the symmetric group S_n. Hence the statement follows from the above theorem. In fact, S_n acts faithfully on the field $\mathbf{F}(t_{ij})$. So we do not need the detail analysis of the exceptions: M is an algebra of $n!$ by $n!$ matrices over the Galois field, and all irreducible right M-modules are of dimension one over $\mathbf{F}(t_{ij})$. □

A similar statement is valid for quantum variables considered by Paregis, see Example 4.5. By definition the quantization parameters defined by the Pareigis quantum variables are related by $p_{it}p_{ti} = \zeta^2$, $1 \leq i \neq j \leq n$, where ζ is a nth primitive root of 1. These relations are invariant with respect to the action of the symmetric group. Moreover, the Pareigis quantum Lie operation given in Example 4.5 is symmetric.

Corollary 5.2 *The total number of linearly independent symmetric multilinear quantum Lie operations for symmetric, but not absolutely symmetric, Pareigis quantum variables is greater than or equal to* $(n-2)!$.

5.11 Chapter Notes

Pareigis quantum Lie operations appeared in [183–185]. The equality

$$\prod_{1\leq i\neq s\leq n} p_{is} = 1$$

as a necessary and sufficient condition for a set of quantum variables to possess a nonzero multilinear quantum Lie operation was established in [125].

In [81, 82], Frønsdal and Galindo determined that the dimension of the space of multilinear constants for differential calculus defined by the diagonal commutations rules, $dx_i \cdot x_s = p_{is}x_s \cdot dx_i$, also equals $(n-2)!$, provided that the above equality holds but that $\prod_{i\neq s,\, i,s\in J} p_{is} \neq 1$ for all proper subsets J of $\{1,2,\ldots,n\}$ containing more than one element. Certainly, this fact implies that the operations and constants are identical in this particular case.

The results concerning symmetric and generic operations are from [126, 130].

Chapter 6
Braided Hopf Algebras

Abstract The main goal of this chapter is a detailed construction of the free braided Hopf algebra $\mathbf{k}\langle V\rangle$ and the shuffle braided Hopf algebra $Sh_\tau(V)$ on the tensor space of a given braided space V. Then we define a Nichols algebra $\mathscr{B}(V)$ as a subalgebra generated by V in $Sh_\tau(V)$ and provide some characterizations of it. Finally we adopt the Radford biproduct and the Majid bozonization to character Hopf algebras. All calculations are done in the braid monoid (not in the braid group), therefore in the constructions there is no need to assume that the braiding is invertible.

The main goal of this chapter is a detailed construction of free braided Hopf algebra $\mathbf{k}\langle V\rangle$ and braided shuffle Hopf algebra $Sh_\tau(V)$ on the tensor space of a given braided space V. We then define a *Nichols algebra* $\mathscr{B}(V)$ as a subalgebra generated by V in $Sh_\tau(V)$ and provide some characterizations of this algebra. Finally we adopt the Radford biproduct decomposition and the Majid bozonization to the class of character Hopf algebras. All calculations are performed in the braid monoid (not in the braid group). Therefore, in the constructions, we are not required to assume that braiding is invertible. In the final section, we discuss when a structure of a braided Hopf algebra on a filtered space R induces that structure on the associated graded space gr R.

6.1 Braided Objects

A linear space V is called a *braided space* if there is fixed a linear map $\tau : V\otimes V \to V\otimes V$ (in general not necessary invertible) that satisfies the *braid relation*:

$$(\tau\otimes \mathrm{id})(\mathrm{id}\otimes\tau)(\tau\otimes \mathrm{id}) = (\mathrm{id}\otimes\tau)(\tau\otimes \mathrm{id})(\mathrm{id}\otimes\tau). \tag{6.1}$$

Example 6.1 If $x_1, x_2, \ldots, x_n$ is the basis of a linear space V, then for arbitrary parameters $q_{is}\in\mathbf{k}$, $1\le i,s\le n$, the map

$$\tau : x_i\otimes x_s \mapsto q_{is}\cdot x_s\otimes x_i$$

satisfies the braid relation. This is the so called *diagonal braiding*.

V. Kharchenko, *Quantum Lie Theory*, Lecture Notes in Mathematics 2150,
DOI 10.1007/978-3-319-22704-7_6

Let V and V' be spaces with braidings τ and τ' respectively. A linear map $\varphi : V \to V'$ is called a *homomorphism of braided spaces* (or *it respects the braidings*) if

$$\tau(\varphi \otimes \varphi) = (\varphi \otimes \varphi)\tau'.$$

Let $(a \otimes b)\tau = \sum b_i \otimes a_i$, and $(a' \otimes b')\tau = \sum b'_i \otimes a'_i$. In this case the definition of the homomorphism takes the form

$$\sum \varphi(b_i) \otimes \varphi(a_i) = \sum \varphi(b)_i \otimes \varphi(a)_i,$$

or, informally, $\varphi(a_i) = \varphi(a)_i$.

Proposition 6.1 *If a linear map $\varphi : V \to V'$ is a homomorphism of braided spaces, then $W = \ker \varphi$ satisfies*

$$(V \otimes W + W \otimes V)\tau \subseteq V \otimes W + W \otimes V. \tag{6.2}$$

Conversely, if a subspace W satisfies (6.2), *then the quotient space V/W has an induced braiding such that the natural homomorphism $\varphi : V \to V/W$ is a homomorphism of braided spaces.*

Proof To prove the statement, we need the following statement on kernels of tensor products of maps.

Lemma 6.1 *If $\varphi : V \to V'$ is a linear map, then*

$$\ker(\varphi \otimes \varphi) = V \otimes \ker \varphi + \ker \varphi \otimes V. \tag{6.3}$$

Proof The required equality follows from (1.49) with $\psi = \varphi$. □

Let φ be a homomorphism of braided spaces. If $w \in W$, $v \in V$, then by definition of the braided homomorphism we have

$$(v \otimes w)\tau(\varphi \otimes \varphi) = (\varphi(v) \otimes \varphi(w))\tau' = 0;$$

that is, $(V \otimes W)\tau \subseteq \ker(\varphi \otimes \varphi)$, and Lemma 6.1 applies. In a perfect analogy, we have $(W \otimes V)\tau \subseteq \ker(\varphi \otimes \varphi)$.

Conversely, assume that W satisfies (6.2). The quotient space V/W is isomorphic to a complement T of W to V. Let us define τ' on T via

$$\tau' = \tau(\pi \otimes \pi),$$

where π is a linear projection $\pi : V \to T$.

We have $(\pi - \mathrm{id})\pi = 0$, and therefore $\mathrm{im}(\pi - \mathrm{id}) \subseteq W$. Due to (6.2), the latter inclusion implies

$$(\mathrm{im}(\pi - \mathrm{id}) \otimes V)\tau(\pi \otimes \pi) = 0,$$

or in the operator form, $((\pi - \mathrm{id}) \otimes \mathrm{id})\tau(\pi \otimes \pi) = 0$, which is equivalent to

$$(\pi \otimes \mathrm{id})\tau(\pi \otimes \pi) = \tau(\pi \otimes \pi). \tag{6.4}$$

In a perfect analogy, we have

$$(\mathrm{id} \otimes \pi)\tau(\pi \otimes \pi) = \tau(\pi \otimes \pi). \tag{6.5}$$

These two equalities imply

$$(\pi \otimes \pi)\tau' = (\pi \otimes \pi)\tau(\pi \otimes \pi) = \tau(\pi \otimes \pi),$$

and for $\tau_1' = \tau' \otimes \mathrm{id}$, $\tau_2' = \mathrm{id} \otimes \tau'$, $\tau_1 = \tau \otimes \mathrm{id}$, $\tau_2 = \mathrm{id} \otimes \tau$, we have

$$\tau_1'\tau_2'\tau_1' = \tau_1\tau_2\tau_1(\pi \otimes \pi \otimes \pi) = \tau_2\tau_1\tau_2(\pi \otimes \pi \otimes \pi) = \tau_2'\tau_1'\tau_2',$$

which is required. □

An algebra R with a multiplication $\mathbf{m} : R \otimes R \to R$ is called a *braided algebra* if it is a braided space and

$$(\mathbf{m} \otimes \mathrm{id})\tau = \tau_2\tau_1(\mathrm{id} \otimes \mathbf{m}), \quad (\mathrm{id} \otimes \mathbf{m})\tau = \tau_1\tau_2(\mathbf{m} \otimes \mathrm{id}). \tag{6.6}$$

In these formulas, as above, we use the so-called "exponential notation" for actions of the operators; that is, the operators in a superposition act from the left to the right. For example, $(\mathbf{m} \otimes \mathrm{id})\tau$ acts on $V \otimes V \otimes V$ via

$$(x \otimes y \otimes z)^{(\mathbf{m}\otimes\mathrm{id})\tau} = (xy \otimes z)^{\tau} = \tau(xy \otimes z),$$

whereas $\tau_2\tau_1(\mathrm{id} \otimes \mathbf{m})$ acts on $V \otimes V \otimes V$ as follows

$$(x\otimes y\otimes z)^{\tau_2\tau_1(\mathrm{id}\otimes\mathbf{m})} = (x\otimes(y\otimes z)^{\tau})^{\tau_1(\mathrm{id}\otimes\mathbf{m})} = \Big(\sum_i (x\otimes z_i)^{\tau}\otimes y_i\Big)^{(\mathrm{id}\otimes\mathbf{m})} = \sum_{i,j} z_{ij}\otimes x_{ij}y_i,$$

where $(y \otimes z)^{\tau} = (y \otimes z)\tau = \sum_i z_i \otimes y_i$ and $(x \otimes z_i)^{\tau} = (x \otimes z_i)\tau = \sum_j z_{ij} \otimes x_{ij}$.

By definition, a *homomorphism of braided algebras* is a linear map that is both a homomorphism of algebras and braided spaces.

A coalgebra $(C, \Delta^b, \varepsilon)$ is called *braided* if it is a braided space and

$$\tau(\varepsilon \otimes \mathrm{id}) = (\mathrm{id} \otimes \varepsilon)\tau, \quad \tau(\mathrm{id} \otimes \varepsilon) = (\varepsilon \otimes \mathrm{id})\tau; \tag{6.7}$$

$$\tau(\mathrm{id} \otimes \Delta^b) = (\Delta^b \otimes \mathrm{id})\tau_2\tau_1, \quad \tau(\Delta^b \otimes \mathrm{id}) = (\mathrm{id} \otimes \Delta^b)\tau_1\tau_2. \tag{6.8}$$

By definition, a *homomorphism of braided coalgebras* $\varphi : V \to V'$ is a homomorphism of coalgebras,

$$\Delta^b(\varphi(a)) = \sum_{(a)} \varphi(a^{(1)}) \otimes \varphi(a^{(2)}), \quad \varepsilon(\varphi(a)) = \varepsilon(a), \tag{6.9}$$

that respects the braidings.

A *braided bialgebra* is an associative braided algebra and a braided coalgebra H (with the same braiding) where the coproduct is an algebra homomorphism

$$\Delta^b : H \to H\underline{\otimes}H. \tag{6.10}$$

Here, $H\underline{\otimes}H$ is the ordinary tensor product of spaces with a new multiplication

$$(a\underline{\otimes}b)(c\underline{\otimes}d) = \sum_i (ac_i\underline{\otimes}b_i d), \text{ where } (b \otimes c)\tau = \sum_i c_i \otimes b_i. \tag{6.11}$$

A *homomorphism of braided bi-algebras* is a homomorphism of coalgebras and braided algebras.

By definition, a *braided Hopf algebra* is a braided bialgebra H with a linear map $\sigma^b : H \to H$ called a *braided antipode* that satisfies the usual identity

$$\Delta^b(a)(\sigma^b\underline{\otimes}\text{id})\mathbf{m} = \Delta^b(\text{id}\underline{\otimes}\sigma^b)\mathbf{m} = \varepsilon(a) \cdot 1. \tag{6.12}$$

A *homomorphism of braided Hopf algebras* is a homomorphism of braided bi-algebras that satisfies

$$\varphi(\sigma^b(a)) = \sigma^b(\varphi(a)). \tag{6.13}$$

Definition 6.1 A subspace $W \subseteq V$ of a braided Hopf algebra V is called a *braided Hopf ideal* if the following conditions are met:

1. W is an ideal of the algebra V;
2. $(V \otimes W + W \otimes V)\tau \subseteq V \otimes W + W \otimes V$;
3. $\varepsilon(W) = 0$;
4. $\Delta^b(W) \subseteq V \otimes W + W \otimes V$;
5. $\sigma^b(W) \subseteq W$.

Lemma 6.2 *If the map $\varphi : V \to V'$ is a homomorphism of braided Hopf algebras, then $W = \ker\varphi$ is a braided Hopf ideal of V. Conversely, if $W \subseteq V$ is a braided Hopf ideal, then the quotient algebra V/W has induced braiding, coproduct, and braided antipode such that the natural homomorphism $\varphi : V \to V/W$ is a homomorphism of braided Hopf algebras.*

Proof We have to check all five conditions of Definition 6.1. The first one is evident. To prove the second one, we may apply Lemma 6.1:

$$(V \otimes W + W \otimes V)\tau(\varphi \otimes \varphi) = [\varphi(V) \otimes \varphi(W) + \varphi(W) \otimes \varphi(V)]\tau' = 0.$$

The third one follows from the second formula of (6.9), whereas the forth one follows from the first formula of (6.9) and Lemma 6.1. Equality (6.13) implies the fifth condition.

Conversely, let W satisfies all conditions of Definition 6.1. By Proposition 6.1 the natural homomorphism of algebras $\varphi : V \to V/W$ is a homomorphism of braided spaces. Formulas (6.9) inspire the definition of a counit and a coproduct on V/W:

$$\Delta^b(v + W) = \sum_{(v)} (a^{(1)} + W) \otimes (v^{(2)} + W), \ \varepsilon(v + W) = \varepsilon(v).$$

In this way the counit and the coproduct are well-defined due to the third and fourth properties. Because $\varphi(v) = v + W$, the map φ is a homomorphism of coalgebras. Due to the fifth condition, the braided antipode is well-defined by the formula

$$\sigma^b(v + W) = \sigma^b(v) + W.$$

Again, the equality $\varphi(v) = v + W$ implies that φ satisfies (6.13). □

6.2 Free Braided Hopf Algebra

Let V be a linear space with a braiding $\tau : V \otimes V \to V \otimes V$. We fix some basis $X = \{x_i, i \in I\}$ of V. The free associative algebra $\mathbf{k}\langle X\rangle$ generated by x_i, $i \in I$ is isomorphic to the tensor algebra $T(V) = \bigoplus_{i=0}^{\infty} V^{\otimes i}$ of a linear space V with the concatenation product $(u\underline{\otimes}v)\mathbf{m} = u \otimes v$. By definition we set $V^{\otimes 0} = \mathbf{k} \cdot 1$, where 1 is the empty word in X, so that $1 \otimes v = v \otimes 1 = v$. Consider the following linear maps

$$\tau_i = \mathrm{id}^{\otimes(i-1)} \otimes \tau \otimes \mathrm{id}^{\otimes(n-i-1)} : V^{\otimes n} \to V^{\otimes n}, \ \ 1 \leq i < n. \tag{6.14}$$

Due to (6.1) the maps τ_i satisfy all defining relations of the braid monoid:

$$\tau_i\tau_{i+1}\tau_i = \tau_{i+1}\tau_i\tau_{i+1}, \ \ 1 \leq i < n-1; \ \ \tau_i\tau_j = \tau_j\tau_i, \ \ |i-j| > 1. \tag{6.15}$$

Therefore, $u \cdot s_i = u\tau_i$ is a well-defined action on $V^{\otimes n}$ of the braid monoid B_n generated by the braids s_i, $1 \leq i < n$. This action is called a *local action.*

Theorem 6.1 *The braiding τ has a unique extension on $\mathbf{k}\langle X\rangle$ so that $\mathbf{k}\langle X\rangle$ is a braided algebra. Normally this extension has the same notation τ.*

Proof Let θ_r, $0 \leq r \leq n$ be the linear map $V^{\otimes n} \to V^{\otimes r} \underline{\otimes} V^{\otimes (n-r)}$ acting as follows

$$(z_1 z_2 \cdots z_n)\,\theta_r = z_1 z_2 \cdots z_r \underline{\otimes} z_{r+1} \cdots z_n, \quad z_i \in X.$$

Consider a map $\nu_r^{k,n} : V^{\otimes n} \to V^{\otimes n}$, $k \leq r < n$ defined as a superposition of the τ_i's:

$$\nu_r^{k,n} = (\tau_r \tau_{r-1} \cdots \tau_k)(\tau_{r+1}\tau_r \cdots \tau_{k+1}) \cdots (\tau_{n-1}\tau_{n-2} \cdots \tau_{n-1-r+k}). \tag{6.16}$$

The operator $\nu_r^{k,n}$ has an alternative representation:

$$\nu_r^{k,n} = (\tau_r \tau_{r+1} \cdots \tau_{n-1})(\tau_{r-1}\tau_r \cdots \tau_{n-2}) \cdots (\tau_k \tau_{k+1} \cdots \tau_{n-1-r+k}). \tag{6.17}$$

Indeed, in (6.16), the first term of each factor commutes with all terms except the first one of the previous factor. Hence, we have

$$\nu_r^{k,n} = (\tau_r \tau_{r+1} \cdots \tau_{n-1}) \cdot (\tau_{r-1} \cdots \tau_k)(\tau_r \cdots \tau_{k+1}) \cdots (\tau_{n-2} \cdots \tau_{n-1-r+k}).$$

Continuation of this process yields (6.17). We extend the braiding on $\mathbf{k}\langle X\rangle$ via

$$(u \underline{\otimes} v)\tau' = (u \otimes v)\nu_r^{1,n}\,\theta_{n-r}, \quad u \in V^{\otimes r},\ v \in V^{\otimes (n-r)}. \tag{6.18}$$

If $r = 0$ or $r = n$, then this definition reads: $(1 \underline{\otimes} v)\tau' = v \underline{\otimes} 1$; $(u \underline{\otimes} 1)\tau' = 1 \underline{\otimes} u$. Let us show that τ' is a braiding of $\mathbf{k}\langle X\rangle$. If $u \in V^{\otimes r}$, $v \in V^{\otimes m}$, $w \in V^{\otimes n-m-r}$, then

$$T_{uvw}(\tau' \underline{\otimes} \mathrm{id}) = (u \otimes v \otimes w)\nu_r^{1,r+m}\theta_m \theta_{m+r},$$
$$T_{uvw}(\mathrm{id} \underline{\otimes} \tau') = (u \otimes v \otimes w)\nu_{r+m}^{r+1,n}\theta_r \theta_{n-m},$$

where $T_{uvw} = (u \underline{\otimes} v \underline{\otimes} w)$. Similarly,

$$T_{uvw}(\tau' \underline{\otimes} \mathrm{id})(\mathrm{id} \underline{\otimes} \tau') = (u \otimes v \otimes w)\nu_r^{1,r+m}\nu_{r+m}^{m+1,n}\theta_m \theta_{n-r},$$
$$T_{uvw}(\mathrm{id} \underline{\otimes} \tau')(\tau' \underline{\otimes} \mathrm{id}) = (u \otimes v \otimes w)\nu_{r+m}^{r+1,n}\nu_r^{1,n-m}\theta_{n-r-m}\theta_{n-m},$$

and

$$T_{uvw}(\tau' \underline{\otimes} \mathrm{id})(\mathrm{id} \underline{\otimes} \tau')(\tau' \underline{\otimes} \mathrm{id}) = (u \otimes v \otimes w)\nu_r^{1,r+m} \cdot \nu_{r+m}^{m+1,n} \cdot \nu_m^{1,n-r}\theta_{n-r-m}\theta_{n-r},$$
$$T_{uvw}(\mathrm{id} \underline{\otimes} \tau')(\tau' \underline{\otimes} \mathrm{id})(\mathrm{id} \underline{\otimes} \tau') = (u \otimes v \otimes w)\nu_{r+m}^{r+1,n} \cdot \nu_r^{1,n-m} \cdot \nu_{n-m}^{n-r-m+1,n}\theta_{n-r-m}\theta_{n-r}.$$

Hence, the braid relation (6.1) for τ' is equivalent to the operator equality

$$\nu_r^{1,r+m} \cdot \nu_{r+m}^{m+1,n} \cdot \nu_m^{1,n-r} = \nu_{r+m}^{r+1,n} \cdot \nu_r^{1,n-m} \cdot \nu_{n-m}^{n-r-m+1,n}$$

•

Taking into account definition (6.18), we obtain

$$\nu_r^{1,r+m}\cdot\nu_{r+m}^{m+1,n}=\nu_r^{1,n},\ \text{and}\ \nu_r^{1,n-m}\cdot\nu_{n-m}^{n-r-m+1,n}=\nu_r^{1,n}. \tag{6.19}$$

Therefore the braid relation for τ' reduces to

$$\nu_r^{1,n}\cdot\nu_m^{1,n-r}=\nu_{r+m}^{r+1,n}\cdot\nu_r^{1,n}. \tag{6.20}$$

We shall prove this equality using commutation rule (1.33) of Lemma 1.13. By (6.17), the operator $\nu_r^{1,n}$ is the following superposition:

$$(\tau_r\tau_{r+1}\cdots\tau_{n-1})(\tau_{r-1}\tau_r\cdots\tau_{n-2})\cdots(\tau_1\tau_2\cdots\tau_{n-r}).$$

At the same time by definition (6.16), we have

$$\nu_m^{1,n-r}=(\tau_m\tau_{m-1}\cdots\tau_1)(\tau_{m+1}\tau_m\cdots\tau_2)\cdots(\tau_{n-r-1}\tau_{n-r-2}\cdots\tau_{n-r-m}).$$

Applying Lemma 1.13 totally $n-r-m$ times, we have

$$(\tau_1\tau_2\cdots\tau_{n-r})\cdot\nu_m^{1,n-r}=\nu_{m+1}^{2,n-r+1}\cdot(\tau_1\tau_2\cdots\tau_{n-r}).$$

Similarly,

$$(\tau_2\tau_3\cdots\tau_{n-r+1})\cdot\nu_{m+1}^{2,n-r+1}=\nu_{m+2}^{3,n-r+2}\cdot(\tau_2\tau_3\cdots\tau_{n-r+1}).$$

Continuation of this process results in (6.20). Thus, the map τ' is a braiding.

Let us check identities of braided algebra (6.6). The concatenation product $\mathbf{m}$ of the free algebra satisfies $(u\underline{\otimes}v)\mathbf{m}=u\otimes v$. Therefore,

$$T_{uvw}(\mathbf{m}\underline{\otimes}\mathrm{id})\tau'=((u\otimes v)\underline{\otimes}w)\tau'=(u\otimes v\otimes w)\nu_{r+m}^{1,n}\cdot\theta_{n-r-m}.$$

At the same time

$$T_{uvw}\tau_2'\tau_1'(\mathrm{id}\otimes\mathbf{m})=(u\otimes v\otimes w)\nu_{r+m}^{r+1,n}\cdot\nu_r^{1,n-m}\cdot\theta_{n-r-m}.$$

To prove the first identity of braided algebra (6.6), it remains to demonstrate the following relation

$$\nu_{r+m}^{r+1,n}\cdot\nu_r^{1,n-m}=\nu_{r+m}^{1,n}. \tag{6.21}$$

We have

$$\begin{aligned}\nu_{r+m}^{r+1,n}\cdot\nu_r^{1,n-m}&=(\tau_{r+m}\tau_{r+m-1}\cdots\tau_{r+1})(\tau_{r+m+1}\tau_{r+m}\cdots\tau_{r+2})\cdots(\tau_{n-1}\tau_{n-2}\cdots\tau_{n-m})\\&\quad\times(\tau_r\tau_{r-1}\cdots\tau_1)(\tau_{r+1}\tau_r\cdots\tau_2)\cdots(\tau_{n-m-1}\tau_{n-m-2}\cdots\tau_{n-r-m}).\end{aligned}$$

Using relations $\tau_i\tau_j = \tau_j\tau_i$, $|i-j| > 1$, we can move the first factor $(\tau_r\tau_{r-1}\cdots\tau_1)$ of $\nu_r^{1,n-m}$ to the left until the first factor of $\nu_{r+m}^{r+1,n}$. Next, we move the second factor of $\nu_r^{1,n-m}$ to the left until the second factor of $\nu_{r+m}^{r+1,n}$, and so on. In this way we obtain the product

$$(\tau_{r+m}\tau_{r+m-1}\cdots\tau_1)(\tau_{r+m+1}\tau_m\cdots\tau_2)\cdots(\tau_{n-1}\tau_{n-2}\cdots\tau_{n-r-m}) = \nu_{r+m}^{1,n}.$$

This completes the proof of (6.21).

In the perfect analogy, the second identity of braided algebra (6.6) reduces to the equality $\nu_r^{1,n} = \nu_r^{1,r+m} \cdot \nu_{r+m}^{m+1,n}$, which was mentioned in (6.19).

Finally, the uniqueness of τ' follows from each one of (6.6) considered as a recurrence relation. □

Theorem 6.2 *The free braided algebra* $\mathbf{k}\langle X\rangle$ *has a natural structure of a braided Hopf algebra where the free generators are primitive with respect to the braided coproduct:*

$$\Delta^b(x_i) = x_i \underline{\otimes} 1 + 1 \underline{\otimes} x_i. \tag{6.22}$$

Proof Because by definition a braided coproduct is a homomorphism of associative algebras, equality (6.22) uniquely defines Δ^b. We have to verify that Δ^b is coassociative, has a counit ε, a braided antipode σ^b, and satisfies identities of braided coalgebra (6.8). Our fundamental idea is to reformulate each of these axioms in terms of the local action of the braid monoid B_n and then to make calculations in the monoid algebra $\mathbf{k}[B_n]$. To this end, we need the braided coproduct in an explicit form in terms of the local action. We fix the following notation

$$\Phi_r^{(t,n)} = \sum_{t\leq k_1<k_2<\ldots<k_{r-t+1}\leq n} [t;k_1][t+1;k_2]\cdots[r;k_{r-t+1}], \tag{6.23}$$

where by definition

$$[k;k] = \text{id};\quad [m;k] = \tau_{k-1}\tau_{k-2}\tau_{k-3}\cdots\tau_{m+1}\tau_m,\quad m<k. \tag{6.24}$$

Lemma 6.3 *In terms of the local action, the braided coproduct has the form*

$$\Delta^b(u) = \sum_{r=0}^{n} \left[u\cdot\Phi_r^{(1,n)}\right]\theta_r,\quad u\in V^{\otimes n}. \tag{6.25}$$

Proof Without loss of generality we may assume that u is a word, $u = z_1z_2\cdots z_n$, $z_i \in V$, or, equivalently, $u = z_1\otimes z_2\otimes\cdots\otimes z_n \in V^{\otimes n}$. Let $A = \{k_1 < k_2 < \ldots < k_r\}$ be an r-element subset of indices. Denote by $\varepsilon(A,i)$ its characteristic function:

$$\varepsilon(A,i) = \begin{cases} + & \text{if } i\in A,\\ - & \text{otherwise.}\end{cases}$$

Let us put $(z_i)^+ = (z_i \underline{\otimes} 1)$ and $(z_i)^- = (1 \underline{\otimes} z_i)$. Then by definition

$$\Delta^b(z_1 z_2 \cdots z_n) = \sum_r \sum_A (z_1)^{\varepsilon(A,1)} (z_2)^{\varepsilon(A,2)} \ldots (z_n)^{\varepsilon(A,n)}.$$

To prove (6.25), it suffices to check that

$$u \cdot [1; k_1][2; k_2] \cdots [r; k_r]\, \theta_r = (z_1)^{\varepsilon(A,1)} (z_2)^{\varepsilon(A,2)} \cdots (z_n)^{\varepsilon(A,n)}. \tag{6.26}$$

We may do it by induction on the lexicographically ordered pairs (r, n). If $r = 0$, then $A = \emptyset$; hence, $\varepsilon(A, i) = -$, whereas (6.26) reduces to

$$1 \underline{\otimes} u = z_1 \cdots z_n\, \theta_0 = (1 \underline{\otimes} z_1) \cdots (1 \underline{\otimes} z_n) = 1 \underline{\otimes} u.$$

Suppose that (6.26) is valid for all pairs $(r_1, n_1) < (r, n)$. If $k_r \neq n$, then $\varepsilon(A, n) = -$, that is $(z_n)^{\varepsilon(A,n)} = 1 \underline{\otimes} z_n$, and we may use the induction supposition:

$$\begin{aligned} u \cdot [1; k_1] \cdots [r; k_r]\, \theta_r &= z_1 \cdots z_{n-1} \cdot [1; k_1] \cdots [r; k_r]\, \theta_r (1 \underline{\otimes} z_n) \\ &= (z_1)^{\varepsilon(A,1)} \cdots (z_{n-1})^{\varepsilon(A,n-1)} (1 \underline{\otimes} z_n), \end{aligned}$$

which is required.

If $k_r = n$, then $(z_n)^{\varepsilon(A,n)} = z_n \underline{\otimes} 1$, and by means of (1.36) we have

$$\begin{aligned} & u \cdot [1; k_1][2; k_2] \cdots [r; k_r]\, \theta_r \\ &= (z_1 \cdots z_{n-1} \cdot [1; k_1] \cdots [r-1; k_{r-1}]\, z_n) \cdot \tau_{n-1} \tau_{n-2} \cdots \tau_r\, \theta_r. \end{aligned} \tag{6.27}$$

Formula (6.11) and the method for extension of the braiding (6.18), (6.16) imply

$$(w\, \theta_{r-1})(z_n \underline{\otimes} 1) = (w z_n) \tau_{n-1} \tau_{n-2} \cdots \tau_r\, \theta_r, \quad w \in V^{\otimes(n-1)}.$$

Hence we may continue (6.27) using the induction supposition

$$\begin{aligned} &= (z_1 \cdots z_{n-1} \cdot [1; k_1] \cdots [r-1; k_{r-1}])\, \theta_{r-1} (z_n \underline{\otimes} 1) \\ &= (z_1)^{\varepsilon(A,1)} \cdots (z_{n-1})^{\varepsilon(A,n-1)} (z_n \underline{\otimes} 1), \end{aligned}$$

which is required. □

Coassociativity By the above lemma, we have

$$\Delta^b(u) = \sum_{(u)} u_{(1)} \underline{\otimes} u_{(2)} = \sum_{r=0}^{n} [u \cdot \Phi_r^{(1,n)}]\, \theta_r.$$

Let us fix nonnegative numbers r, m, such that $r + m \leq n$. The sum of all terms of $\Delta^b(u)$ that belong to $V^{\otimes(r+m)}\underline{\otimes}V^{\otimes(n-r-m)}$ takes the form $u \cdot \Phi_{r+m}^{(1,n)}\theta_{r+m}$, whereas the sum $R_{k,m}$ of all terms of $\sum_{(u)} \Delta^b(u_{(1)})\underline{\otimes}u_{(2)}$ that belong to $V^{\otimes r}\underline{\otimes}V^{\otimes m}\underline{\otimes}V^{\otimes n-r-m}$ is

$$R_{k,m} = \left[u \cdot \Phi_{r+m}^{(1,n)} \cdot \Phi_r^{(1,r+m)}\right] \theta_{r+m}\theta_r.$$

Similarly, the sum of all terms of $\Delta^b(u)$ that belong to $V^{\otimes r}\underline{\otimes}V^{\otimes(n-r)}$ takes the form $u \cdot \Phi_r^{(1,n)}\theta_r$, and the sum $R'_{k,m}$ of all terms of $\sum_{(u)} u_{(1)}\underline{\otimes}\Delta^b(u_{(2)})$ that belong to $V^{\otimes r}\underline{\otimes}V^{\otimes m}\underline{\otimes}V^{\otimes n-r-m}$ takes the form

$$R'_{k,m} = \left[u \cdot \Phi_r^{(1,n)} \cdot \Phi_{r+m}^{(r+1,n)}\right] \theta_r\theta_{r+m}.$$

It remains to check that $R_{k,m} = R'_{k,m}$.

Lemma 6.4 *In* $\mathbf{k}[B_n]$ *the following equality is valid*

$$\Phi_{r+m}^{(1,n)} \cdot \Phi_r^{(1,r+m)} = \Phi_r^{(1,n)} \cdot \Phi_{r+m}^{(r+1,n)}. \tag{6.28}$$

Proof We have

$$\Phi_{r+m}^{(1,n)} \cdot \Phi_r^{(1,r+m)} = \left(\sum_{1\leq k_1<k_2<\ldots<k_{r+m}\leq n} \prod_{i=1}^{r+m}[i;k_i]\right) \cdot \left(\sum_{1\leq t_1<t_2<\ldots<t_r\leq r+m} \prod_{i=1}^{r}[i;t_i]\right).$$

Let us analyze an arbitrary term of the above product

$$[1;k_1][2;k_2]\cdots[r+m;k_{r+m}] \cdot [1;t_1][2;t_2]\cdots[r;t_r]. \tag{6.29}$$

The relations $\tau_i\tau_j = \tau_j\tau_i$, $i > j+1$ imply that $[i;k_i][1;t_1] = [1;t_1][i;k_i]$ provided that $i > t_1$, whereas for $i = t_1$, we have $[t_1;k_{t_1}][1;t_1] = [1;k_{t_1}]$. This allows us to remove the factor $[1;t_1]$ from (6.29) replacing $[t_1;k_{t_1}]$ with $[1;k_{t_1}]$. Further, the commutation rule (1.34) under substitution $t \leftarrow k_{t_1} - 1$, $k \leftarrow 1$, $r \leftarrow k_i$, $m \leftarrow i+1$ demonstrates that

$$[i;k_i][1;k_{t_1}] = [1;k_{t_1}][i+1;k_i+1], \quad 1 \leq i < t_1$$

because $1 \leq i \leq k_i \leq k_{t_1} - 1$ holds (we stress that $1 \leq k_1 < k_2 < \ldots < k_{t_1}$ implies $i \leq k_i$, and $i < t_1$ implies $k_i < k_{t_1}$). Hence, we may move the new factor $[1;k_{t_1}]$ to the left margin position of (6.29) replacing each factor $[i;k_i]$, $1 \leq i < t_1$ by $[i+1;k_i+1]$.

If after that we do the same with the factor $[2;t_2]$, then in the left margin position of (6.29) appears to be $[1;k_{t_1}][2;k_{t_1}]$, and each factor $[i;k_i]$, $1 \leq i < t_1$ is replaced with $[i+2;k_i+2]$, whereas each factor $[i;k_i]$, $t_1 < i < t_2$ is replaced with $[i+1;k_i+1]$.

Applying this procedure further to the factors $[3; t_3]$, $[4; t_4]$, $\ldots$, $[r; t_r]$, we transform (6.29) thus:

$$[1; k_{t_1}][2; k_{t_2}] \cdots [r; k_{t_r}] \cdot \prod_{s=0}^{r} \Big(\prod_{i=t_s+1}^{t_{s+1}-1} [i + r - s,\ k_i + r - s] \Big), \tag{6.30}$$

where we postulate $t_0 = 0$, $t_{r+1} = r + m + 1$. Let us replace the summation index i of the elementary product Π_s in the parenthesis with $j = i + r - s$. We have

$$\Pi_s \stackrel{\text{df}}{=} \prod_{i=t_s+1}^{t_{s+1}-1} [i + r - s,\ k_i + r - s] = \prod_{j=t_s+r-s+1}^{t_{s+1}+r-s-1} [j,\ k_{j-r+s} + r - s].$$

The upper limit, $j = t_{s+1} + r - s - 1$, of Π_s and the lower limit, $j = t_{s+1} + r - (s + 1) + 1$, of Π_{s+1} are consecutive integer numbers. The smallest value of j is $t_0 + r - 0 + 1 = 1$, whereas the biggest one is $t_{r+1} + r - (r + 1) + 1 = r + m$. Therefore, if we set $k'_j = k_{j-r+s} + r - s$, then

$$\prod_{s=0}^{r} \Big(\prod_{i=t_s+1}^{t_{s+1}-1} [i + r - s,\ k_i + r - s] \Big) = \prod_{j=r+1}^{r+m} [j; k'_j].$$

In which case, we have $r < k'_1 < k'_2 < \ldots < k'_m \leq n$, so that the above term occurs in

$$\Phi_{r+m}^{(r+1,n)} = \sum_{r<s_1<s_2<\ldots<s_m\leq n} \prod_{i=r+1}^{r+m} [i; s_i].$$

In formula (6.30), we have $1 \leq k_{t_1} < k_{t_2} < \ldots < k_{t_r} \leq n$. In particular, the product $[1; k_{t_1}][2; k_{t_2}] \cdots [r; k_{t_r}]$ occurs in

$$\Phi_r^{(1,n)} = \sum_{1\leq s_1<s_2<\ldots s_r\leq n} \prod_{i=1}^{r} [i; s_i].$$

Thus, each transformed term of $\Phi_{r+m}^{(1,n)} \cdot \Phi_r^{(1,r+m)}$ occurs in $\Phi_r^{(1,n)} \cdot \Phi_{r+m}^{(r+1,n)}$. Because different terms are transformed to different ones (the above described transformation is invertible) and both sums have the same number of elements,

$$\binom{n}{r+m}\binom{r+m}{r} = \binom{n}{r}\binom{n-r}{m},$$

we have $\Phi_{r+m}^{(1,n)} \cdot \Phi_r^{(1,r+m)} = \Phi_r^{(1,n)} \cdot \Phi_{r+m}^{(r+1,n)}$. □

Counit The counit is defined in a standard way as a homomorphism $\varepsilon(x_i) = 0$, $\varepsilon(\alpha \cdot 1) = \alpha$. The kernel of ε is $\Lambda = \sum_{i=1}^{\infty} V^{\otimes^i}$. Evidently Λ satisfies (6.2) so that ε is a homomorphism of braided algebras. Further,

$$\Phi_r^{(1,n)}\theta_r \cdot (\varepsilon \underline{\otimes} \mathrm{id}) = \Phi_r^{(1,n)}\theta_r \cdot (\mathrm{id} \underline{\otimes} \varepsilon) = 0, \quad 0 < r < n,$$

whereas $\Phi_0^{(1,n)} = \Phi_n^{(1,n)} = \mathrm{id}$. This implies the counit properties (1.53):

$$\sum_{(u)} \varepsilon(u_{(1)})u_{(2)} = \sum_{(u)} u_{(1)}\varepsilon(u_{(2)}) = u.$$

Similarly, we have

$$\nu_r^{(1,n)}\theta_{n-r} \cdot (\varepsilon \underline{\otimes} \mathrm{id}) = \nu_{n-r}^{(1,n)}\theta_r \cdot (\mathrm{id} \underline{\otimes} \varepsilon) = 0, \quad 0 \leq r < n,$$

and $\nu_0^{(1,n)} = \nu_n^{(1,n)} = \mathrm{id}$. This implies (6.7) connecting the braiding and counit:

$$\tau'(\varepsilon \underline{\otimes} \mathrm{id}) = (\mathrm{id} \underline{\otimes} \varepsilon)\tau', \quad \tau'(\mathrm{id} \underline{\otimes} \varepsilon) = (\varepsilon \underline{\otimes} \mathrm{id})\tau'.$$

Identities of Braided Coalgebra Let us check (6.8) by connecting the braiding and coproduct:

$$\tau'(\Delta^b \underline{\otimes} \mathrm{id}) = (\mathrm{id} \underline{\otimes} \Delta^b)\tau_1'\tau_2', \quad \tau'(\mathrm{id} \underline{\otimes} \Delta^b) = (\Delta^b \underline{\otimes} \mathrm{id})\tau_2'\tau_1'. \tag{6.31}$$

If $u \in V^{\otimes r}$, $v \in V^{\otimes n-r}$, then by definition, we have

$$(u \underline{\otimes} v)\tau'(\Delta^b \underline{\otimes} \mathrm{id}) = (u \otimes v)\nu_r^{1,n}\theta_{n-r}(\Delta^b \underline{\otimes} \mathrm{id}) = (u \otimes v)\nu_r^{1,n} \sum_{m=0}^{n-r} \Phi_m^{(1,n-r)}\theta_m\theta_{n-r},$$

whereas

$$\begin{aligned}(u \underline{\otimes} v)(\mathrm{id} \underline{\otimes} \Delta^b)\tau_1'\tau_2' &= (u \otimes v) \sum_{m=0}^{n-r} \Phi_{r+m}^{(r+1,n)}\, \theta_{r+m}\theta_r\tau_1'\tau_2' \\ &= (u \otimes v) \sum_{m=0}^{n-r} \Phi_{r+m}^{(r+1,n)}\, \nu_r^{1,r+m}\, \theta_m\theta_{r+m}\tau_2' \\ &= (u \otimes v) \sum_{m=0}^{n-r} \Phi_{r+m}^{(r+1,n)}\, \nu_r^{1,r+m}\, \nu_{m+r}^{m+1,n}\, \theta_m\theta_{n-r}.\end{aligned}$$

Due to (6.19), we have $\nu_r^{1,r+m}\nu_{m+r}^{m+1,n} = \nu_r^{1,n}$. Therefore, the result follows from the next lemma.

Lemma 6.5 *In* $\mathbf{k}[B_n]$, *the following commutation rule holds*

$$\nu_r^{1,n}\,\Phi_m^{(1,n-r)} = \Phi_{r+m}^{(r+1,n)}\,\nu_r^{1,n}. \tag{6.32}$$

Proof By (6.17), we have

$$\nu_r^{1,n} = (\tau_r\tau_{r+1}\cdots\tau_{n-1})(\tau_{r-1}\tau_r\cdots\tau_{n-2})\cdots(\tau_1\tau_2\cdots\tau_{n-r}), \tag{6.33}$$

and definition (6.23) reads:

$$\Phi_m^{(1,n-r)} = \sum_{1\leq k_1<k_2<\ldots<k_m\leq n-r}\left(\prod_{i=1}^{m}[i;k_i]\right).$$

Conditions $1 \leq k_1 < k_2 < \ldots < k_m$ imply $i \leq k_i$. In particular, the chain of inequalities $1 \leq i \leq k_i - 1 < n - r$ holds unless $k_i = i$. Because $[i;i] = \mathrm{id}$, the latter chain of inequalities allows us to apply commutation rule (1.33) under the substitution $k \leftarrow 1,\, t \leftarrow n-r,\, r \leftarrow k_i - 1,\, m \leftarrow i$:

$$(\tau_1\tau_2\cdots\tau_{n-r})[i;k_i] = [i+1;k_i+1](\tau_1\tau_2\cdots\tau_{n-r}).$$

This implies

$$(\tau_1\tau_2\cdots\tau_{n-r})\Phi_m^{(1,n-r)} = \Phi_{m+1}^{(2,n-r+1)}(\tau_1\tau_2\cdots\tau_{n-r}).$$

In the same way, we have

$$(\tau_2\tau_3\cdots\tau_{n-r+1})\Phi_{m+1}^{(2,n-r+1)} = \Phi_{m+2}^{(3,n-r+2)}\Phi_{m+1}^{(2,n-r+1)}.$$

Continuation of this process ends with the required commutation formula (6.32). □

Similarly we shall check the second of identities (6.31). We have,

$$\begin{aligned}(u\underline{\otimes}v)\tau'(\mathrm{id}\underline{\otimes}\Delta^b) &= (u\otimes v)\nu_r^{1,n}\theta_{n-r}(\mathrm{id}\underline{\otimes}\Delta^b)\\ &= (u\otimes v)\nu_r^{1,n}\sum_{m=0}^{r}\Phi_{n-r+m}^{(n-r+1,n)}\theta_{n-r+m}\theta_{n-r},\end{aligned}$$

and

$$
\begin{aligned}
(u\underline{\otimes}v)(\Delta^b\underline{\otimes}\mathrm{id})\tau_2'\tau_1' &= (u\otimes v)\sum_{m=0}^{r}\Phi_m^{(1,r)}\,\theta_m\theta_r\tau_2'\tau_1' \\
&= (u\otimes v)\sum_{m=0}^{r}\Phi_m^{(1,r)}\,\nu_r^{m+1,n}\,\theta_{n-r+m}\theta_m\tau_1' \\
&= (u\otimes v)\sum_{m=0}^{r}\Phi_m^{(1,r)}\,\nu_r^{m+1,n}\,\nu_m^{1,n-r+m}\,\theta_{n-r}\theta_{n-r+m}.
\end{aligned}
$$

Taking into account representation (6.17), we obtain $\nu_r^{m+1,n}\,\nu_m^{1,n-r+m}=\nu_r^{1,n}$. Hence, it remains to apply the commutation rules of the next lemma.

Lemma 6.6 *In* $\mathbf{k}[B_n]$, *the following commutation rules hold*

$$
\nu_r^{1,n}\,\Phi_{n-r+m}^{(n-r+1,n)}=\Phi_m^{(1,r)}\,\nu_r^{1,n}. \tag{6.34}
$$

Proof By definition (6.16), we have

$$
\nu_r^{1,n}=(\tau_r\tau_{r-1}\cdots\tau_1)(\tau_{r+1}\tau_r\cdots\tau_2)\cdots(\tau_{n-1}\tau_{n-2}\cdots\tau_{n-r}),
$$

whereas the definition (6.23) states

$$
\Phi_{n-r+m}^{(n-r+1,n)}=\sum_{n-r+1\le k_1<k_2<\ldots<k_m\le n}\left(\prod_{i=1}^{m}[n-r+i;k_i]\right).
$$

Conditions $n-r+1\le k_1<k_2<\ldots<k_m$ imply $n-r+i\le k_i$. In particular, the following chain of inequalities holds: $n-r\le(n-r+i)-1\le k_i-1\le n-1$. This chain allows us to apply commutation rule (1.34) proven in Lemma 1.13 under the substitution $t\leftarrow n-1,\ k\leftarrow n-r,\ r\leftarrow k_i-1,\ m\leftarrow n-r+i:$

$$
(\tau_{n-1}\tau_{n-2}\cdots\tau_{n-r})[n-r+i;k_i]=[n-r+i-1;k_i-1](\tau_{n-1}\tau_{n-2}\cdots\tau_{n-r}).
$$

This implies

$$
(\tau_{n-1}\tau_{n-2}\cdots\tau_{n-r})\Phi_{n-r+m}^{(n-r+1,n)}=\Phi_{n-r+m-1}^{(n-r,n-1)}(\tau_{n-1}\tau_{n-2}\cdots\tau_{n-r}).
$$

Similarly,

$$
(\tau_{n-2}\tau_{n-3}\cdots\tau_{n-r-1})\Phi_{n-r+m-1}^{(n-r,n-1)}=\Phi_{n-r+m-2}^{(n-r-1,n-2)}(\tau_{n-2}\tau_{n-3}\cdots\tau_{n-r-1}).
$$

In this way, after $n-r$ steps, we obtain the required commutation rule. □

Braided Antipode We define $\sigma^b(u) = (-1)^n u \cdot \mu_n^1$, $u \in V^{\otimes n}$, where the *mirror operators* μ_n^t, $1 \le t < n$ are set up thus:

$$\mu_n^t = (\tau_t\tau_{t+1} \dots \tau_{n-1})(\tau_t\tau_{t+1} \dots \tau_{n-2}) \cdots (\tau_t\tau_{t+1})\tau_t. \tag{6.35}$$

If $u = \alpha \cdot 1 \in V^0$, then $\sigma^b(u) = u$; that is, the restriction of σ^b on $V^{\otimes 0}$ is the identity.

Lemma 6.7 *The mirror operator has another representation:*

$$\mu_n^t = (\tau_{n-1}\tau_{n-2} \cdots \tau_{t+1}\tau_t)(\tau_{n-1}\tau_{n-2} \cdots \tau_{t+2}\tau_{t+1}) \cdots (\tau_{n-1}\tau_{n-2}) \cdot \tau_{n-1} = (\mu_{n-t+1}^1)^\iota,$$

where $\iota : \tau_i \mapsto \tau_{n-i}$ *is an automorphism of* B_n; *see Sect.* 1.4.

Proof The latter equality follows from definitions of the mirror operator and ι. To check the former, we use induction on n :

$$\mu_n^t = (\tau_t\tau_{t+1} \dots \tau_{n-1})\mu_{n-1}^t = (\tau_t\tau_{t+1} \dots \tau_{n-1}) \prod_{i=t}^{n-2} [i; n-1], \tag{6.36}$$

where, as above, $[i; k] = \tau_{k-1}\tau_{k-2} \cdots \tau_i$. By definition, $\tau_{n-1}[i; n-1] = [i; n]$, whereas the commutation rule (1.34) under the substitution $t \leftarrow n-1$, $k \leftarrow i$, $r \leftarrow s+1$, $m \leftarrow s+1$ reads: $\tau_s[i; n] = [i; n]\tau_{s+1}$, provided that $i \le s \le n-2$. These relations allow us to continue (6.36):

$$= (\tau_t\tau_{t+1} \dots \tau_{n-2})\tau_{n-1}[t; n-1] \prod_{i=t+1}^{n-2} [i; n-1]$$

$$= (\tau_t\tau_{t+1} \dots \tau_{n-3})[t; n]\tau_{n-1}[t+1; n-1] \prod_{i=t+2}^{n-2} [i; n-1]$$

$$= (\tau_t\tau_{t+1} \dots \tau_{n-4})[t; n][t+1; n]\tau_{n-1}[t+2; n-1] \prod_{i=t+3}^{n-2} [i; n-1] = \dots = \prod_{i=t}^{n-1} [i; n].$$

□

Let us check the properties of the antipode, (6.12). If $u \in V^{\otimes 0}$, the properties are clear. Because $(u \underline{\otimes} v)\mathbf{m} = u \otimes v$, the sum $\sum_{(u)} \sigma^b(u_{(1)})u_{(2)}$ with $u \in V^{\otimes n}$, $n > 0$ takes the following form:

$$u \cdot \sum_{r=0}^{n} (-1)^r \Phi_r^{(1,n)} \theta_r (\sigma^b \underline{\otimes} \mathrm{id})\mathbf{m} = u \cdot \sum_{r=0}^{n} (-1)^r \Phi_r^{(1,n)} \mu_r^1,$$

whereas $\sum_{(u)} u_{(1)}\sigma^b(u_{(2)})$ reduces to

$$u \cdot \sum_{r=0}^{n} (-1)^r \Phi_r^{(1,n)} \theta_r(\mathrm{id} \underline{\otimes} \sigma^b)\mathbf{m} = u \cdot \sum_{r=0}^{n} (-1)^{n-r} \Phi_r^{(1,n)} \mu_n^{r+1}.$$

Hence, the required equalities follow from the lemma below. □

Lemma 6.8 *If $n > 0$, then*

$$\sum_{r=0}^{n} (-1)^r \Phi_r^{(1,r)} \mu_r^1 = 0 = \sum_{r=0}^{n} (-1)^{n-r} \Phi_r^{(1,r)} \mu_n^{r+1}.$$

Proof Let $T_0 = \mathrm{id}$, and for $1 \leq r \leq n$ define

$$T_r = \Phi_r^{(1,n)} \mu_r^1 = \sum_{1 \leq k_1 < k_2 < \ldots < k_r \leq n} [1; k_1][2; k_2] \cdots [r; k_r] \cdot \prod_{i=1}^{r-1} (\tau_1 \tau_2 \ldots \tau_{r-i}).$$

The operator τ_j commutes with all operators $[s; k_s]$, $s > j + 1$. Therefore

$$[2; k_2] \cdots [r; k_r] \cdot \prod_{i=1}^{r-1} (\tau_1 \tau_2 \ldots \tau_{r-i}) = [1; k_2][2; k_3] \cdots [r-1; k_r] \cdot \prod_{i=2}^{r-1} (\tau_1 \tau_2 \ldots \tau_{r-i})$$

$$= [1; k_2][1; k_3] \cdots [r-2; k_r] \cdot \prod_{i=3}^{r-1} (\tau_1 \tau_2 \ldots \tau_{r-i}) = \ldots = [1; k_2][1; k_3] \cdots [1; k_r],$$

which implies

$$T_r = \sum_{1 \leq k_1 < k_2 < \ldots < k_r \leq n} [1; k_1][1; k_2] \cdots [1; k_r].$$

We shall prove by induction on s the following equality

$$\sum_{r=0}^{s} (-1)^r T_r = (-1)^s \sum_{1 < k_1 < k_2 < \ldots < k_s \leq n} [1; k_1][1; k_2] \cdots [1; k_s]. \tag{6.37}$$

If $s = 0$, then $T_s = T_0 = \mathrm{id}$, whereas the set of sequences of length $s = 0$ contains the only sequence—the empty one. At the same time, the empty product of operators by definition is the identity operator. This explains (6.37) with $s = 0$.

If $s \geq 1$, we have the following partition:

$$\begin{aligned} &\{1 \leq k_1 < k_2 < \ldots < k_s \leq n\} \\ &= \{1 = k_1 < k_2 < \ldots < k_s \leq n\} \cup \{1 < k_1 < k_2 < \ldots < k_s \leq n\}. \end{aligned}$$

Because $[1;1] = \mathrm{id}$, this implies

$$T_s = \sum_{1<k_2<\ldots<k_s\le n} [1;k_2]\cdots[1;k_s] + \sum_{1<k_1<k_2<\ldots<k_s\le n} [1;k_1][1;k_2]\cdots[1;k_s].$$

By the induction supposition the first sum equals $(-1)^{s-1}\sum_{r=0}^{s-1}(-1)^r T_r$, whereas

$$(-1)^s \sum_{1<k_1<k_2<\ldots<k_s\le n} [1;k_1][1;k_2]\cdots[1;k_s]$$
$$= (-1)^s T_s - (-1)^s(-1)^{s-1}\sum_{r=0}^{s-1}(-1)^r T_r = \sum_{r=0}^{s}(-1)^r T_r,$$

which completes the proof of (6.37).

If $n \ge 1$, then the set $\{1 < k_1 < k_2 < \ldots < k_n \le n\}$ is empty; hence, (6.37) implies

$$\sum_{r=0}^{n}(-1)^r \Phi_r^{(1,r)}\mu_r^1 = \sum_{r=0}^{n}(-1)^r T_r = 0. \tag{6.38}$$

Further, by Lemma 6.7, we have $\mu_n^{r+1} = (\mu_{n-r}^1)^\iota$, where by definition $\tau_i^\iota = \tau_{n-i}$. Lemma 1.14 claims that

$$[1;k_1][2;k_2]\cdots[r;k_r] = [n;i_{n-r}][n-1;i_{n-r-1}]\cdots[r+2;i_2][r+1;i_1],$$

where $\{i_1 < i_2 < \ldots < i_{n-r}\}$ is the complement of $\{k_1 < k_2 < \ldots < k_r\}$ to the set $\{1,2,\ldots,n\}$. Considering that $[k;m]^\iota = [n-k+1;n-m+1]$, we have $\Phi_r^{(1,r)} = (\Phi_{n-r}^{(1,n-r)})^\iota$. This implies

$$\sum_{r=0}^{n}(-1)^{n-r}\Phi_r^{(1,r)}\mu_n^{r+1} = (\sum_{r=0}^{n}(-1)^{n-r}\Phi_{n-r}^{(1,n-r)}\mu_{n-r}^1)^\iota = 0$$

due to (6.38) with $n \leftarrow n-r$. Theorem 6.2 is completely proved. □

Proposition 6.2 *The braided antipode σ^b of the free braided Hopf algebra $\mathbf{k}\langle X\rangle$ is a braided antihomomorphism:*

$$\mathbf{m}\,\sigma^b = \tau(\sigma^b \underline{\otimes} \sigma^b)\,\mathbf{m}. \tag{6.39}$$

Proof Let $u \in V^{\otimes r}$, $v \in V^{\otimes(n-r)}$. Using (6.18) and the representation of Lemma 6.7, we have

$$(u\underline{\otimes}v)\tau\,(\sigma^b\underline{\otimes}\sigma^b)\,\mathbf{m} = (u\otimes v)\nu_r^{1,n}(-1)^{n-r}\prod_{i=1}^{n-r-1}[i;n-r] \times (-1)^r \prod_{i=n-r+1}^{n-1}[i;n].$$

By definition, $[n-r;n][i;n-r] = [i;n]$, $1 \le i \le n-r$, whereas the commutation rule (1.34) under the substitution $t \leftarrow n-1$, $k \leftarrow i$, $r \leftarrow j+r$, $m \leftarrow j+1$ takes the form $[j;r+j][i;n] = [i;n][j+1;r+j+1]$ provided that $i \le j \le n-r-1$. Applying these relations and definition (6.16) of $v_r^{1,n}$, we obtain

$$v_r^{1,n} \prod_{i=1}^{n-r-1} [i;n-r] = \prod_{j=1}^{n-r}[j;r+j] \times \prod_{i=1}^{n-r-1} [i;n-r]$$

$$= \prod_{j=1}^{n-r-1} [j;r+j] \times [n-r;n][1;n-r] \times \prod_{i=2}^{n-r-1} [i;n-r]$$

$$= \prod_{j=1}^{n-r-2} [j;r+j] \times [1;n][n-r;n][2;n-r] \times \prod_{i=3}^{n-r-1} [i;n-r]$$

$$= \prod_{j=1}^{n-r-3} [j;r+j] \times [1;n][2;n][n-r;n][3;n-r]$$

$$\times \prod_{i=4}^{n-r-1} [i;n-r] = \ldots = \prod_{i=1}^{n-r}[i;n].$$

This implies

$$(u\underline{\otimes}v)\tau\,(\sigma^b\underline{\otimes}\sigma^b)\,\mathbf{m} = (u\otimes v)(-1)^n \prod_{i=1}^{n-r}[i;n] \times \prod_{i=n-r+1}^{n-1} [i;n] = (u\underline{\otimes}v)\,\mathbf{m}\,\sigma^b,$$

which is required. □

6.3 Differential Calculi and Constants

Consider the following commutation rules for differentials

$$x_i dx_k = \sum_{s,t} \alpha_{ik}^{st} dx_s x_t, \quad \text{where} \quad (x_i \otimes x_k)\tau = \sum_{s,t} \alpha_{ik}^{st} x_s \otimes x_t, \ \ \alpha_{ik}^{st} \in \mathbf{k}.$$

In other words, the operators A_k^s are defined on generators by

$$A(x_i)_k^s = \sum_t \alpha_{ik}^{st} x_t,$$

and they are extended on $\mathbf{k}\langle X\rangle$ by formula (1.78):

$$A(vw)^i_k = \sum_s A(v)^i_s A(w)^s_k. \tag{6.40}$$

Proposition 6.3 *The above commutation rules define a right coordinate differential calculus on* $\mathbf{k}\langle X\rangle$, *which is connected with the coproduct of* $\mathbf{k}\langle X\rangle$ *as follows:*

$$\Delta^b(u) \equiv 1 \underline{\otimes} u + \sum_i x_i \underline{\otimes} \frac{\partial u}{\partial x_i} \pmod{\Lambda^2 \underline{\otimes} \mathbf{k}\langle X\rangle}, \tag{6.41}$$

where as above $\Lambda = \ker \varepsilon$ *is the ideal generated by* x_i, $i \in I$. *The partial derivatives are connected with the coproduct* $\Delta^b(u) = \sum_{(u)} u_{(1)} \underline{\otimes} u_{(2)}$ *as follows:*

$$\Delta^b\left(\frac{\partial u}{\partial x_i}\right) = \sum_{(u)} \frac{\partial u_{(1)}}{\partial x_i} \underline{\otimes} u_{(2)}. \tag{6.42}$$

Proof Let us demonstrate that the operators A^i_k are related to the braiding thus:

$$(u \underline{\otimes} x_k)\tau = \sum_i x_i \underline{\otimes} A(u)^i_k. \tag{6.43}$$

Because by definition A^i_k are linear maps, it suffices to prove this equality when u is a word in the x_i's. We perform induction on the length of u. If $u = x_s$, then

$$(x_s \underline{\otimes} x_k)\tau = \sum_{i,t} \alpha^{it}_{sk} x_i \underline{\otimes} x_t = \sum_i x_i \underline{\otimes} A(x_s)^i_k.$$

Let $u = vw$, where v, w are nonempty subwords. Using the axioms of braided algebra, the induction supposition, and (6.40), we have

$$\begin{aligned}(vw \underline{\otimes} x_k)\tau &= (v \underline{\otimes} w \underline{\otimes} x_k)(\mathbf{m} \underline{\otimes} \mathrm{id})\tau = (v \underline{\otimes} w \underline{\otimes} x_k)\tau_2\tau_1(\mathrm{id} \underline{\otimes} \mathbf{m})\\ &= (v \underline{\otimes} \sum_s x_s \underline{\otimes} A(w)^s_k)\tau_1(\mathrm{id} \underline{\otimes} \mathbf{m}) = (\sum_{s,i} x_i \underline{\otimes} A(v)^i_s \underline{\otimes} A(w)^s_k)(\mathrm{id} \underline{\otimes} \mathbf{m})\\ &= \sum_i x_i \underline{\otimes} (\sum_s A(v)^i_s A(w)^s_k) = \sum_i x_i \underline{\otimes} A(vw)^i_k.\end{aligned}$$

This completes the proof of (6.43).

Because all monomials are linearly independent in $\mathbf{k}\langle X\rangle$, for each u the elements $\varphi^i(u)$ such that $\Delta^b(u) \equiv 1 \underline{\otimes} u + \sum_i x_i \underline{\otimes} \varphi^i(u) \pmod{\Lambda^2 \underline{\otimes} \mathbf{k}\langle X\rangle}$ are uniquely defined.

Of course $\varphi^i(x_s) = \delta^i_s$. We have

$$\begin{aligned}
1 \underline{\otimes} uv + \sum_i x_i \underline{\otimes} \varphi^i(uv) &\equiv \Delta^b(uv) \\
&\equiv (1 \underline{\otimes} u + \sum_i x_i \underline{\otimes} \varphi^i(u))(1 \underline{\otimes} v + \sum_k x_k \underline{\otimes} \varphi^k(v)) \\
&\equiv 1 \underline{\otimes} uv + \sum_i x_i \underline{\otimes} \varphi^i(u)v + \sum_k \tau(u \underline{\otimes} x_k)(1 \underline{\otimes} \varphi^k(v)) \\
&= 1 \underline{\otimes} uv + \sum_i x_i \underline{\otimes} (\varphi^i(u)v + \sum_k A^i_k(u)\varphi^k(v)).
\end{aligned}$$

Hence $\varphi^i(uv) = \varphi^i(u)v + \sum_k A^i_k(u)\varphi^k(v)$; that is, the Leibniz formula (1.79) holds and φ^i are precisely the partial derivatives. This completes the proof of (6.41).

Thus we have

$$\Delta^b(u) = 1 \underline{\otimes} u + \sum_i x_i \underline{\otimes} \frac{\partial u}{\partial x_i} + \sum_{u_{(1)} \in \Lambda^2} u_{(1)} \underline{\otimes} u_{(2)}.$$

This implies

$$\begin{aligned}
(\Delta^b(u))^2(u) = &\, 1 \underline{\otimes} 1 \underline{\otimes} u + \sum_i x_i \underline{\otimes} 1 \underline{\otimes} \frac{\partial u}{\partial x_i} + \sum_i 1 \underline{\otimes} x_i \underline{\otimes} \frac{\partial u}{\partial x_i} \\
&+ \sum_{u_{(1)} \in \Lambda^2} \Delta^b(u_{(1)}) \underline{\otimes} u_{(2)}.
\end{aligned}$$

Since the coproduct is coassociative, it follows that

$$\begin{aligned}
(\Delta^b(u))^2(u) = &\, 1 \underline{\otimes} 1 \underline{\otimes} u + \sum_i 1 \underline{\otimes} x_i \underline{\otimes} \frac{\partial u}{\partial x_i} + 1 \underline{\otimes} \sum_{u_{(1)} \in \Lambda^2} u_{(1)} \underline{\otimes} u_{(2)}. \\
&+ \sum_i x_i \underline{\otimes} \Delta^b\left(\frac{\partial u}{\partial x_i}\right) + \sum_{u_{(1)} \in \Lambda^2} u_{(1)} \underline{\otimes} u_{(2)} \underline{\otimes} u_{(3)}.
\end{aligned}$$

Considering that (6.41) is already proved, we have

$$\Delta^b(u_{(1)}) \equiv 1 \underline{\otimes} u_{(1)} + \sum_i x_i \underline{\otimes} \frac{\partial u_{(1)}}{\partial x_i} \pmod{\Lambda^2 \underline{\otimes} \mathbf{k}\langle X \rangle}.$$

Therefore, the coassociativity implies

$$\Delta^b\left(\frac{\partial u}{\partial x_i}\right) = 1 \underline{\otimes} \frac{\partial u}{\partial x_i} + \sum_{u_{(1)} \in \Lambda^2} \frac{\partial u_{(1)}}{\partial x_i} \underline{\otimes}\, u_{(2)} = \sum_{(u)} \frac{\partial u_{(1)}}{\partial x_i} \underline{\otimes}\, u_{(2)}.$$

□

In a perfect analogy, there exists a left coordinate differential calculi on $\mathbf{k}\langle X\rangle$ defined by the commutation rules

$$d^* x_k \cdot v = \sum_s B(v)_k^s \cdot d^* x_s,$$

where $B(x_i)_k^s = \sum_t \alpha_{ki}^{ts}$ and $B(uv)_k^s = \sum_t B(u)_k^t B(v)_t^s$. This calculus is connected with the coproduct similarly:

$$\Delta^b(u) \equiv u \underline{\otimes} 1 + \sum_i \frac{\partial^* u}{\partial x_i} \underline{\otimes}\, x_i \ (\mathrm{mod}\ \mathbf{k}\langle X\rangle \underline{\otimes} \Lambda^2); \tag{6.44}$$

$$\Delta^b\left(\frac{\partial^* u}{\partial x_i}\right) = \sum_{(u)} u_{(1)} \underline{\otimes} \frac{\partial^* u_{(2)}}{\partial x_i}. \tag{6.45}$$

Recall that a polynomial $u \in \mathbf{k}\langle X\rangle$ is a d-constant (d^*-constant) if $\partial u/\partial x_i = 0$, $i \in I$ (respectively, $\partial^* u/\partial x_i = 0$, $i \in I$); see Definition 1.16 .

Corollary 6.1 *All primitive elements from Λ^2 are constants for both calculi.*

Proof We have $u \underline{\otimes} 1 + 1 \underline{\otimes} u \equiv 1 \underline{\otimes} u \ (\mathrm{mod}\ \Lambda^2 \underline{\otimes} \mathbf{k}\langle X\rangle)$; hence, (6.41) implies $\partial u/\partial x_i = 0$. Similarly, (6.44) implies $\partial^* u/\partial x_i = 0$. □

Corollary 6.2 *The algebra C of all d-constants is a right coideal; that is, $\Delta^b(C) \subseteq C \underline{\otimes} \mathbf{k}\langle X\rangle$. The algebra C^* of all d^*-constants is a left coideal: $\Delta^b(C^*) \subseteq \mathbf{k}\langle X\rangle \underline{\otimes} C^*$.*

Proof If $\partial u/\partial x_i = 0$, then by (6.42) we have $\sum_{(u)} (\partial u_{(1)}/\partial x_i) \underline{\otimes}\, u_{(2)} = 0$. This implies $\partial u_{(1)}/\partial x_i = 0$ because without loss of generality one may suppose that the set $\{u_{(2)}\}$ is linearly independent. In a perfect analogy, $\partial^* u/\partial x_i = 0$ implies $\partial^* u_{(2)}/\partial x_i = 0$. □

In view of the fact that the space of all constants is a subalgebra, the following note demonstrates that the constants are far from always being primitive

Lemma 6.9 *A product uv of two primitives is primitive only if $(u \otimes v)\tau = -u \otimes v$.*

Proof $\Delta^b(uv) - uv \underline{\otimes} 1 - 1 \underline{\otimes} uv = u \underline{\otimes} v + (u \underline{\otimes} v)\tau$. □

6.4 Categorical Subspaces

The free braided Hopf algebra allows one to construct braided Hopf algebras as its quotients. To this end, we have to find a way to construct braided Hopf ideals, see Lemma 6.2.

Lemma 6.10 *If a subspace $W \subseteq \mathbf{k}\langle X\rangle$ satisfies conditions 2, 3, 4, and 5 of Definition 6.1, then the ideal $I(W)$ generated by W is a braided Hopf ideal.*

Proof We have to demonstrate that $I(W)$ satisfies conditions $2-5$. For short, let S denote the space $\mathbf{k}\langle X\rangle$. Using identities of braided algebra (6.6), we obtain

$$\begin{aligned}(S\underline{\otimes}SW)\tau &= (S\underline{\otimes}S\underline{\otimes}W)(\mathrm{id}\underline{\otimes}\mathbf{m})\tau\\ &= (S\underline{\otimes}S\underline{\otimes}W)\tau_1\tau_2(\mathbf{m}\underline{\otimes}\mathrm{id}) \subseteq (S\underline{\otimes}S\underline{\otimes}W)\tau_2(\mathbf{m}\underline{\otimes}\mathrm{id})\\ &\subseteq (S\underline{\otimes}S\underline{\otimes}W + S\underline{\otimes}W\underline{\otimes}S)(\mathbf{m}\underline{\otimes}\mathrm{id}) \subseteq S\underline{\otimes}W + SW\underline{\otimes}S.\end{aligned}$$

Applying the obtained inclusion, we have

$$\begin{aligned}(S\underline{\otimes}SWS)\tau &= (S\underline{\otimes}SW\underline{\otimes}S)(\mathrm{id}\underline{\otimes}\mathbf{m})\tau = (S\underline{\otimes}SW\underline{\otimes}S)\tau_1\tau_2(\mathbf{m}\underline{\otimes}\mathrm{id})\\ &\subseteq (S\underline{\otimes}W\underline{\otimes}S + SW\underline{\otimes}S\underline{\otimes}S)\tau_2(\mathbf{m}\underline{\otimes}\mathrm{id}) \subseteq (SW\underline{\otimes}S + S\underline{\otimes}W) + SWS\underline{\otimes}S.\end{aligned}$$

The resulting inclusion takes the form

$$(S\underline{\otimes}I(W))\tau \subseteq I(W)\underline{\otimes}S + S\underline{\otimes}W. \tag{6.46}$$

Similarly,

$$\begin{aligned}(SW\underline{\otimes}S)\tau &= (S\underline{\otimes}W\underline{\otimes}S)(\mathbf{m}\underline{\otimes}\mathrm{id})\tau = (S\underline{\otimes}W\underline{\otimes}S)\tau_2\tau_1(\mathrm{id}\underline{\otimes}\mathbf{m})\\ &= (S\underline{\otimes}(W\underline{\otimes}S)\tau)\tau_1(\mathrm{id}\underline{\otimes}\mathbf{m})\\ &\subseteq (S\underline{\otimes}S\underline{\otimes}W + S\underline{\otimes}W\underline{\otimes}S)\tau_1(\mathrm{id}\underline{\otimes}\mathbf{m}) \subseteq S\underline{\otimes}SW + S\underline{\otimes}WS + W\underline{\otimes}S,\end{aligned}$$

and

$$\begin{aligned}(SWS\underline{\otimes}S)\tau &= (SW\underline{\otimes}S\underline{\otimes}S)(\mathbf{m}\underline{\otimes}\mathrm{id})\tau = (SW\underline{\otimes}S\underline{\otimes}S)\tau_2\tau_1(\mathrm{id}\underline{\otimes}\mathbf{m}) \subseteq (SW\underline{\otimes}S\underline{\otimes}S)\tau_1(\mathrm{id}\underline{\otimes}\mathbf{m})\\ &\subseteq ((S\underline{\otimes}SW + S\underline{\otimes}WS + W\underline{\otimes}S)\underline{\otimes}S)(\mathrm{id}\underline{\otimes}\mathbf{m}) \subseteq S\underline{\otimes}SWS + S\underline{\otimes}WS + W\underline{\otimes}S.\end{aligned}$$

Therefore

$$(I(W)\underline{\otimes}S)\tau \subseteq S\underline{\otimes}I(W) + W\underline{\otimes}S. \tag{6.47}$$

This completes the proof of the second property.

The third one is evident, for ε is a homomorphism of algebras. To check the fourth one, we remember that Δ^b is also a homomorphism of algebras:

$$\Delta^b(SWS) \subseteq (S\underline{\otimes}S)(S\underline{\otimes}W + W\underline{\otimes}S)(S\underline{\otimes}S),$$

where the product on $S\underline{\otimes}S$ is defined via (6.11). Using (6.46) and (6.47), we obtain the required inclusion,

$$\Delta^b(I(W)) \subseteq I(W)\underline{\otimes}S + S\underline{\otimes}I(W). \tag{6.48}$$

The fifth condition follows from the fact that σ^b is a braided anti-homomorphism; see Proposition 6.2:

$$\begin{aligned}(SW + WS)\sigma^b &= (S\underline{\otimes}W + W\underline{\otimes}S)\,\mathbf{m}\,\sigma^b = (S\underline{\otimes}W + W\underline{\otimes}S)\tau(\sigma^b\underline{\otimes}\sigma^b)\,\mathbf{m}\\ &\subseteq (S\underline{\otimes}W + W\underline{\otimes}S)(\sigma^b\underline{\otimes}\sigma^b)\,\mathbf{m} \subseteq (S\underline{\otimes}W + W\underline{\otimes}S)\,\mathbf{m} \subseteq SW + WS,\end{aligned}$$

and

$$\begin{aligned}(SWS)\sigma^b &= (SW\underline{\otimes}S)\,\mathbf{m}\,\sigma^b = (SW\underline{\otimes}S)\tau(\sigma^b\underline{\otimes}\sigma^b)\,\mathbf{m}\\ &\subseteq (S\underline{\otimes}SW + S\underline{\otimes}WS + W\underline{\otimes}S)(\sigma^b\underline{\otimes}\sigma^b)\,\mathbf{m}\\ &\subseteq (S\underline{\otimes}(SW + WS) + S\underline{\otimes}(SW + WS) + W\underline{\otimes}S)\,\mathbf{m} \subseteq SWS.\end{aligned}$$

□

Definition 6.2 A subspace W of a braided space V is called *right categorical* if $(V \otimes W)\tau \subseteq W \otimes V$. It is *left categorical* if $(W \otimes V)\tau \subseteq V \otimes W$. A left and right categorical subspace is called *categorical*.

Every categorical subspace is a braided subspace, $\tau(W \otimes W) \subseteq W \otimes W$, but not vice versa: A sum of two (right) categorical subspaces is (right) categorical, but a sum of two braided subspaces is not necessary braided. If W is a categorical subspace, than it satisfies (6.2), and by Proposition 6.1, the quotient space V/W has induced braiding.

The simplest examples of categorical subspaces in $\mathbf{k}\langle X\rangle$ are $V^{\otimes n}$. The local action provides more examples.

Lemma 6.11 *Let R be an arbitrary subset of the monoid algebra $\mathbf{k}[B_r]$ of the braid monoid B_r. If $A \subseteq V^{\otimes r}$ is a (right) categorical subspace, then so is $A \cdot R$.*

Proof Because a sum of (right) categorical subspaces is (right) categorical, we may suppose that R has just one element, Ξ. Let $v \in V^{\otimes m}$, $a \in A$. We have

$$(v\underline{\otimes}a)\tau = \sum_j a_j\underline{\otimes}v_j, \quad a_j \in A.$$

We identify the braid monoid B_r with the submonoid of B_{r+m} generated by the braids $s_1, s_2, \ldots, s_{r-1}$, whereas by B'_r we denote the submonoid generated by $s_{m+1}, s_{m+2}, \ldots, s_{m+r-1}$. Certainly, the map $\phi : s_i \to s_{m+i}$ defines an isomorphism between B_r and B'_r. We have

$$v\underline{\otimes}(a \cdot \Xi) = (v \otimes a) \cdot \phi(\Xi)\, \theta_m, \text{ and } (a_j \cdot \Xi)\underline{\otimes} v_j = (a_j \otimes v_j) \cdot \Xi\, \theta_r.$$

Consider in B_{r+m} the following element, cf. (6.16), (6.17):

$$\nu_m^{1,m+r} = \prod_{i=1}^{r} [i; m+i] = \prod_{j=m}^{1} (s_j s_{j+1} \cdots s_{j+r-1}). \tag{6.49}$$

Commutation rules (1.33) under the substitution

$$k \leftarrow j,\ t \leftarrow j+r-1,\ r \leftarrow i+j-1,\ m \leftarrow i+j-1$$

hold provided that $j \leq i+j-1 \leq i+j-1 < j+r-1$. The latter inequalities are equivalent to $1 \leq i < r$. Hence,

$$(s_j s_{j+1} \cdots s_{j+r-1}) s_{i+j-1} = s_{i+j}(s_j s_{j+1} \cdots s_{j+r-1}),\ \ 1 \leq i < r,\ 1 \leq j \leq m.$$

Therefore, decomposition (6.49) implies the commutation rule

$$\nu_m^{1,m+r} s_i = s_{m+i} \nu_m^{1,m+r} = \phi(s_i)\, \nu_m^{1,m+r}, \quad 1 \leq i < r.$$

In particular $\phi(\Xi)\, \nu_m^{1,m+r} = \nu_m^{1,m+r}\, \Xi$. Thus we obtain

$$\begin{aligned}(v\underline{\otimes}(a \cdot \Xi))\tau &= (v \otimes a) \cdot \phi(\Xi) \nu_m^{1,m+r}\, \theta_r \\ &= (v \otimes a) \cdot \nu_m^{1,m+r}\, \Xi\, \theta_r = \sum_j a_j v_j \cdot \Xi\, \theta_r = \sum_j (a_j \cdot \Xi)\underline{\otimes} v_j.\end{aligned} \tag{6.50}$$

Hence, $A \cdot \Xi$ is right categorical. In a perfect analogy, one may demonstrate that

$$((a \cdot \Xi)\underline{\otimes} v)\tau = \sum_k v_k \underline{\otimes}(a_k \cdot \Xi), \tag{6.51}$$

where $(a\underline{\otimes} v)\tau = \sum_k v_k \underline{\otimes} a_k$. Thus, $A \cdot \Xi$ is left categorical as soon as A is. □

Lemma 6.12 *The left annihilator in $V^{\otimes n}$ of any subset $R \subseteq \mathbf{k}[B_n]$ is categorical.*

Proof This follows from equalities (6.50) and (6.51). □

Corollary 6.3 *The space C of all d-constants and the space C^* of all d^*-constants are categorical.*

Proof Comparing the coproduct formula (6.25) and decomposition (6.41), we see that $u \in V^{\otimes n}$ is a constant for the right calculus if and only if $u \cdot \Phi_1^{(1,n)} = 0$. By the above lemma, each homogeneous component $C_n = C \cap V^{\otimes n}$ is categorical. Hence, so is C. Similarly, $u \in V^{\otimes n}$ is a constant for the left calculus if and only if $u \cdot \Phi_{n-1}^{(1,n)} = 0$. □

6.5 Combinatorial Rank

In this section, we adopt the concept of the combinatorial rank—see Sect. 1.5.6—to braided Hopf algebras.

Theorem 6.3 *Let H be a braided Hopf algebra generated by a braided subspace V of primitive elements. Every nonzero coideal C of H contains a nonzero primitive element.*

Proof Let $\{a_i \mid i \in I\}$ be a basis of V. Consider the free braided Hopf algebra $\mathbf{k}\langle X\rangle$ introduced in Theorem 6.2, and let us fix a natural homomorphism

$$\xi : \mathbf{k}\langle X\rangle \to H, \quad \xi(x_i) = a_i, \quad i \in I.$$

We are reminded that a constitution (multidegree) of a word u in $X = \{x_i \mid i \in I\}$ is a family $\{m_i \mid i \in I\}$ such that u has m_i occurrences of x_i. A *total degree* of u is $\sum_i m_i$. A total degree $d_t(f)$ of a polynomial f is the maximum of total degrees of its monomials.

Let us choose a polynomial $f \in \mathbf{k}\langle X\rangle$ of minimal total degree such that $\xi(f) \in C$, $\xi(f) \neq 0$. We claim that $\xi(f)$ is a primitive element.

The coproduct of a monomial has a decomposition $\Delta^b(u) = \sum u_{(1)} \otimes u_{(2)}$ where $u_{(1)}, u_{(2)}$ are monomials such that $d_t(u) = d_t(u_{(1)}) + d_t(u_{(2)})$. This implies that $\Delta^b(f) - f \underline{\otimes} 1 - 1 \underline{\otimes} f$ has a decomposition $\sum_i f_1^i \otimes f_2^i$, where the total degree of each f_1^i, f_2^i is less than the total degree of f. Our aim is to show that

$$\sum_i f_1^i \otimes f_2^i \in \mathbf{k}\langle X\rangle \underline{\otimes} \ker \xi + \ker \xi \underline{\otimes} \mathbf{k}\langle X\rangle.$$

Therefore, we may assume that the f_2^i's are linearly independent modulo $\ker \xi$. Let $\tilde{C} = \xi^{-1}(C)$. As ξ is a homomorphism of coalgebras, $\tilde{C}$ is a coideal of $\mathbf{k}\langle X\rangle$. Hence,

$$\sum_i f_1^i \otimes f_2^i \in \tilde{C} \underline{\otimes} \mathbf{k}\langle X\rangle + \mathbf{k}\langle X\rangle \underline{\otimes} \tilde{C}. \tag{6.52}$$

Consider an arbitrary linear map $\pi : \mathbf{k}\langle X\rangle \to \mathbf{k}$ such that $\pi(\tilde{C}) = 0$. Applying $\pi \underline{\otimes} \mathrm{id}$ to (6.52), we obtain $\sum_i \pi(f_1^i) f_2^i \in \tilde{C}$. The total degree of the latter sum is less than that of f. Hence $\sum_i \pi(f_1^i) f_2^i \in \ker \xi$, which implies $\pi(f_1^i) = 0$ for all i.

Inasmuch as π is arbitrary, this implies $f_i \in \tilde{C}$. Because the total degree of f_i is less than the total degree of f, we have the required inclusion $f_i \in \ker \xi$. □

Denote by J the kernel of $\xi : G\langle X\rangle \to H$. By Theorem 6.3, J contains nonzero primitive elements. Let J_1 be an ideal generated by all primitive elements of J. Clearly J_1 is a Hopf ideal. Consider the Hopf ideal J/J_1 in the quotient Hopf algebra $\mathbf{k}\langle X\rangle/J_1$. By Theorem 6.3, either $J_1 = J$ or J/J_1 has nonzero primitive elements. Let J_2/J_1 be an ideal generated by all primitive elements of J/J_1, and J_2 be its pre-image with respect to the natural homomorphism $\mathbf{k}\langle X\rangle \to \mathbf{k}\langle X\rangle/J_1$. In this way, we find a strictly increasing chain of Hopf ideals

$$0 = J_0 \subset J_1 \subset J_2 \subset \ldots \subset J_s \subset \ldots. \qquad (6.53)$$

In this chain, the ideal J_s/J_{s-1} of $\mathbf{k}\langle X\rangle/J_{s-1}$ is generated by primitive elements.

Definition 6.3 The length $\kappa(H)$ of chain (6.53) is called a *combinatorial rank* of the braided Hopf algebra H with respect to the primitive generators a_i, $i \in I$.

Lemma 6.13 *Let $\mathbf{k}\langle X\rangle_s$ be a subspace spanned by all polynomials of total degree less than or equal to s. The following inclusion holds: $\mathbf{k}\langle X\rangle_s \cap J \subseteq J_{s+1}$.*

Proof We use induction on s. If $s = 0$, the required inclusion is evident.

Assume that $\mathbf{k}\langle X\rangle_s \cap J \subseteq J_{s+1}$ for a given s. Consider the natural homomorphism $\xi : \mathbf{k}\langle X\rangle \to \mathbf{k}\langle X\rangle/J_{s+1}$. For each polynomial f of total degree $\leq s$ such that $\xi(f) \in J/J_{s+1}$ we have $\xi(f) = 0$. This means that all elements from $\mathbf{k}\langle X\rangle_{s+1} \cap J/J_{s+1}$ are primitive. In particular, $\mathbf{k}\langle X\rangle_{s+1} \cap J_{s+1} \subseteq J_{s+2}$, which is required. □

In view of the fact that $J = \bigcup_{s=1}^{\infty}(\mathbf{k}\langle X\rangle_s \cap J)$, the above lemma implies $\bigcup_{s=1}^{\infty} J_s = J$.

6.6 Braided Shuffle Hopf Algebra

In this section, as above V is a braided space with a basis $X = \{x_i, i \in I\}$. The tensor space $T(V) = \sum_{n=0}^{\infty} V^{\otimes n}$ has another structure of braided Hopf algebra, $Sh_\tau(V)$, called a *quantum shuffle algebra* or a *braided shuffle Hopf algebra*. To distinguish between elements of $Sh_\tau(V)$ and $\mathbf{k}\langle X\rangle$, a word $u = x_{i_1}x_{i_2}\cdots x_{i_n}$ considered as an element of $Sh_\tau(V)$ is designated by $(u) = (x_{i_1}x_{i_2}\cdots x_{i_n})$. The expression $(x_{i_1}x_{i_2}\cdots x_{i_n})$ is called a *co-monomial*. We extend the map $u \mapsto (u)$ to an isomorphism of linear spaces $() : \mathbf{k}\langle X\rangle \to Sh_\tau(V)$. The coproduct on the co-monomials is the *co-concatenation*:

$$\Delta^b((u)) = \sum_{u=vw} (v) \underline{\otimes} (w) = (\sum_{s=0}^{n} u \cdot \theta_s), \qquad (6.54)$$

where, as above,

$$(x_{i_1}x_{i_2}\cdots x_{i_n}\cdot\theta_s) = (x_{i_1}x_{i_2}\cdots x_{i_s})\,\underline{\otimes}\,(x_{i_{s+1}}x_{i_{s+2}}\cdots x_{i_n}).$$

We stress that both the co-concatenation coproduct and the concatenation product are independent of the braiding.

The product of co-monomials is defined as the *shuffle product*:

$$(u)(v) = \left(u\otimes v\cdot(\Phi_r^{(1,n)})^*\right),\quad u\in V^{\otimes r},\ v\in V^{\otimes(n-r)},\tag{6.55}$$

where the operators $\Phi_r^{(t,n)}$, $1\leq t\leq r\leq n$ are defined in (6.23), and $*:\mathbf{k}[B_n]\to\mathbf{k}[B_n]$ is the involution of the monoid algebra $\mathbf{k}[B_n]$ such that

$$(\tau_{i_1}\tau_{i_2}\cdots\tau_{i_s})^* = \tau_{i_s}\cdots\tau_{i_2}\tau_{i_1}.$$

In particular, the operator $(\Phi_r^{(1,n)})^*$ takes the following explicit form

$$(\Phi_r^{(1,n)})^* = \sum_{1\leq k_1<k_2<\ldots<k_r\leq n}[k_r;r][k_{r-1};r-1]\ldots[k_2;2][k_1;1],$$

where due to (6.24), we have

$$[k;k]=\text{id};\quad [k;m]=[m;k]^*=\tau_m\tau_{m+1}\cdots\tau_{k-3}\tau_{k-2}\tau_{k-1},\quad m<k.\tag{6.56}$$

Due to equality (1.43) proven in Lemma 1.14, there is another representation

$$(\Phi_r^{(1,n)})^* = \sum_{1\leq i_1<i_2<\ldots<i_{n-r}\leq n}[i_1;r+1][i_2;r+2]\cdots[i_{n-r};n].\tag{6.57}$$

The braiding is extended on co-monomials by the same formula (6.18):

$$(u)\underline{\otimes}(v)\cdot\tau = (u\otimes v\cdot\nu_r^{1,n}\cdot\theta_{n-r}),\quad u\in V^{\otimes r},\ v\in V^{\otimes(n-r)},\tag{6.58}$$

where due to (6.16), (6.17), we have

$$\begin{aligned}\nu_r^{1,n} &= (\tau_r\tau_{r-1}\cdots\tau_1)(\tau_{r+1}\tau_r\cdots\tau_2)\cdots(\tau_{n-1}\tau_{n-2}\cdots\tau_{n-r})\\ &= (\tau_r\tau_{r+1}\cdots\tau_{n-1})(\tau_{r-1}\tau_r\cdots\tau_{n-2})\cdots(\tau_1\tau_2\cdots\tau_{n-r}) = (\nu_{n-r}^{1,n})^*.\end{aligned}\tag{6.59}$$

The counit and the braided antipode on co-monomials remain unchanged also:

$$\varepsilon((u)) = \varepsilon(u),\quad \sigma^b((u)) = (\sigma^b(u)) = (u\cdot\mu_n^1),\quad u\in V^{\otimes n}.$$

In view of (6.55), it looks much more natural to define the antipode by the equality $\sigma^b((u)) = (u\cdot(\mu_n^1)^*)$, $u\in V^{\otimes n}$. In fact, it leads to the same definition.

Lemma 6.14 $(\mu_n^t)^* = \mu_n^t$, $1 \le t \le n$.

Proof We shall perform induction on n. By definition we have

$$(\mu_n^t)^* = \underline{\tau_t}(\underline{\tau_{t+1}}\tau_t)(\underline{\tau_{t+2}}\tau_{t+1}\tau_t)\cdots(\underline{\tau_{n-2}}\tau_{n-3}\cdots\tau_{t+1}\tau_t)(\underline{\tau_{n-1}}\tau_{n-2}\cdots\tau_{t+1}\tau_t).$$

Using relations $\tau_a\tau_b = \tau_b\tau_a$, $|a-b|>1$, we may move all underlined operators to the left margin position. This transformation and the induction supposition imply

$$(\mu_n^t)^* = (\tau_t\tau_{t+1}\cdots\tau_{n-2}\tau_{n-1})(\mu_{n-1}^t)^* = (\tau_t\tau_{t+1}\cdots\tau_{n-2}\tau_{n-1})(\mu_{n-1}^t) = \mu_n^t.$$

□

Theorem 6.4 *The tensor space $T(V) = Sh_\tau(V)$ with the braiding* (6.58), *the co-concatenation coproduct* (6.54), *the shuffle product* (6.55), *the counit ε, and the braided antipode σ^b is a braided Hopf algebra.*

Proof We are going to verify all axioms step by step.

1. *The co-concatenation coproduct is coassociative.* Evident.
2. *The counit properties* (1.53) are evident.
3. *Identities of braided coalgebra* (6.8). Let $u \in V^{\otimes r}$, $v \in V^{\otimes(n-r)}$. We have

$$(u)\underline{\otimes}(v)\cdot\tau(\mathrm{id}\,\underline{\otimes}\,\Delta^b) = (\sum_{i=n-r}^{n} u\otimes v\cdot\nu_r^{1,n}\cdot\theta_{n-r}\,\theta_i) = (\sum_{i=0}^{r} u\otimes v\cdot\nu_r^{1,n}\cdot\theta_{n-r}\,\theta_{n-r+i}),$$

and

$$\begin{aligned}(u)\underline{\otimes}(v)\cdot(\Delta^b\,\underline{\otimes}\,\mathrm{id})\tau_2\tau_1 &= \sum_{i=0}^{r}(u\cdot\theta_i)\underline{\otimes}(v)\cdot\tau_2\tau_1\\ &= (\sum_{i=0}^{r} u\otimes v\cdot\nu_r^{i+1,n}\cdot\theta_i\theta_{n-r+i})\tau_1\\ &= (\sum_{i=0}^{r} u\otimes v\cdot\nu_r^{i+1,n}\,\nu_i^{1,n-r+i}\cdot\theta_{n-r}\,\theta_{n-r+i}).\end{aligned}$$

The equality (6.21) proven in Theorem 6.1 reads: $\nu_{r+m}^{r+1,n}\cdot\nu_r^{1,n-m} = \nu_{r+m}^{1,n}$. Under the substitutions $r \leftarrow i$, $m \leftarrow r-i$, it reduces to $\nu_r^{1,n} = \nu_r^{i+1,n}\,\nu_i^{1,n-r+i}$. This completes the proof of the first of braided coalgebra equalities (6.8).

Similarly,

$$(u)\underline{\otimes}(v)\cdot\tau(\Delta^b\,\underline{\otimes}\,\mathrm{id}) = (\sum_{m=0}^{n-r} u\otimes v\cdot\nu_r^{1,n}\cdot\theta_{n-r}\,\theta_m),$$

and

$$(u)\underline{\otimes}(v)\cdot(\mathrm{id}\underline{\otimes}\Delta^b)\tau_1\tau_2 = (u)\underline{\otimes}\sum_{m=0}^{n-r}(v\cdot\theta_m)\tau_1\tau_2$$

$$= (\sum_{m=0}^{n-r} u\otimes v\cdot\nu_r^{1,r+m}\theta_m\theta_{r+m})\tau_2$$

$$= (\sum_{m=0}^{n-r} u\otimes v\cdot\nu_r^{1,r+m}\,\nu_{r+m}^{m+1,n}\cdot\theta_m\,\theta_{n-r})$$

It remains to apply the already proven equality $\nu_r^{1,n} = \nu_r^{1,r+m}\cdot\nu_{r+m}^{m+1,n}$, see (6.19).

4. *The shuffle product is associative*. If $u\in V^{\otimes r}$, $v\in V^{\otimes m}$, $w\in V^{\otimes(n-r-m)}$, then

$$[(u)(v)](w) = \left(u\otimes v\cdot(\Phi_r^{(1,r+m)})^*\right)(w) = \left(u\otimes v\otimes w\cdot(\Phi_r^{(1,r+m)})^*(\Phi_{r+m}^{(1,n)})^*\right),$$

whereas

$$(u)[(v)(w)] = (u)\left(v\otimes w\cdot(\Phi_{r+m}^{(r+1,n)})^*\right) = \left(u\otimes v\otimes w\cdot(\Phi_{r+m}^{(r+1,n)})^*(\Phi_r^{(1,n)})^*\right).$$

The equality $(\Phi_r^{(1,r+m)})^*(\Phi_{r+m}^{(1,n)})^* = (\Phi_{r+m}^{(r+1,n)})^*(\Phi_r^{(1,n)})^*$ follows from (6.28) stated in Lemma 6.4 applying the involution $*$.

5. *Identities of braided algebra* (6.6). Let $u\in V^{\otimes r}$, $v\in V^{\otimes m}$, $w\in V^{\otimes(n-m-r)}$. We have

$$(u)\underline{\otimes}(v)\underline{\otimes}(w)\cdot(\mathbf{m}\underline{\otimes}\mathrm{id})\tau = [(u\otimes v\cdot(\Phi_r^{(1,r+m)})^*)\underline{\otimes}(w)]\tau$$

$$= \left(u\otimes v\otimes w\cdot(\Phi_r^{(1,r+m)})^*\,\nu_{r+m}^{1,n}\,\theta_{n-m-r}\right),$$

whereas

$$(u)\underline{\otimes}(v)\underline{\otimes}(w)\cdot\tau_2\tau_1(\mathrm{id}\underline{\otimes}\mathbf{m}) = (u)\underline{\otimes}(v\otimes w\cdot\nu_m^{1,n-r}\,\theta_{n-m-r})\cdot\tau_1(\mathrm{id}\underline{\otimes}\mathbf{m})$$

$$= \left(u\otimes v\otimes w\cdot\nu_{m+r}^{r+1,n}\,\theta_{n-m}\right)\cdot\tau_1(\mathrm{id}\underline{\otimes}\mathbf{m})$$

$$= \left(u\otimes v\otimes w\cdot\nu_{m+r}^{r+1,n}\,\nu_r^{1,n-m}\theta_{n-m-r}\theta_{n-m}\right)\cdot(\mathrm{id}\underline{\otimes}\mathbf{m})$$

$$= \left(u\otimes v\otimes w\cdot\nu_{m+r}^{r+1,n}\,\nu_r^{1,n-m}(\Phi_{n-m}^{n-m-r+1,n})^*\,\theta_{n-m-r}\right).$$

Equality (6.21) reads: $\nu_{m+r}^{r+1,n}\,\nu_r^{1,n-m} = \nu_{r+m}^{1,n}$. Therefore, it remains to check that

$$(\Phi_r^{(1,r+m)})^*\,\nu_{r+m}^{1,n} = \nu_{r+m}^{1,n}\,(\Phi_{n-m}^{n-m-r+1,n})^*.$$

The commutation rule $\nu_r^{1,n}\,\Phi_m^{(1,n-r)} = \Phi_{r+m}^{(r+1,n)}\,\nu_r^{1,n}$ proven in Lemma 6.5 reduces to the required form after the replacements $r \leftarrow n-r-m$, $m \leftarrow r$ and application of the involution $*$, taking into account (6.59).

In a perfect analogy, we analyze the second equality of (6.6). We have

$$\begin{aligned}(u)\,\underline{\otimes}\,(v)\,\underline{\otimes}\,(w)\cdot(\mathrm{id}\,\underline{\otimes}\,\mathbf{m})\tau &= (u)\,\underline{\otimes}\,(v\otimes w\cdot(\Phi_m^{(1,n-r)})^*)\cdot\tau\\ &= (u\otimes v\otimes w\cdot(\Phi_{m+r}^{(r+1,n)})^*\theta_r)\cdot\tau\\ &= (u\otimes v\otimes w\cdot(\Phi_{m+r}^{(r+1,n)})^*\,\nu_r^{1,n}\,\theta_{n-r}),\end{aligned}$$

and

$$\begin{aligned}(u)\,\underline{\otimes}\,(v)\,\underline{\otimes}\,(w)\cdot\tau_1\tau_2(\mathbf{m}\,\underline{\otimes}\,\mathrm{id}) &= (u\otimes v\cdot\nu_r^{1,r+m}\,\theta_m)\,\underline{\otimes}\,(w)\cdot\tau_2(\mathbf{m}\,\underline{\otimes}\,\mathrm{id})\\ &= \left(u\otimes v\otimes w\cdot\nu_r^{1,r+m}\,\nu_{r+m}^{m+1,n}\,\theta_m\theta_{n-r}\right)\cdot(\mathbf{m}\,\underline{\otimes}\,\mathrm{id})\\ &= \left(u\otimes v\otimes w\cdot\nu_r^{1,r+m}\,\nu_{r+m}^{m+1,n}\,(\Phi_m^{1,n-r})^*\,\theta_{n-r}\right).\end{aligned}$$

Equality (6.19) states that $\nu_r^{1,r+m}\,\nu_{r+m}^{m+1,n} = \nu_r^{1,n}$; hence, it suffices to check that

$$(\Phi_{m+r}^{(r+1,n)})^*\,\nu_r^{1,n} = \nu_r^{1,n}\,(\Phi_m^{1,n-r})^*.$$

The commutation rule $\nu_r^{1,n}\,\Phi_{n-r+m}^{(n-r+1,n)} = \Phi_m^{(1,r)}\,\nu_r^{1,n}$ proven in Lemma 6.6 reduces to the required form after the replacement $r \leftarrow n-r$ and application of the involution $*$, taking into account (6.59).

6. *Antipode*. If $u \in V^{\otimes n}$, $n > 0$, then Lemmas 6.14 and 6.8 imply

$$\sum_{((u))}\sigma^b((u_{(1)}))\cdot(u_{(2)}) = \sum_{r=0}^{n} u\cdot\mu_r^1(\Phi_r^{(1,r)})^* = (u\cdot(\sum_{r=0}^{n}\Phi_r^{(1,r)}\mu_r^1)^*) = 0.$$

Similarly,

$$\sum_{((u))}(u_{(1)})\cdot\sigma^b((u_{(2)})) = \sum_{r=0}^{n} u\cdot\mu_n^{r+1}(\Phi_r^{(1,r)})^* = (u\cdot(\sum_{r=0}^{n}\Phi_r^{(1,r)}\mu_n^{r+1})^*) = 0.$$

□

Remark 6.1 Let V be finite dimensional. On the dual space $V^* = \mathrm{Hom}(V, \mathbf{k})$, one may define a braiding $x_*^i \otimes x_*^s \to \sum_{k,r}\alpha_{i,s}^{k,r}x_*^k \otimes x_*^r$ where $x_i \otimes x_s \to \sum_{k,r}\alpha_{i,s}^{k,r}x_k \otimes x_r$ is the initial braiding of V, and $\{x_*^i \mid i \in I\}$ is the dual basis of V^*, so that $x_*^i(x_s) = \delta_s^i$. In this case, the map

$$\Delta_{k,s}^* : V^{\otimes(k+m)} \to V^{\otimes k} \otimes V^{\otimes s}$$

dual to the (k,s)-component of the coproduct of the free braided Hopf algebra $\mathbf{k}\langle V^*\rangle$,

$$\Delta_{k,s} : (V^*)^{\otimes(k+s)} \to (V^*)^{\otimes k} \otimes (V^*)^{\otimes s},$$

is precisely the (k,s)-component of the shuffle product of $Sh_\tau(V)$. Similarly, the map

$$\mathbf{m}^*_{k,s} : V^{\otimes k} \otimes V^{\otimes s} \to V^{\otimes(k+m)}$$

dual to the (k,s)-component of the concatenation product,

$$\mathbf{m}_{k,s} : (V^*)^{\otimes k} \otimes (V^*)^{\otimes s} \to (V^*)^{\otimes(k+s)},$$

is precisely the (k,s)-component of the co-concatenation coproduct of $Sh_\tau(V)$.

This clearly explains why relations of $\mathbf{k}[B_n]$ that are responsible for $\mathbf{k}\langle X\rangle$ being a braided Hopf algebra are responsible for Sh_τ being a braided Hopf algebra too.

6.7 Nichols Algebra

In this section, we consider a very important construction of the Nichols algebra (or, equivalently, *quantum symmetric algebra* or *braided symmetric algebra*) $\mathcal{B}(V)$ related to a braided space V with a basis $\{x_i \mid i \in I\}$. This braided Hopf algebra appeared independently in various articles with different definitions. In fact, each rediscovering may be considered a demonstration of a new property of that object. According to one of a myriad of characterizations, this is a subalgebra of the quantum shuffle algebra.

Definition 6.4 A subalgebra $\mathcal{B}(V)$ generated by V in $Sh_\tau(V)$ is called a *Nichols algebra* related to a braided space V.

The Nichols algebra is a braided Hopf subalgebra of $Sh_\tau(V)$ and a homomorphic image of the braided Hopf algebra $\mathbf{k}\langle V\rangle$. The epimorphism

$$\Omega : \mathbf{k}\langle V\rangle \to \mathcal{B}(V)$$

has a lot of nice characterizations. We consider just two of them. Additionally we prove fundamental properties of the Hopf ideal $\ker \Omega$.

For each permutation $\pi \in S_n$, we fix an element $\pi^b \in B_n$ as follows. If $\pi \in S_0 = \{1\}$, then $\pi^b = 1$. Assume that π^b is already defined for all $\pi \in S_{n-1}$. If $\pi \in S_n \setminus S_{n-1}$, then the permutation $\pi t_{\pi(n)} t_{\pi(n)+1} \cdots t_{n-2} t_{n-1}$ belongs to S_{n-1}, where as usual t_i is the elementary transposition $i \leftrightarrow i+1$. We put

$$\pi^b = (\pi t_{\pi(n)} t_{\pi(n)+1} \cdots t_{n-2} t_{n-1})^b\, \tau_{n-1} \tau_{n-2} \cdots \tau_{\pi(n)+1} \tau_{\pi(n)}. \tag{6.60}$$

Proposition 6.4 *If* $u \in V^{\otimes n}$, *then*

$$\Omega(u) = (\sum_{\pi \in S_n} u \cdot \pi^b). \tag{6.61}$$

Proof We use induction on n. If $n = 1$, the equality takes a form $\Omega(x) = (x), x \in V$. Let $u = v \otimes x$, $v \in V^{\otimes(n-1)}$, $x \in V$. Using the induction supposition, representation (6.57), and definition (6.60), we have

$$\Omega(u) = \Omega(v)\Omega(x) = (\sum_{\pi \in S_{n-1}} v \cdot \pi^b)(x) = (\sum_{\pi \in S_{n-1}} v \cdot \pi^b \otimes x \cdot \sum_{1 \le i \le n} [i;n])$$

$$= \sum_{\pi \in S_{n-1}} \sum_{1 \le i \le n} u \cdot \pi^b \tau_{n-1}\tau_{n-2} \cdots \tau_i = \sum_{\pi \in S_{n-1}} \sum_{1 \le i \le n} u \cdot (\pi t_{n-1} t_{n-2} \cdots t_i)^b.$$

For a given i the set $\Sigma_i = S_{n-1} t_{n-1} t_{n-2} \cdots t_i$ consists of all permutations $\nu \in S_n$ such that $\nu(n) = i$. Hence the union of Σ_i, $1 \le i \le n$ equals S_n. □

Lemma 6.15 *The Hopf ideal* $\ker \Omega$ *is equal to the sum of all coideals C such that* $C \subseteq \Lambda^2$, *where* $\Lambda = \ker \varepsilon$ *is the ideal of* $\mathbf{k}\langle X \rangle$ *generated by* x_i, $i \in I$. *In particular,* $\ker \Omega$ *is the biggest Hopf ideal contained in* Λ^2.

Proof If $f = \alpha_0 + \sum_i \alpha_i x_i + a$, $a \in \Lambda^2$, then $\Omega(f) = \alpha_0 + \sum \alpha_i (x_i) + \Omega(a)$, and $\Omega(a) \in (\Omega(\Lambda))^2 \subseteq (\Lambda^2)$. By definition of Sh_τ, the elements 1, (x_i), $i \in I$ are linearly independent modulo the space (Λ^2) spanned by all (u), where u is a word of length greater than 1. Therefore if $f \in \ker \Omega$, then $\alpha_0 = \alpha_i = 0$, $i \in I$; that is, $\ker \Omega \subseteq \Lambda^2$.

If $C \subseteq \Lambda^2$ is a coideal, then $\Omega(C)$ is a coideal of $Sh_\tau(V)$ such that $\Omega(C) \subseteq (\Lambda^2)$. If $\Omega(C) \neq 0$, then by Theorem 6.3 there exists a nonzero primitive element $a \in \Omega(C)$. However, the definition of the co-concatenation coproduct shows that no one element from (Λ^2) is primitive. Hence $\Omega(C) = 0$, and $C \subseteq \ker \Omega$. □

The next characterization of Ω is related to the coordinate differential calculi.

Theorem 6.5 *The Nichols algebra has a right differential calculus such that* Ω *is a homomorphism of differential algebras. The partial derivatives with respect to* (x_i), $i \in I$ *connect the calculus on* $\mathcal{B}(V)$ *with the coproduct via*

$$\Delta^b((u)) \equiv 1 \underline{\otimes} (u) + \sum_i (x_i) \underline{\otimes} \frac{\partial(u)}{\partial(x_i)} \pmod{(\Lambda)^2 \underline{\otimes} \mathcal{B}(V)}, \quad (u) \in \mathcal{B}(V), \tag{6.62}$$

where, as usual, $\Lambda = \ker \varepsilon$ *is an ideal generated by* x_i, $i \in I$.

Proof By Lemma 1.25 we have to show that the kernel J of Ω is a differential ideal. Let $u \in J$. Because J is a Hopf ideal, we have

$$\Delta^b(u) = 1 \underline{\otimes} u + \sum_i x_i \underline{\otimes} \frac{\partial u}{\partial x_i} + \cdots \in J \underline{\otimes} \mathbf{k}\langle V \rangle + \mathbf{k}\langle V \rangle \underline{\otimes} J. \tag{6.63}$$

Let π^s be a linear map $\pi^s : \mathbf{k}\langle V\rangle \to \mathbf{k}$, such that $\pi(1) = 0$, $\pi^s(x_i) = \delta_i^s$, $\pi(\Lambda^2) = 0$. Applying $\pi^s \underline{\otimes}\,\mathrm{id}$ to both sides of (6.63), we obtain $\partial u/\partial x_s \in J$ because due to Lemma 6.15 the inclusion $J \subseteq \Lambda^2$ holds. □

Proposition 6.5 (M. Graña) *The homomorphism* $\Omega : \mathbf{k}\langle V\rangle \to \mathscr{B}(V)$, $x_i \mapsto (x_i)$ *has the following representation in terms of the above defined differential calculus:*

$$\Omega(u) = \sum_{i_1, i_2, \ldots, i_n} \frac{\partial^n u}{\partial x_{i_1} \partial x_{i_2} \ldots \partial x_{i_n}} (x_{i_1} x_{i_2} \ldots x_{i_n}), \quad u \in V^{\otimes n}. \tag{6.64}$$

Proof Let us define linear maps $D^i : \mathbf{k}\langle V\rangle \to \mathbf{k}\langle V\rangle$, $i \in I$ such that $D^i(1) = 0$, $D^i(x_s u) = \delta_s^i u$. The equality

$$u = \sum_{i_1, i_2, \ldots, i_n} u D^{i_1} D^{i_2} \cdots D^{i_n} x_{i_1} x_{i_2} \cdots x_{i_{n-1}} x_{i_n}.$$

is evident if u is a monomial of length $n > 0$, and by linearity it is valid for arbitrary $u \in V^{\otimes n}$, $n > 0$. Considering that $u D^{i_1} D^{i_2} \cdots D^{i_n}$ are scalars, we obtain

$$(u) = \sum_{i_1, i_2, \ldots, i_n} u D^{i_1} D^{i_2} \cdots D^{i_n} (x_{i_1} x_{i_2} \cdots x_{i_{n-1}} x_{i_n}) \tag{6.65}$$

Definition of the co-concatenation coproduct and decomposition (6.62) imply

$$\frac{\partial(w)}{\partial(x_i)} = (u D^i), \qquad (w) \in \mathscr{B}(V). \tag{6.66}$$

Let $\Omega(v) = (w)$, $w \in \mathbf{k}\langle V\rangle$. By Theorem 6.5 the map Ω is a homomorphism of differential algebras. Therefore it commutes with partial derivatives. Using (6.65) and (6.66), we have

$$\begin{aligned}
\Omega(v) = (w) &= \sum_{i_1, i_2, \ldots, i_n} w D^{i_1} D^{i_2} \cdots D^{i_n} (x_{i_1} x_{i_2} \cdots x_{i_{n-1}} x_{i_n}) \\
&= \sum_{i_1, i_2, \ldots, i_n} \frac{\partial^n (w)}{\partial(x_{i_1}) \partial(x_{i_2}) \ldots \partial(x_{i_n})} (x_{i_1} x_{i_2} \ldots x_{i_n}) \\
&= \sum_{i_1, i_2, \ldots, i_n} \frac{\partial^n \,\Omega(v)}{\partial(x_{i_1}) \partial(x_{i_2}) \ldots \partial(x_{i_n})} (x_{i_1} x_{i_2} \ldots x_{i_n}) \\
&= \sum_{i_1, i_2, \ldots, i_n} \Omega\left(\frac{\partial^n u}{\partial x_{i_1} \partial x_{i_2} \ldots \partial x_{i_n}} \right) (x_{i_1} x_{i_2} \ldots x_{i_n}),
\end{aligned}$$

which proves (6.64) because in the latter expression Ω acts on scalars. □

In a perfect analogy, there exists a left coordinate differential calculus on $\mathscr{B}(V)$ defined by the commutation rules

$$d^*x_k \cdot v = \sum_s B(v)_k^s \cdot d^*x_s,$$

where $B(x_i)_k^s = \sum_t \alpha_{ki}^{ts}$ and $B(uv)_k^s = \sum_t B(u)_k^t B(v)_t^s$. This calculus is connected with the coproduct in a similar way; see (6.44):

$$\Delta^b((u)) \equiv (u) \underline{\otimes} 1 + \sum_i \frac{\partial^*(u)}{\partial(x_i)} \underline{\otimes} (x_i) \pmod{\mathscr{B}(V) \underline{\otimes} (\Lambda)^2}, \tag{6.67}$$

where, as above, $\Lambda = \ker \varepsilon$. The same reasoning shows that the homomorphism Ω has a representation in terms of d^* as well:

$$\Omega(u) = \sum_{i_1, i_2, \ldots, i_n} \frac{(\partial^*)^n u}{\partial^* x_{i_1} \partial^* x_{i_2} \ldots \partial^* x_{i_n}} (x_{i_n} x_{i_{n-1}} \ldots x_{i_1}), \quad u \in V^{\otimes n}. \tag{6.68}$$

The combinatorial rank of a Nichols algebra is an invariant related to the braiding τ. By Lemma 6.15 it has a maximal value among all Hopf homomorphic images of $\mathbf{k}\langle X\rangle$ when the images of the generators x_i, $i \in I$ are linearly independent modulo the space spanned by values of words of length > 1. The following statements shows that the combinatorial rank is finite if $\mathscr{B}(V)$ has a finite dimension.

Proposition 6.6 (A. Ardizzoni) *The combinatorial rank of a finite dimensional Nichols algebra is finite.*

Proof If dimension of $\mathscr{B}(V)$ is finite, then, of course, so is the dimension of V. Because $\mathscr{B}(V)$ is a homogeneous subalgebra of Sh_τ, the ideal Λ generated by V in $\mathscr{B}(V)$ is nilpotent; that is, $V^{\otimes s} \subseteq \ker \Omega$ for a suitable $s > 0$. Using Lemma 6.13, we have $V^{\otimes s} \subseteq \mathbf{k}\langle X\rangle_s \cap J \subseteq J_{s+1}$, where $J = \ker \Omega$. Therefore $\mathbf{k}\langle X\rangle / J_{s+1}$ is a finite dimensional algebra. This implies that the chain $J_{s+1} \subset J_{s+2} \subset \ldots$ is not infinite. □

6.8 Radford Biproduct

In this section, we focus on the relations between character Hopf algebras and braided Hopf algebras. Consider first the free character Hopf algebra $G\langle Y\rangle$ generated by skew-primitive variables y_i, $i \in I$:

$$\Delta(y_i) = y_i \otimes h_i + f_i \otimes y_i, \quad y_i g = \chi^i(g) g y_i, \quad h_i, f_i, g \in G, \ i \in I, \tag{6.69}$$

see Sect. 1.5.3. Let $x_i = h_i^{-1} y_i$, $i \in I$ be the normalized skew-primitive variables, and let V be a linear space spanned by the x_i's equipped by a diagonal braiding

$$(x_i \otimes x_s)\tau = p_{si}^{-1}(x_s \otimes x_i).$$

Consider the free subalgebra $\mathbf{k}\langle X\rangle$ as the free braided Hopf algebra defined by the braided space V in Sect. 6.2.

Lemma 6.16 *If u, v are polynomials in X homogeneous in each x_i, $i \in I$, then*

$$(u \otimes v)\tau = \chi^v(g_u)^{-1}(v \otimes u),$$

where as usual $g_u = \mathrm{gr}(u)$ is a group-like element that appears from each monomial of u under the substitutions $x_i \leftarrow g_i = h_i^{-1} f_i$, $i \in I$, and χ^v is a character that appears from each monomial of v under the substitutions $x_i \leftarrow \chi^i$, $i \in I$.

Proof It suffices to demonstrate the formula for monomials u, v. We perform induction on the sum of lengths of u and v. If u, v are the generators, there is nothing to prove. Applying the axioms of braided algebra (6.6), we have

$$\begin{aligned}(ux_s \otimes x_i)\tau &= (u \otimes x_s \otimes x_i)(\mathbf{m} \otimes \mathrm{id})\tau = (u \otimes x_s \otimes x_i)\tau_2\tau_1(\mathrm{id} \otimes \mathbf{m})\\ &= \chi^i(g_s)^{-1}(u \otimes x_i \otimes x_s)\tau_1(\mathrm{id} \otimes \mathbf{m})\\ &= \chi^i(g_s)^{-1}\chi^i(g_u)^{-1}(x_i \otimes u \otimes x_s)(\mathrm{id} \otimes \mathbf{m})\\ &= \chi^i(g_{ux_s})^{-1}(x_i \otimes ux_s).\end{aligned}$$

Similarly

$$\begin{aligned}(u \otimes vx_i)\tau &= (u \otimes v \otimes x_i)(\mathrm{id} \otimes \mathbf{m})\tau = (u \otimes v \otimes x_i)\tau_1\tau_2(\mathbf{m} \otimes \mathrm{id})\\ &= \chi^v(g_u)^{-1}(v \otimes u \otimes x_i)\tau_2(\mathbf{m} \otimes \mathrm{id})\\ &= \chi^v(g_u)^{-1}\chi^i(g_u)^{-1}(v \otimes x_i \otimes u)(\mathbf{m} \otimes \mathrm{id})\\ &= \chi^{vx_i}(g_u)^{-1}(vx_i \otimes u).\end{aligned}$$

□

Theorem 6.6 *If u is a polynomial in X homogeneous in each x_i, $i \in I$ and $\Delta(u) = \sum_{(u)} u_{(1)} \otimes u_{(2)}$ is the coproduct of u in $G\langle X\rangle$ with homogeneous $u_{(1)}, u_{(2)}$, then*

$$\Delta^b(u) = \sum_{(u)} u_{(1)}\mathrm{gr}(u_{(2)})^{-1} \underline{\otimes}\, u_{(2)}. \tag{6.70}$$

In other words, $u^b_{(1)} = u_{(1)}\mathrm{gr}(u_{(2)})^{-1}$ and $u^b_{(2)} = u_{(2)}$. The antipode σ of H and the braided antipode σ^b of $\mathbf{k}\langle X\rangle$ are related by $\sigma^b(u) = g_u\sigma(u)$.

Proof It suffices to check the formulas for monomials. We use induction on the length of u. If $u = x_i$, the formulas are clear. We have

$$\Delta(ux_i) = (\sum_{(u)} u_{(1)} \otimes u_{(2)})(x_i \otimes 1 + g_i \otimes x_i) = \sum_{(u)} u_{(1)}x_i \otimes u_{(2)} + \sum u_{(1)}g_i \otimes u_{(2)}x_i.$$

Using induction supposition, we obtain

$$\begin{aligned}\Delta^b(ux_i) &= (\sum_{(u)} u_{(1)}\mathrm{gr}(u_{(2)})^{-1} \underline{\otimes}\, u_{(2)})(x_i \underline{\otimes}\, 1 + 1 \underline{\otimes}\, x_i)\\ &= \sum_{(u)} \chi^i(\mathrm{gr}(u_{(2)}))^{-1} u_{(1)}\mathrm{gr}(u_{(2)})^{-1}x_i \underline{\otimes}\, u_{(2)} + \sum_{(u)} u_{(1)}\mathrm{gr}(u_{(2)})^{-1} \underline{\otimes}\, u_{(2)}x_i\\ &= \sum_{(u)} u_{(1)}\, x_i\, \mathrm{gr}(u_{(2)})^{-1} \underline{\otimes}\, u_{(2)} + \sum_{(u)} u_{(1)}\, g_i\, \mathrm{gr}(u_{(2)}x_i)^{-1} \underline{\otimes}\, u_{(2)}x_i,\end{aligned}$$

which is required. Similarly, we have $\sigma(x_i) = -g_ix_i$, $\sigma^b(x_i) = -x_i$. By Proposition 6.2, the braided antipode σ^b is a braided anti-homomorphism. Therefore, using induction supposition, we have

$$\begin{aligned}\sigma^b(ux_i) &= (u \otimes x_i)\mathbf{m}\sigma^b = (u \otimes x_i)\tau(\sigma^b \underline{\otimes} \sigma^b)\mathbf{m}\\ &= \chi^i(g_u)^{-1}(x_i \otimes u)(\sigma^b \underline{\otimes} \sigma^b)\mathbf{m} = \chi^i(g_u)^{-1}(-x_i)\sigma^b(u)\\ &= -\chi^i(g_u)^{-1}x_ig_u\sigma(u).\end{aligned}$$

Considering that the antipode σ is an anti-homomorphism, we may develop

$$g_{ux_i}\sigma(ux_i) = -g_ux_i\sigma(u) = -\chi^i(g_u)^{-1}x_ig_u\sigma(u) = \sigma^b(ux_i).$$

□

Proposition 6.7 *The partial derivatives* $\partial_i(u)$ *and* $\partial_i^*(u)$ *defined in Proposition* 1.8 *and Proposition* 1.9 *are related to the partial derivatives defined on the free braided Hopf algebra* $\mathbf{k}\langle X\rangle$ *as follows:*

$$\partial_i(u) = p(u, x_i)p_{ii}^{-1}\frac{\partial^* u}{\partial x_i}, \quad \partial_i^*(u) = p(x_i, u)p_{ii}^{-1}\frac{\partial u}{\partial x_i}.$$

Proof Using relation (6.70) between coproduct Δ of $G\langle X\rangle$ and braided coproduct Δ^b of $\mathbf{k}\langle X\rangle$, we may compare decompositions (1.80) and (6.44). We obtain

$$\partial_i(u) = g_i^{-1}\frac{\partial^* u}{\partial x_i}g_i = p(u, x_i)p_{ii}^{-1}\frac{\partial^* u}{\partial x_i}.$$

In the same way, we may compare decompositions (1.85) and (6.41). This yields

$$g_u g_i^{-1} x_i \, \mathrm{gr}(\partial_i^*(u))^{-1} \otimes \partial_i^*(u) = x_i \otimes \frac{\partial^* u}{\partial x_i}.$$

We have $\mathrm{gr}(\partial_i^*(u)) = \mathrm{gr}(u) g_i^{-1}$ provided that u depends on x_i. Otherwise, both the required relations reduce to $0 = 0$. □

Let H be an arbitrary character Hopf algebra generated by normalized skew-primitive semi-invariants a_i, $i \in I$ and group G. There exists a Hopf algebra homomorphism

$$\xi : G\langle X\rangle \to H, \ \ \xi(x_i) = a_i, \ \ \xi(g) = g, \ \ i \in I, \ \ g \in G. \tag{6.71}$$

Let A be a subalgebra of H generated by the elements a_i, $i \in I$. Recall that H has a grading by the character group $\hat{G}$; see Lemma 1.17. At the same time, the grading by the group G defined on the free character Hopf algebra by (1.92) not always retains on the Hopf algebra H. For example, the relations $[x_i, x_i^-] = 1 - g_i g_i^-$ connecting positive and negative components of a quantization are not homogeneous with respect to the grading by G, so that the quantizations do have grading by $\hat{G}$ but do not have grading by G.

Lemma 6.17 *If* $\ker \xi \subseteq G\Lambda$, *then both H and A are bigraded algebras; that is, the gradings by G and $\hat{G}$ retain on H and A. Here, as above, Λ is the ideal of $\mathbf{k}\langle X\rangle$ generated by x_i, $i \in I$.*

Proof The gradings retain on H due to Lemmas 1.17 and 1.26. The subalgebra A generated by a_i, $i \in I$ is homogeneous because all generators are. □

We stress that each word in the a_i's is homogeneous with respect to both gradings. Consequently, all expressions homogeneous in each a_i are homogeneous with respect to both gradings but not vice versa in general.

Theorem 6.7 *If* $\ker \xi \subseteq G\Lambda$, *then subalgebra A generated by a_i, $i \in I$ has a structure of a braided Hopf algebra such that the restriction of ξ on $\mathbf{k}\langle X\rangle$ is a homomorphism of braided Hopf algebras. In particular, the braided coproduct and the braided antipode on A and the coproduct and the antipode on H are related by*

$$\Delta^b(a) = \sum_{(a)} a_{(1)} \mathrm{gr}(a_{(2)})^{-1} \underline{\otimes}\, a_{(2)}, \ \ \sigma^b(a) = g_a \sigma(a), \tag{6.72}$$

where $a \in A$ is an arbitrary homogeneous element with respect to both gradings.

Proof By Lemma 6.2, it suffices to demonstrate that $W = \ker \xi \cap \mathbf{k}\langle X\rangle$ is a braided Hopf ideal of $\mathbf{k}\langle X\rangle$. Let us check all five conditions of Definition 6.1.

1. W is an ideal of the algebra $\mathbf{k}\langle X\rangle$ because $\ker \xi$ is an ideal of $G\langle X\rangle$.

2. $(\mathbf{k}\langle X\rangle \underline{\otimes} W + W \underline{\otimes} \mathbf{k}\langle X\rangle)\tau \subseteq \mathbf{k}\langle X\rangle \underline{\otimes} W + W \underline{\otimes} \mathbf{k}\langle X\rangle$ as W is homogeneous and due to Lemma 6.16, we have $(\mathbf{k}\langle X\rangle \underline{\otimes} W)\tau \subseteq W \underline{\otimes} \mathbf{k}\langle X\rangle$, $(W \underline{\otimes} \mathbf{k}\langle X\rangle)\tau \subseteq \mathbf{k}\langle X\rangle \underline{\otimes} W$.
3. $\varepsilon(W) = 0$ is evident.
4. $\Delta^b(W) \subseteq \mathbf{k}\langle X\rangle \underline{\otimes} W + W \underline{\otimes} \mathbf{k}\langle X\rangle$. We have

$$\Delta^b(W) \subseteq \ker\xi \underline{\otimes} G\langle X\rangle + G\langle X\rangle \underline{\otimes} \ker\xi$$

by virtue of the fact that $\ker\xi$ is a coideal of $G\langle X\rangle$. If T is a complement of the linear space W to $\ker\xi$, then $T \cap \mathbf{k}\langle X\rangle = 0$. Let U be a complement of $T \oplus \mathbf{k}\langle X\rangle$ to $G\langle X\rangle$. Consider a linear map $\pi : G\langle X\rangle \to G\langle X\rangle$ such that

$$\pi(U \oplus T) = 0, \quad \pi|_{\mathbf{k}\langle X\rangle} = \text{id}.$$

We have $\pi(G\langle X\rangle) \subseteq \mathbf{k}\langle X\rangle$, and $\pi(\ker\xi) \subseteq W$ because $G\langle X\rangle = U \oplus T \oplus \mathbf{k}\langle X\rangle$, and $T \oplus W = \ker\xi$. Considering that by definition $\Delta^b(W) \subseteq \mathbf{k}\langle X\rangle \underline{\otimes} \mathbf{k}\langle X\rangle$, we obtain the required inclusion:

$$\begin{aligned}\Delta^b(W) &= \Delta^b(W)(\pi \underline{\otimes} \pi) \subseteq \pi(\ker\xi) \underline{\otimes} \pi(G\langle X\rangle) + \pi(G\langle X\rangle) \underline{\otimes} \pi(\ker\xi)\\ &= \mathbf{k}\langle X\rangle \underline{\otimes} W + W \underline{\otimes} \mathbf{k}\langle X\rangle,\end{aligned}$$

5. $\sigma^b(W) \subseteq W$ because $\sigma^b(W) \subseteq G \ker\xi = \ker\xi$, and $\sigma^b(\mathbf{k}\langle X\rangle) \subseteq \mathbf{k}\langle X\rangle$. □

The proven theorem and Lemma 6.17 demonstrate that A is a bigraded braided Hopf algebra in the sense of the definition below.

Definition 6.5 A braided Hopf algebra L is said to be a *bigraded braided Hopf algebra* with respect to an Abelian group G if L is graded by $G \times \hat{G}$,

$$L = \bigoplus_{g\in G,\ \chi\in\hat{G}} L_g^\chi, \quad L_g^\chi L_h^{\chi'} \subseteq L_{gh}^{\chi\chi'}, \quad \Delta^b(L_g^\chi) \subseteq \bigoplus_{fh=g,\ \chi'\chi''=\chi} L_f^{\chi'} \underline{\otimes} L_h^{\chi''},$$

and the braiding is defined by the grading as follows:

$$\tau : u \otimes v \mapsto (\chi(g))^{-1} \cdot (v \otimes u), \quad u \in L^\chi,\ v \in L_g,\ g \in G,\ \chi \in \hat{G}.$$

In the above theorem, one may reconstruct H from A as a *Radford biproduct* $H = \mathbf{k}[G] \star A$. Let L be an arbitrary bigraded braided Hopf algebra generated by homogeneous primitive elements a_i, $\Delta^b(a_i) = a_i \underline{\otimes} 1 + 1 \underline{\otimes} a_i$, $i \in I$. As an algebra the Radford biproduct $H = \mathbf{k}[G] \star L$ is a skew group ring with commutation rules

$$ag = \chi(g)ga, \quad a \in L^\chi, \quad g \in G, \tag{6.73}$$

whereas the coproduct, the antipode, and the counit are defined by

$$\Delta(ga) = \sum_{(a)} ga^b_{(1)}\mathrm{gr}(a^b_{(2)}) \otimes ga^b_{(2)},\ \sigma(ga) = g_u^{-1}g^{-1}\sigma^b(a),\ \varepsilon(ga) = \varepsilon^b(a), \tag{6.74}$$

where a is an arbitrary homogeneous element with respect to both gradings, $a \in L_h^\chi$, and $\Delta^b(a) = \sum_{(a)} a^b_{(1)} \underline{\otimes} a^b_{(2)}$ with $a^b_{(1)} \in L_f^{\chi'}$, $a^b_{(2)} \in L_h^{\chi''}$, $h = \mathrm{gr}(a^b_{(2)})$.

Theorem 6.7 in terms of the Radford biproduct states that $H = \mathbf{k}[G] \star A$ provided that $\ker \xi \subseteq G\Lambda$. The converse statement is also valid.

Proposition 6.8 *If L is a bigraded braided Hopf algebra generated by homogeneous primitive elements a_i, $i \in I$, then $H = \mathbf{k}[G] \star L$ is a character Hopf algebra and $\ker \xi \subseteq G\Lambda$, where $\xi : G\langle X\rangle \to H$ is a natural homomorphism $\xi(x_i) = a_i$, $i \in I$.*

Proof Consider a space spanned by x_i, $i \in I$ as a braided space with a braiding

$$\tau(x_i \otimes x_s) = (\chi^s(g_i))^{-1}(x_s \otimes x_i), \quad \text{where} \quad a_i \in L_{g_i},\ a_s \in L^{\chi^s}.$$

By Theorem 6.7, we have $G\langle X\rangle = \mathbf{k}[G] \star \mathbf{k}\langle X\rangle$. The map $x_i \mapsto a_i$ defines an isomorphism between V and a braided space spanned by the a_i's. Therefore, it has an extension to a homomorphism of braided algebras $\varphi : \mathbf{k}\langle X\rangle \to L$. Let $\xi = \mathrm{id} \star \varphi$ be a linear map such that $\xi(gu) = g\varphi(u)$, $g \in G$, $u \in L$. In this case, $\ker \xi = G \ker \varphi$ is a Hopf ideal of $G\langle X\rangle$ because $\ker \varphi$ is a braided Hopf ideal of $\mathbf{k}\langle X\rangle$. Consequently, there is a Hopf algebra structure on $H = \mathbf{k}[G] \star L$ such that ξ is a homomorphism of Hopf algebras. Comparing the braided Hopf algebra structure on $\mathbf{k}\langle X\rangle$ defined in Theorem 6.6 with (6.72), we see that the induced coproduct, counit and antipode coincide with that given in (6.74). This proves that $\mathbf{k}[G] \star L$ is a character Hopf algebra.

As $\ker \varphi \subseteq \ker \varepsilon^b = \Lambda$, we have $\ker \xi = G \ker \varphi \subseteq G\Lambda$, which is required. □

Remark 6.2 In the general case, the decomposition of a Hopf algebra H in a Radford biproduct $H = F \star A$, where F is a Hopf subalgebra and A is a braided Hopf algebra, exists if and only if there is a Hopf algebra homomorphism (projection) $\pi : H \to F$, $\pi|_F = \mathrm{id}$. In our case $F = \mathbf{k}[G]$, and the map $\pi : a_i \mapsto 0$, $\pi : g \mapsto g$ is a Hopf algebra projection if and only if $\ker \xi \subseteq G\Lambda$. If $G\Lambda$ does not contain $\ker \xi$ but there exists a Hopf algebra projection $\nu : H \to \mathbf{k}[G]$, then we may replace the generators a_i by $a_i' = a_i - \nu(a_i)$. In this case $\ker \xi' \subseteq G\Lambda$, where $\xi'(x_i) = a_i'$.

Lemma 6.18 *The space* Prim (L) *of all primitive elements of a bigraded braided Hopf algebra L and a space* Prin $(G \star L)$ *spanned by all normalized skew-primitive elements (see* (1.59)*) of the character Hopf algebra $G \star L$ are related as follows:*

$$\mathrm{Prin}\,(G \star L) = \mathrm{Prim}\,(L) \oplus \bigoplus_{g \in G} (1 - g)\mathbf{k}.$$

Proof The proof follows from the coproduct formula (6.72) and Lemma 1.19. □

We conclude this section with a useful particular formula for products of comonomials. Let, as above, $G\langle X\rangle$ be the free character Hopf algebra with commutation rules $ug_v = \chi^u(g_v)g_v u$, $p(u,v) = \chi^u(g_v)$. Then, we have a decomposition in Radford biproduct $G\langle X\rangle = \mathbf{k}[G] \star \mathbf{k}\langle X\rangle$, where $\mathbf{k}\langle X\rangle$ is the free braided Hopf algebra with the braiding defined on X via $(x_i \otimes x_s)\tau = p_{si}^{-1} x_s \otimes x_i$. In this particular case the product of comonomials in the shuffle algebra $Sh(X)$ takes the form

$$(w)(x_i) = \sum_{uv=w} p(x_i, v)^{-1}(ux_iv), \quad (x_i)(w) = \sum_{uv=w} p(u, x_i)^{-1}(ux_iv). \tag{6.75}$$

6.9 Filtrations and Subalgebras of the Free Braided Hopf Algebra

We are reminded that a filtration on a linear space R is an increasing chain

$$\{0\} = R_{(-\infty)} \subseteq R_{(0)} \subseteq R_{(1)} \subseteq \ldots \subseteq R_{(n)} \subseteq \ldots,$$

of linear subspaces such that $\bigcup_i R_{(i)} = R$. The related degree function is defined by $d(x) = \min\{i \mid x \in R_{(i)}\}$. A tensor product $R \otimes T$ of two filtered spaces is a filtered space with a filtration $(R \otimes T)_{(n)} = \sum_{i+s=n} R_{(i)} \otimes T_{(s)}$. A linear map $\varphi : R \to T$ is filtered if $\varphi(R_{(n)}) \subseteq T_{(n)}$; see details in Sect. 1.6.

Definition 6.6 A braided Hopf algebra H is said to be *filtered* if on the space H a filtration is fixed so that the braiding, the product, the unit map $\alpha \to \alpha \cdot 1$, the coproduct, and the braided antipode are filtered linear maps. We always consider the ground field as a filtered space with $\mathbf{k}_{(n)} = \mathbf{k}$, $n \geq 0$. In particular, the counit is automatically a filtered map, whereas the rest of the maps are filtered if and only if:

1. $(H_{(i)} \otimes H_{(s)})\tau \subseteq \sum_{k+m=i+s} H_{(k)} \otimes H_{(m)}, \quad i,\ s \geq 0;$
2. $H_{(i)}H_{(s)} \subseteq H_{(i+s)}, \quad 1 \in H_{(0)}, \qquad i,\ s \geq 0;$
3. $\Delta^b(H_{(n)}) \subseteq \sum_{s+i=n} H_{(i)} \otimes H_{(s)}, \ n \geq 0;$
4. $\sigma^b(H_{(n)}) \subseteq H_{(n)}, \qquad n \geq 0.$

With each filtered space R a graded space $\mathrm{gr}\, R$ is associated as follows:

$$\mathrm{gr}\, R = \bigoplus_{i=0}^{\infty} \mathrm{gr}_i\, R,$$

where $\mathrm{gr}_i\, R$, $i \geq 0$ is the quotient space $R_{(i)}/R_{(i-1)}$. A filtered linear map $\varphi : R \to T$ induces a linear map $\mathrm{gr}\, \varphi$ of associated graded spaces:

$$\mathrm{gr}\, \varphi : u + R_{(n-1)} \mapsto \varphi(u) + T_{(n-1)}, \quad u \in R_{(n)}.$$

In particular, if $(H, \tau, \mathbf{m}, \Delta^b, \varepsilon, \sigma^b)$ is a filtered braided Hopf algebra, then there are defined linear maps $\operatorname{gr}\tau$, $\operatorname{gr}\mathbf{m}$, $\operatorname{gr}\Delta^b$, $\operatorname{gr}\varepsilon$, and $\operatorname{gr}\sigma^b$ related to the associated graded space $\operatorname{gr} H$.

Theorem 6.8 *If H is a filtered braided Hopf algebra, then $\operatorname{gr} H$ with the braiding $\operatorname{gr}\tau$, the product $\operatorname{gr}\mathbf{m}$, the coproduct $\operatorname{gr}\Delta$, the counit $\operatorname{gr}\varepsilon$, and the antipode $\operatorname{gr}\sigma$ is a braided Hopf algebra.*

Proof Recall that the operator gr is a functor of tensor categories; that is, it satisfies

1. $\operatorname{gr}(\varphi\cdot\xi) = \operatorname{gr}\varphi\cdot\operatorname{gr}\xi$ (Lemma 1.29);
2. $\operatorname{gr}(R\otimes T) = \operatorname{gr}R\otimes\operatorname{gr}T$ (Lemma 1.32);
3. $\operatorname{gr}(\varphi\otimes\xi) = \operatorname{gr}\varphi\otimes\operatorname{gr}\xi$ (Lemma 1.33).

Therefore, to check that $\operatorname{gr} H$ satisfies the braided Hopf algebra axioms, it suffices to write down those axioms as operator equalities and apply the functor gr. □

Proposition 6.9 *Let R, T be filtered braided Hopf algebras. If $\varphi : R \to T$ is a filtered homomorphism of braided Hopf algebras, then $\operatorname{gr}\varphi : \operatorname{gr}R \to \operatorname{gr}T$ is a homomorphism of braided Hopf algebras.*

Proof The condition that φ is a homomorphism of braided Hopf algebras can be written in the operator form as follows: $\tau\cdot(\varphi\otimes\varphi) = (\varphi\otimes\varphi)\cdot\tau$; $\mathbf{m}\,\varphi = (\varphi\otimes\varphi)\,\mathbf{m}$; $\Delta^b\cdot(\varphi\otimes\varphi) = \varphi\cdot\Delta^b$. It remains to apply the functor gr. □

We conclude this section by considering subalgebras of the free braided Hopf algebra.

Theorem 6.9 *If a subalgebra U of a free braided Hopf algebra $\mathbf{k}\langle X\rangle$ is a right categorical right coideal, that is*

$$\Delta^b(U) \subseteq U \underline{\otimes}\, \mathbf{k}\langle X\rangle, \qquad (\mathbf{k}\langle X\rangle \otimes U)\tau \subseteq U \otimes \mathbf{k}\langle X\rangle,$$

then U is a free subalgebra.

Proof Let Sh denote the tensor algebra $T(V)$ of the space spanned by X considered as a coalgebra with co-concatenation coproduct

$$\Delta(x_{i_1}\otimes\cdots\otimes x_{i_n}) = \sum_{k=0}^{n} x_{i_1}\otimes\cdots\otimes x_{i_k}\underline{\otimes} x_{i_{k+1}}\otimes\ldots\otimes x_{i_n}. \tag{6.76}$$

We fix a natural non-degenerate paring $\langle -,-\rangle$ on $\mathbf{k}\langle X\rangle \times Sh$:

$$\langle x_{i_1}x_{i_2}\cdots x_{i_n},\, x_{s_1}\otimes x_{s_2}\otimes\cdots\otimes x_{s_m}\rangle = \delta^n_m\cdot\delta^{s_1}_{i_1}\delta^{s_2}_{i_2}\cdots\delta^{s_n}_{i_n}.$$

In this case, the following relation is valid

$$\langle uv, h\rangle = \sum_{(h)} \langle u, h^{(1)}\rangle\langle v, h^{(2)}\rangle, \quad u, v \in \mathbf{k}\langle X\rangle, \ h \in Sh. \tag{6.77}$$

Define a measure (action) of Sh on $\mathbf{k}\langle X\rangle$ as follows:

$$u \leftharpoonup h = \sum_{(u)} \langle u_b^{(1)}, h\rangle u_b^{(2)}, \quad u \in \mathbf{k}\langle X\rangle, \ h \in Sh. \tag{6.78}$$

Let U be a right categorical right coideal subalgebra of $\mathbf{k}\langle X\rangle$. According to Theorem 1.9, to demonstrate that U is a free subalgebra, it suffices to check that U has the weak algorithm with respect to the formal degree $d(x_i) = 1$. Suppose first that U is homogeneous, $U = \operatorname{gr} U$.

Let us show that U is left closed in $\mathbf{k}\langle X\rangle$; see Definition 1.22. Consider a left linearly independent over $\mathbf{k}\langle X\rangle$ homogeneous elements $u_1, u_2, \ldots, u_n \in U$. Suppose that

$$\sum_{i=1}^{n} r_i u_i = u \in U, \quad r_i \in \mathbf{k}\langle X\rangle. \tag{6.79}$$

We have $\Delta^b(u_i) = \sum u_i^{(1)} \underline{\otimes} u_i^{(2)}$, $\Delta^b(u) = \sum u^{(1)} \underline{\otimes} u^{(2)}$ with $u^{(1)}, u_i^{(1)} \in U$, and $\Delta^b(r_i) = \sum r_i^{(1)} \underline{\otimes} r_i^{(2)}$. Since U is right categorical, it follows that

$$(r_i^{(2)} \otimes u_i^{(1)})\tau = \sum_s u_{is} \otimes r_{is}, \quad u_{is} \in U. \tag{6.80}$$

Without loss of generality, we may suppose that all elements $u_{is}, u_i^{(1)}, u^{(1)}$ are homogeneous. Obviously, this is a finite set; hence, the space

$$D = \sum u_{is}\mathbf{k} + \sum u_i^{(1)}\mathbf{k} + \sum u^{(1)}\mathbf{k}$$

spanned by these elements has a finite dimension. Let T be a subalgebra generated by D, whereas T_k, $k \geq 0$ are its homogeneous components.

Denote $J_k = \{h \in V^{\otimes k} \mid \langle T_k, h\rangle = 0\} \stackrel{df}{=} T_k^{\perp}$. Because $\dim(T_k) < \infty$, we may write

$$J_k^{\perp} \stackrel{df}{=} \{u \in V^{\otimes k} | \ \langle u, J_k\rangle = 0\} = T_k.$$

By definition, T is a subalgebra. Therefore, its annihilator with respect to the pairing, $\sum_{i\geq 0} J_i$, is a coideal. Thus for every $h \in J_{k+1}$ we have the following inclusion:

$$\Delta^b(h) - 1\underline{\otimes} h - h\underline{\otimes} 1 \in (\sum_{i=0}^{k} J_i)\underline{\otimes} Sh + Sh\underline{\otimes}(\sum_{i=0}^{k} J_i). \tag{6.81}$$

Let us prove by induction on k, starting with $k = 0$, that $r_i \leftharpoondown J_k = 0, 1 \leq i \leq n$; that is (see (6.78)), if $\deg r_i^{(1)} = k$, then $r_i^{(1)} \in J_k^{\perp} = T_k$. We shall show first that under the induction supposition for k and smaller values the following equality is valid

$$r_i u_i \leftharpoondown h = (r_i \leftharpoondown h) u_i, \quad h \in J_{k+1}. \tag{6.82}$$

Indeed, by means of (6.80) we have

$$\begin{aligned} r_i u_i \leftharpoondown h &= \sum \langle (r_i u_i)^{(1)}, h \rangle (r_i u_i)^{(2)} = \sum \langle r_i^{(1)} u_{is}, h \rangle r_{is} u_i^{(2)} \\ &= \sum \langle r_i^{(1)}, h^{(1)} \rangle \langle u_{is}, h^{(2)} \rangle r_{is} u_i^{(2)}. \end{aligned}$$

Formula (6.81) and the induction suppositions (with the definition of T) demonstrate that in the above sum, all summands are equal to zero with the exception of two types, where $h^{(1)} = 1, h^{(2)} = h$, or $h^{(1)} = h$, $h^{(2)} = 1$. Moreover $\langle u_{is}, h \rangle = 0$ because either $u_{is} \in T_{k+1}$ or $\deg u_{is} \neq k+1$. Thus, we have got just the sum

$$\sum \langle r_i^{(1)}, h \rangle \langle u_{is}, 1 \rangle r_{is} u_i^{(2)}.$$

Again $\langle u_{is}, 1 \rangle$ is not zero only if $\deg u_{is} = 0$. By (6.80) this is equivalent to $\deg u_1^{(1)} = 0$; hence, $u_i^{(2)} = u_i$, $r_{is} = r_i^{(2)}$, which proves (6.82).

Let us apply $\leftharpoondown h$ to both sides of (6.79). We have $u \leftharpoondown h = 0$ because all $u^{(1)}$ of degree s by definition belong to $T_s = J_s^{\perp}$. Therefore (6.82) implies

$$\sum_i (r_i \leftharpoondown h) u_i = 0.$$

Because $u_1, u_2, \ldots, u_n$ are left linearly independent over $\mathbf{k}\langle X \rangle$, we get $r_i \leftharpoondown h = 0$, which completes the induction step.

In particular we have proved that $r_i \leftharpoondown J_m = 0$ with $m = \deg r_i$. This implies $\langle r_i, J_m \rangle = 0$; that is $r_i \in J_m^{\perp} = T_m \subseteq U$. Thus U is left closed in $\mathbf{k}\langle X \rangle$.

If U is not necessarily homogeneous, we consider associated graded algebra gr U with respect to the induced filtration. In this case, gr U is a homogeneous subalgebra of gr $\mathbf{k}\langle X \rangle = \mathbf{k}\langle X \rangle$. Moreover, gr U is a right categorical right coideal:

$$\Delta^b(\text{gr}\, U) = \text{gr}\,(\Delta^b(U)) \subseteq \text{gr}\,(U \underline{\otimes} \mathbf{k}\langle X \rangle) = \text{gr}\, U \underline{\otimes} \mathbf{k}\langle X \rangle;$$
$$(\mathbf{k}\langle X \rangle \otimes \text{gr}\, U)\text{gr}\, \tau = \text{gr}\,[(\mathbf{k}\langle X \rangle \otimes U)\tau] \subseteq \text{gr}\,(U \otimes \mathbf{k}\langle X \rangle) = \text{gr}\, U \otimes \mathbf{k}\langle X \rangle.$$

By the above arguments, the space gr U as a homogeneous subalgebra is left closed in $\mathbf{k}\langle X \rangle = \text{gr}\, \mathbf{k}\langle X \rangle$. It remains to apply Proposition 1.14. □

Corollary 6.4 *The subalgebra C of all constants for the right calculus d and the subalgebra C^* of all constants for the left calculus d^* are free.*

Proof By Corollary 6.3, the spaces C and C^* are categorical. By Corollary 6.2, the spaces C and C^* are left and right coideals, respectively. Hence the above theorem applies. □

In the proof of the above theorem, we see that an arbitrary weak basis of U is a set of free generators. The construction of the weak algebra basis given in Lemma 1.38 has an important freedom in choosing a transversal. Sometimes it is possible to use this freedom to make the generators primitive.

Definition 6.7 A subcoalgebra A of a coalgebra C with a distinguished element 1 is said to be *conservative* in C if for $u \in C$ the inclusion

$$\Delta^o(u) \stackrel{df}{=} \Delta^b(u) - 1\underline{\otimes}u - u\underline{\otimes}1 \in A\underline{\otimes}A \tag{6.83}$$

implies that there exists $a \in A$ such that $u-a$ is a primitive element; that is, $\Delta^o(u) = \Delta^o(a)$.

Lemma 6.19 *Let A be a subcoalgebra of $\mathbf{k}\langle X\rangle$. If $\mathrm{gr}\,A$ is conservative in $\mathrm{gr}\,\mathbf{k}\langle X\rangle = \mathbf{k}\langle X\rangle$, then A is conservative in $\mathbf{k}\langle X\rangle$, and the element $a \in A$ from the above definition can be chosen so that $d(a) \leq d(u)$, where d is the formal degree $d(x_i) = 1$.*

Proof Let $\Delta^o(u) \in A\underline{\otimes}A$. We shall use induction on $d(u)$. By definition of the coproduct in the associated graded algebra we have

$$\Delta^o(\bar{u}) \in \mathrm{gr}A \underline{\otimes} \mathrm{gr}A,$$

where $\bar{u} = u + \mathbf{k}\langle X\rangle_{(d(u)-1)}$ may be identified with the leading component of u. Because $\mathrm{gr}A$ is conservative, we may choose $\bar{a} \in \mathrm{gr}A$ such that $\bar{u} - \bar{a}$ is primitive. Let $a \in A$ be such that $\bar{a}$ is the leading component of a.

If $d(u) = d(a)$, then $d(u - \bar{u} - a + \bar{a}) < d(u)$, and

$$\Delta^o(u - a - \bar{u} + \bar{a}) = \Delta^o(u - a) \in A \underline{\otimes} A.$$

By the induction supposition there exists $a_1 \in A$ such that $v = u - a - \bar{u} + \bar{a} - a_1$ is primitive and $d(a_1) < d(u)$. Therefore $u - a - a_1 = v + (\bar{u} - \bar{a})$ is primitive as a sum of two primitives, and $d(a + a_1) = d(u)$.

If $d(u) \neq d(a)$, then $\bar{u}$ itself is primitive (because $\mathbf{k}\langle X\rangle$ is graded), hence, we may repeat the above argument with $a = 0$. □

Lemma 6.20 *Let A be a braided subbialgebra of the free braided Hopf algebra $\mathbf{k}\langle X\rangle$ equipped with filtration $d(x_i) = 1$. If for each primitively generated braided subbialgebra $A_1 \subseteq A$ the subcoalgebra $\mathrm{gr}\,A_1$ is conservative in $\mathbf{k}\langle X\rangle$, then A has a weak algebra basis of primitive elements.*

Proof Recall that the weak algebra basis may be constructed as follows. For each $n > 0$ denote by $R'_{(n)}$ the subspace of $R_{(n)}$ spanned by the products ab, where $a, b \in$

$R_{(n-1)}$ and $d(a) + d(b) \leq n$. Choose a minimal set X_n spanning $R_{(n)}$ (mod$R'_{(n)}$) over **k**, and put $X = \cup X_n$.

Because $1 \in A$, we may choose $X_1 \subseteq V$. Suppose that each of X_k, $k \leq m$ consists of primitive elements. The subalgebra U generated by all X_k, $k \leq m$ equals the subalgebra generated by $A_{(m)}$. Since the filtration is compatible with the braiding, it follows that the subalgebra U is a braided subbialgebra, and thus gr U is conservative in $\mathbf{k}\langle X\rangle$. We replace X_{m+1} with a set X'_{m+1} of the primitive elements in the following way.

For each $w \in X_{m+1}$, we have $\Delta^o(w) \in A_{(m)}\underline{\otimes}A_{(m)} \subseteq U\underline{\otimes}U$. By Lemma 6.19 there exists $u_1 \in U_{(m+1)}$ such that $u - u_1$ is primitive. In X_{m+1} we replace u with $u - u_1$. By induction on m, the lemma is proved. □

Remark 6.3 We do not claim that the weak algebra basis is a braided subspace.

Definition 6.8 A coideal J of a coalgebra C with a distinguished element 1 is said to be *conservative* in C if for each $u \in C$ the inclusion

$$\Delta^o(u) \in J\underline{\otimes}C + C\underline{\otimes}J \tag{6.84}$$

implies that there exists $a \in J$, such that $u - a$ is primitive.

Lemma 6.21 *Let J be a coideal of $\mathbf{k}\langle X\rangle$. If* gr J *is conservative in* gr $\mathbf{k}\langle X\rangle = \mathbf{k}\langle X\rangle$, *then J is conservative in $\mathbf{k}\langle X\rangle$.*

Proof The proof will literally coincide with that of Lemma 6.19, if one replaces $A \underline{\otimes} A$ by $J \underline{\otimes} \mathbf{k}\langle X\rangle + \mathbf{k}\langle X\rangle \underline{\otimes} J$ and gr $A \underline{\otimes}$ gr A by (gr J) $\underline{\otimes}\, \mathbf{k}\langle X\rangle + \mathbf{k}\langle X\rangle \underline{\otimes}$ (gr J). □

6.10 Chapter Notes

The notion of a braided Hopf algebra is one of the basic features of braided monoidal categories defined by Joyal and Street in [109], although braided Hopf algebras appeared first, before their formalization within category theory, in the famous paper by Milnor and Moore [175], as graded Hopf algebras. They appeared also as universal enveloping algebras of color Lie algebras introduced by Scheunert [203, 204]. A standard method of obtaining a braided monoidal category is to consider all modules over a quasitriangular Hopf algebra or all comodules over a coquasitriangular Hopf algebra: Lyubashenko [156], Drinfeld [66, 67], Larson and Towber [145], Schauenberg [201]. In the book [142, Chap. 10], Klimik and Schmüdgen expound this approach to the construction of braided Hopf algebras.

A slightly different but essentially equivalent approach is to consider categories of Yetter–Drinfeld modules over Hopf algebras. This approach was proposed by Andruskiewitsch and Graña [2] and Andruskiewitsch and Schneider [3]. It provides an effective tool in the classification of pointed Hopf algebras using the lifting method theorized by Andruskiewitsch and Schneider [3, 4] because the main invariant, the diagram of a pointed Hopf algebra, is a Hopf algebra in a Yetter–Drinfeld category.

In [221], Takeuchi surveys the progress of braided Hopf algebra theory using a noncategorical framework in which braided bialgebras are formulate as algebras and coalgebras with a Yang–Baxter operator (alongside compatibility conditions). This approach is actually more general and convenient than the others approaches, which why we adopt it here.

In [8], Ardizzoni considered the combinatorial rank of graded braided bialgebras by investigating conditions guaranteeing that the combinatorial rank is finite. However the following question remains unresolved: is the combinatorial rank of a finitely generated braided bialgebra is finite? Whether the combinatorial rank of a Nichols algebra defined by a finite dimensional braided space is finite is also unknown.

In constructing of the Nichols algebra, we mainly follow Rosso [196, 197]. Using a similar construction, one of the *braided bitensor algebra* has been introduced independently by Schauenberg [202] in terms of braided categories. Before this, Nichols [181] used a similar construction applied to Hopf bimodules over a bialgebra to provide examples of *bialgebras of type one*. S.L. Woronowicz used (6.61) up to changing τ by $-\tau$ to define the external algebra [230, pp. 154–155], whereas Schauenberg [201] proved (6.61) to demonstrate that the external algebra is a braided bitensor algebra [201, Theorem 2.9]. Andruskiewitsch and Graña [2] proposed another approach to constructing the Nichols algebra based on braided pairings. The same object appeared as an *optimal algebra* for noncommutative differential calculi conditioned by Yang–Baxter commutation rules [129]. Some additional general properties of the Nichols algebras appear in [77, 78, 222]. Finite-dimensional Nichols algebras have been widely studied over the past several years, with approximately 100 preprints on the arXiv concerning this topic.

The decomposition of a Hopf algebra with a projection in the biproduct as a purely mathematical statement was discovered by Radford [190]. Majid [159, 160] subsequently advanced a physical interpretation of the inverse process as a bosonization of fermions.

The filtrations undoubtedly provide a fundamental tool for all modern mathematics. In the area of Hopf algebras and quantum groups, N. Andruskiewitsch lifting classification method includes the classification of associated graded braided Hopf algebras as a first step. Recently, Ardizzoni and Menini [11] investigated the notion of associated graded (co)algebra within the framework of abelian braided monoidal categories.

Chapter 7
Binary Structures

Abstract In this chapter, we consider binary generalizations of Lie algebras appeared in modern mathematics and mathematical physics. We consider recent developments and remaining problems on the subject. The chapter discusses Lie superalgebras, color Lie algebras, and Lie algebras in symmetric categories, free Lie τ-algebras.

In this chapter, we consider binary generalizations of Lie algebras appeared in modern mathematics and mathematical physics. We consider recent developments and remaining problems on the subject. The chapter discusses Lie superalgebras, color Lie algebras, and Lie algebras in symmetric categories, free Lie τ-algebras.

7.1 Lie Superalgebras

By definition, a *Lie superalgebra* is a graded linear space $L = L_0 \oplus L_1$ endowed with a bilinear operation $[\,,\,] : L^{\otimes 2} \to L$ that satisfies the following graded versions of the antisymmetry and the Jacobi identity:

$$[u, v] = -(-1)^{|u||v|}[v, u], \tag{7.1}$$

$$[[u, v], w] + (-1)^{|u|(|v|+|w|)}[[v, w], u] + (-1)^{(|u|+|v|)|w|}[[w, u], v] = 0, \tag{7.2}$$

where u, v, w are homogeneous elements $u \in L_{|u|}$, $v \in L_{|v|}$, $w \in L_{|w|}$, while $|u||v|$ means the product of integer numbers $|u|$ and $|v|$.

A fundamental example of a Lie superalgebra appears from an associative superalgebra (that is, graded associative algebra $R = R_0 \oplus R_1$) when in place of the bilinear operation one considers the superbracket:

$$[u, v] = u \cdot v - (-1)^{|u||v|} v \cdot u, \tag{7.3}$$

where, as above, $u \in L_{|u|}$, $v \in L_{|v|}$, $|u|, |v| \in \{0, 1\}$. In line with the Lie algebra theory tradition, this Lie superalgebra is denoted by $R^{(-)}$. The super version of the Poincaré-Birkhoff-Witt Theorem holds: every Lie superalgebra L is a subalgebra of

V. Kharchenko, *Quantum Lie Theory*, Lecture Notes in Mathematics 2150,
DOI 10.1007/978-3-319-22704-7_7

the universal enveloping associative superalgebra $U(L)$. Here, the algebra $U(L)$ may be defined in perfect analogy with the classical case

$$U(L) = \mathbf{k}\langle L\rangle/\{u\cdot v = [u, v] + (-1)^{|u||v|} v\cdot u\}, \tag{7.4}$$

where u, v run through a fixed homogeneous basis of L, and $[u, v]$ is a linear combination of that basis elements, whereas $\mathbf{k}\langle L\rangle$ is a free associative algebra generated by that basis (in the invariant form, this is the tensor algebra of the linear space L). The super structure on $\mathbf{k}\langle L\rangle$, and hence on $U(L)$, is defined in a natural way via

$$|u\cdot v\cdot\ldots\cdot w| = |u| + |v| + \cdots + |w|.$$

In particular, $U(L)$ is a *quadratic algebra*; that is, it is defined by relations of degree two.

Recall that if L is an ordinary Lie algebra, then $U(L)$ has a structure of a Hopf algebra with the coproduct $\Delta(u) = 1\otimes u + u\otimes 1$, $u\in L$. In general case, the same formula defines a structure of Hopf superalgebra on $U(L)$. By definition a *Hopf superalgebra* is the braided Hopf algebra related to the following diagonal braiding:

$$\tau(u\otimes v) = (-1)^{|u||v|}(v\otimes u),$$

see Example 6.1. More precisely, we may define a braided coproduct Δ^b as a homomorphism of associative algebras

$$\Delta^b : U(L) \to U(L)\underline{\otimes} U(L)$$

setting $\Delta^b(u) = 1\underline{\otimes} u + u\underline{\otimes} 1$, $u\in L$. Here $U(L)\underline{\otimes} U(L)$ is the space $U(L)\otimes U(L)$ with the product

$$(u\underline{\otimes} v)\cdot(w\underline{\otimes} t) = (-1)^{|v||w|}(uw\underline{\otimes} vt).$$

In this case, the Radford bi-product construction $G\star U(L)$ is an ordinary Hopf algebra, see Sect. 6.8. Here G is a two element group $G = \{1, g \mid g^2 = 1\}$ with the action $u^g = (-1)^{|u|}u$, $u\in L_{|u|}$. Respectively, in $G\star U(L)$ the following commutation rules holds: $ug = (-1)^{|u|}gu$, $u\in L_{|u|}$, whereas the coproduct is defined via

$$\Delta(u) = u\otimes 1 + g^{|u|}\otimes u.$$

Of course, $G\star U(L)$ is a character Hopf algebra. Let $x_1, x_2, \ldots, x_k$ be a basis of L_0 and let $x_{k+1}, x_{k+2}, \ldots, x_n$ be a basis of L_1. Then the parameters p_{ij} related to this character Hopf algebra are

$$p_{ij} = \begin{cases} -1, & \text{if } k < i, j \le n;\\ 1, & \text{otherwise.}\end{cases}$$

In particular, $p_{ij}p_{ji} = 1, 1 \leq i,j \leq n$. By this reason the principle bilinear quantum Lie operation $[\![-,-]\!]$ introduced in (4.25) is almost well-defined on the set of normalized skew-primitive elements Prin $U(L)$ of $G \star U(L)$, see Sect. 4.3 for details. More precisely, if characteristic of the ground field is zero, then the general formula of Lemma 6.18 takes the form

$$\text{Prin}\, U(L) = (1-g)\mathbf{k} \oplus L,$$

and the operation [-, -] of Lie superalgebra given on L coincides with $[\![-,-]\!]$. We shall prove this statement in a more general context of color Lie algebras. Now we only stress that the whole of Prin $U(L)$ with the brackets $[\![-,-]\!]$ is not a Lie superalgebra. Indeed, let $x_0 = 1 - g$. Then $x_0 g = gx_0$, so that $p_{0n} = 1$ if $n > k$. The equalities $\Delta(x_0) = x_0 \otimes 1 + g \otimes x_0$ and $x_n g = -gx_n$ imply $p_{0n}p_{n0} = -1 \neq 1$. In particular $[\![x_0, x_n]\!]$ is undefined provided that $n > k$.

7.2 Color Lie Algebras

Let G be an Abelian group and let $\alpha : G \times G \to \mathbf{k}^*$ be its bicharacter:

$$\alpha(f \cdot g, h) = \alpha(f,h)\alpha(g,h), \quad \alpha(f, g \cdot h) = \alpha(f,g)\alpha(f,h),$$

where, as usual, $\mathbf{k}^*$ denotes the multiplicative group of nonzero elements of the ground field $\mathbf{k}$. Suppose additionally that α is multiplicatively antisymmetric:

$$\alpha(g,h) = \alpha(h,g)^{-1}.$$

In this case, a linear space $L = \oplus_{g\in G} L_g$ graded by G is said to be a *color Lie algebra* if it is endowed with a bilinear operation $[\ ,\ ] : L^{\otimes 2} \to L$ which satisfies the color versions of the antisymmetry and Jacobi identities:

$$[u,v] = -\alpha(f,g)[v,u], \tag{7.5}$$

$$[[u,v],w] + \alpha(f,gh)[[v,w],u] + \alpha(fg,h)[[w,u],v] = 0, \tag{7.6}$$

where u, v, w are homogeneous elements $u \in L_f$, $v \in L_g$, $w \in L_h$, $f, g, h \in G$.

If $G = \{0, 1\}$ is the two-element group, then there exists just one nontrivial bicharacter: $\alpha(0,0) = \alpha(0,1) = \alpha(1,0) = 1, \alpha(1,1) = -1$; that is,

$$\alpha(|u|, |v|) = (-1)^{|u||v|}.$$

Hence, the Lie superalgebras are precisely the color Lie algebras when the group G (of colors) has just two elements.

It is very important that any associative G-graded algebra

$$R = \bigoplus_{g \in G} R_g, \qquad R_g \cdot R_h \subseteq R_{gh},$$

defines a color Lie algebra when in place of the bilinear operation we consider the following *color bracket*:

$$[u, v] = u \cdot v - \alpha(f, g) v \cdot u, \tag{7.7}$$

where, as usual, $u \in L_f$, $v \in L_g, f, g \in G$. This color Lie algebra is denoted by $R^{(-)}$ as well.

The G-graded universal enveloping algebra $U(L)$ is defined quite similarly to the "super" case

$$U(L) = \mathbf{k}\langle L \rangle / \{u \cdot v = [u, v] + \alpha(f, g) v \cdot u\}, \tag{7.8}$$

where again $u \in L_f$ and $v \in L_g$ run through a fixed homogeneous basis of L, and $[u, v] = \sum \beta_i u_i$ is a linear combination of basis elements u_i. The grading on the tensor (free) algebra $\mathbf{k}\langle L \rangle$, and on $U(L)$, is defined in a similar way

$$|u \cdot v \cdot \ldots \cdot w| = f \cdot g \cdot \ldots \cdot h.$$

In particular, $U(L)$ is still a quadratic algebra.

The "color" version of the Poincaré-Birkhoff-Witt Theorem was proven by Scheunert in [202].

Theorem 7.1 (M. Scheunert) *If $x_i \in L_{g_i}$, $i \geq 1$ is a basis of a color Lie algebra $L = \oplus_{g \in G} L_g$, then the following products form a basis of $U(L)$:*

$$x_1^{n_1} x_2^{n_2} \cdots x_m^{n_m},$$

where n_i, $1 \leq i \leq m$ are nonnegative integer numbers with $n_i \leq 1$ provided that $\alpha(g_i, g_i) = -1 \neq 1$. In particular every color Lie algebra L is a subalgebra of the color Lie algebra $U(L)^-$.

This theorem may be easily proved using the Composition Lemma (Theorem 1.2).

The formula $\Delta^b(u) = 1 \underline{\otimes} u + u \underline{\otimes} 1$, $u \in L$ defines a structure of a color Hopf algebra on $U(L)$. By definition a *color Hopf algebra* is the braided Hopf algebra related to the following diagonal braiding

$$\tau(u \otimes v) = \alpha(f, g) v \otimes u, \quad u \in L_f, \ v \in L_g.$$

The braided coproduct Δ^b is a homomorphism of associative algebras

$$\Delta^b : U(L) \to U(L) \underline{\otimes} U(L),$$

where $U(L)\underline{\otimes}U(L)$ is the space $U(L)\otimes U(L)$ with the product

$$(u\underline{\otimes}v)\cdot(w\underline{\otimes}t)=\alpha(g,h)(uw\underline{\otimes}vt),\quad v\in L_g,\ w\in L_h. \tag{7.9}$$

Lemma 7.1 *If the characteristic of the ground field is zero, then* Prim $(U(L))=L$.

Proof By M. Scheunert theorem each element $a\in U(L)$ has a unique representation

$$a=\sum_{i=1}^{M}\beta_i x_1^{n_1^i}x_2^{n_2^i}\cdots x_m^{n_m^i},\quad \beta_i\in\mathbf{k}. \tag{7.10}$$

Without loss of generality we may suppose that $n_1^1>0$, $\beta_1\neq 0$. We have to show that if a is primitive, then $n_j^i=0$ unless $i=j=1$. By definition of the braided coproduct we have

$$\Delta^b(a)=\sum_{i=1}^{M}\beta_i(x_1\underline{\otimes}1+1\underline{\otimes}x_1)^{n_1^i}(x_2\underline{\otimes}1+1\underline{\otimes}x_2)^{n_2^i}\cdots(x_m\underline{\otimes}1+1\underline{\otimes}x_m)^{n_m^i}. \tag{7.11}$$

The latter formula and definition (7.9) imply that the braided coproduct of the summand $w=x_1^{n_1^i}x_2^{n_2^i}\cdots x_m^{n_m^i}$ is a linear combination of tensors $w^A\underline{\otimes}w^{\bar{A}}$, where A appears from w by deleting of some letters, while $w^{\bar{A}}$ appears from w by deleting of all letters remaining in w^A. According to M. Scheunert theorem all different words that appear from summands of a given in (7.10) deleting some letters are linearly independent in $U(L)$. In particular, the resulting coefficient of the linear combination of all tensors of the form $x_1\underline{\otimes}x_1^{n_1^1-1}x_2^{n_2^i}\cdots x_m^{n_m^i}$ that appear under developing the product in (7.11) has to be zero. Applying (7.9), we see that this coefficient equals

$$1+\alpha(g_1,g_1)+\alpha(g_1,g_1)^2+\cdots+\alpha(g_1,g_1)^{n_1-1}\stackrel{df}{=}\alpha(g_1,g_1)^{[n_1]}.$$

Because α is multiplicatively skew symmetric, we have $\alpha(g_1,g_1)^2=1$; that is, $\alpha(g_1,g_1)=\pm 1$. If $\alpha(g_1,g_1)=1$, then $\alpha(g_1,g_1)^{[n_1]}=n_1\neq 0$ in $\mathbf{k}$. If $\alpha(g_1,g_1)=-1$, then by M. Scheunert theorem $n_1=1$, and $\alpha(g_1,g_1)^{[n_1]}=1\neq 0$. □

The Radford bi-product $G\star U(L)$ is an ordinary Hopf algebra, see Sect. 6.8. In $G\star U(L)$ the following commutation rules hold:

$$ug=\alpha(f,g)gu,\quad u\in L_f,\quad g\in G,$$

whereas the coproduct is defined via

$$\Delta(u)=u\otimes 1+g\otimes u,\quad u\in L_g.$$

In particular all elements of L are normalized skew-primitive in $G \star U(L)$, and $G \star U(L)$ is a character Hopf algebra. Let $\{x_i \mid 1 \leq i \leq n\}$ be a homogeneous basis of L, $x_i \in L_{g_i}$, $1 \leq i \leq n$. The parameters p_{ij} related to this basis are

$$p_{ij} = \alpha(g_i, g_j),$$

where $x_i \in L_{g_i}$, $x_j \in L_{g_j}$. Since the form α is multiplicatively skew symmetric, we have $p_{ij}p_{ji} = 1$, $1 \leq i, j \leq n$. Further, due to (6.18), we have

$$\text{Prin}\,(G \star U(L)) = L \oplus \bigoplus_{g \in G} (1 - g)\mathbf{k}, \tag{7.12}$$

and the operation [-, -] of color Lie algebra given on L coincides with the principle bilinear quantum Lie operation $[\![-, -]\!]$, see (4.25) of Sect. 4.3.

7.3 Lie Algebras in Symmetric Categories

A more general concept of a "Lie τ-algebra" related to a symmetry τ (a braiding such that $\tau^2 = \text{id}$) was introduced by Gurevich [88].

We are reminded that a linear space V is a *braided space* if there is fixed a linear map $\tau : V \otimes V \to V \otimes V$ which satisfies a *braid relation* $\tau_1\tau_2\tau_1 = \tau_2\tau_1\tau_2$, where

$$\tau_i = \text{id}^{\otimes(i-1)} \otimes \tau \otimes \text{id}^{\otimes(n-i-1)} : V^{\otimes n} \to V^{\otimes n}, \quad 1 \leq i < n.$$

An algebra R (associative or non associative) with a multiplication $\mathbf{m} : R \otimes R \to R$ is a *braided algebra* if it is a braided space and

$$(\mathbf{m} \otimes \text{id})\tau = \tau_2\tau_1(\text{id} \otimes \mathbf{m}), \qquad (\text{id} \otimes \mathbf{m})\tau = \tau_1\tau_2(\mathbf{m} \otimes \text{id}), \tag{7.13}$$

where as above we use the exponential notation; that is, the operators act from the left to the right: $(u \otimes v \otimes w) \cdot (\mathbf{m} \otimes \text{id})\tau = (uv \otimes w)^\tau = \tau(uv \otimes w)$.

A braided algebra L is said to be a *Lie τ-algebra* if the braiding is involutive ($\tau^2 = \text{id}$) and if it is connected with the multiplication $\mathbf{m} : L \otimes L \to L$ thus:

$$\begin{aligned} &\mathbf{m} + \tau\mathbf{m} = 0, && \text{antisymmetry;} \\ &(\text{id} + \tau_1\tau_2 + \tau_2\tau_1)(\mathbf{m} \otimes \text{id})\mathbf{m} = 0, && \text{Jacobi identity.} \end{aligned} \tag{7.14}$$

First of all, we note that this concept generalizes the above notion of the color Lie algebra. Indeed, if L is a color Lie algebra with the multiplication $\mathbf{m} = [\ ,\ ]$, and with the bicharacter α, then we define a symmetry τ as follows:

$$(u \otimes v)\tau = \alpha(f, g)v \otimes u, \quad u \in L_f,\ v \in L_g. \tag{7.15}$$

Since α is multiplicatively skew symmetric, we have

$$(u \otimes v)\tau^2 = (\alpha(f,g)v \otimes u)\tau = \alpha(f,g)\alpha(g,f)(u \otimes v) = u \otimes v;$$

that is, τ is involutive. To check (7.13), we have

$$(u \otimes v \otimes w)(\mathbf{m} \otimes \mathrm{id})\tau = ([u,v] \otimes w)\tau = \alpha(fg,h)w \otimes [u,v].$$

At the same time,

$$\begin{aligned}(u \otimes v \otimes w)\tau_2\tau_1(\mathrm{id} \otimes \mathbf{m}) &= \alpha(g,h)(u \otimes w \otimes v)\tau_1(\mathrm{id} \otimes \mathbf{m}) \\ &= \alpha(g,h)\alpha(f,h)(w \otimes u \otimes v)(\mathrm{id} \otimes \mathbf{m}) \\ &= \alpha(gf,h)w \otimes [u,v],\end{aligned}$$

which proves the first of (7.13). The second of (7.13) is quite similar. Hence $\langle L, [\,,\,]\rangle$ is a braided algebra.

The antisymmetry identity (7.14) applied to L coincides with the antisymmetry identity (7.5). To check the Jacobi identity (7.14), we note that

$$\begin{aligned}(u \otimes v \otimes w)(\mathbf{m} \otimes \mathrm{id})\mathbf{m} &= [[u,v],w], \\ (u \otimes v \otimes w)\tau_1\tau_2(\mathbf{m} \otimes \mathrm{id})\mathbf{m} &= \alpha(f,gh)[[v,w],u], \\ (u \otimes v \otimes w)\tau_2\tau_1(\mathbf{m} \otimes \mathrm{id})\mathbf{m} &= \alpha(gf,h)[[w,u],v],\end{aligned}$$

hence (7.14) follows from (7.6).

Lemma 7.2 *If A is an associative braided algebra with a multiplication*

$$\mathbf{m} : u \otimes v \to uv,$$

then the braided space A with a new multiplication, [-, -] $= \mathbf{m} - \tau\mathbf{m}$,

$$[u,v] = uv - \sum_i v_i u_i, \quad \textit{where } (u \otimes v)\tau = \sum_i v_i \otimes u_i, \tag{7.16}$$

is a Lie τ-algebra provided that $\tau^2 = \mathrm{id}$.

Proof We have to check the axioms of braided algebra (7.13) and the axioms of a Lie τ-algebra (7.14) with $\mathbf{m} - \tau\mathbf{m}$ in place of $\mathbf{m}$. Applying the first of the axioms (7.13), we have

$$((\mathbf{m} - \tau\mathbf{m}) \otimes \mathrm{id})\tau = (\mathbf{m} \otimes id)\tau - \tau_1(\mathbf{m} \otimes \mathrm{id})\tau = \tau_2\tau_1(\mathrm{id} \otimes \mathbf{m}) - \tau_1\tau_2\tau_1(\mathrm{id} \otimes \mathbf{m}).$$

This implies the first of (7.13) with [-, -] in place of $\mathbf{m}$ because $\tau_1\tau_2\tau_1 = \tau_2\tau_1\tau_2$ and $\tau_2(\mathrm{id} \otimes \mathbf{m}) = id \otimes \tau\mathbf{m}$. The second of (7.13) with $\mathbf{m} \leftarrow$ [-, -] can be checked quite similar.

The antisymmetry identity takes the form

$$\mathbf{m} - \tau\mathbf{m} + \tau(\mathbf{m} - \tau\mathbf{m}) = \mathbf{m} - \tau^2\mathbf{m} = 0.$$

To check the Jacoby identity, we note that the associativity of the algebra A may be written as an equality:

$$(\mathrm{id} \otimes \mathbf{m})\mathbf{m} = (\mathbf{m} \otimes \mathrm{id})\mathbf{m}. \tag{7.17}$$

The Jacobi identity with $\mathbf{m} \leftarrow [\text{-}, \text{-}]$ reads:

$$(\mathrm{id} + \tau_1\tau_2 + \tau_2\tau_1)([\text{-}, \text{-}] \otimes \mathrm{id})[\text{-}, \text{-}] = 0.$$

We have

$$\begin{aligned}([\text{-}, \text{-}] \otimes \mathrm{id})[\text{-}, \text{-}] &= ((\mathbf{m} - \tau\mathbf{m}) \otimes \mathrm{id})(\mathbf{m} - \tau\mathbf{m}) = (\mathbf{m} \otimes \mathrm{id} - \tau_1(\mathbf{m} \otimes \mathrm{id}))(\mathbf{m} - \tau\mathbf{m}) \\ &= (\mathbf{m} \otimes \mathrm{id})\mathbf{m} - \tau_1(\mathbf{m} \otimes \mathrm{id})\mathbf{m} - (\mathbf{m} \otimes \mathrm{id})\tau\mathbf{m} + \tau_1(\mathbf{m} \otimes \mathrm{id})\tau\mathbf{m}.\end{aligned}$$

If we apply $(\mathbf{m} \otimes \mathrm{id})\tau = \tau_2\tau_1(\mathrm{id} \otimes \mathbf{m})$ and the associativity (7.17), then the latter expression reduces to

$$(\mathrm{id} - \tau_1 - \tau_2\tau_1 + \tau_1\tau_2\tau_1)(\mathbf{m} \otimes \mathrm{id})\mathbf{m}.$$

It remains to note that the equalities $\tau_1^2 = \tau_2^2 = \mathrm{id}$ and $\tau_1\tau_2\tau_1 = \tau_2\tau_1\tau_2$ imply

$$\mathrm{id} + \tau_1\tau_2 + \tau_2\tau_1 = (\mathrm{id} + \tau_1\tau_2 + \tau_2\tau_1)\tau_2\tau_1 = (\mathrm{id} + \tau_1\tau_2 + \tau_2\tau_1)\tau_1\tau_2,$$

and therefore $(\mathrm{id} + \tau_1\tau_2 + \tau_2\tau_1)(\mathrm{id} - \tau_1 - \tau_2\tau_1 + \tau_1\tau_2\tau_1) = 0$. □

The Lie τ-algebra constructed in the above lemma has a standard notation: $A_\tau^{(-)}$.

7.3.1 Universal Enveloping Algebra

Definition 7.1 Given a Lie τ-algebra L, one constructs the *universal enveloping algebra* as follows: Let $\mathbf{k}\langle L\rangle = \bigoplus_{n=0}^{\infty} L^{\otimes n}$ be the free associative algebra. Let $J \subseteq \mathbf{k}\langle L\rangle$ be the ideal generated by

$$\{a \otimes b - (a \otimes b)\tau - [a, b] \mid a, b \in L\}, \tag{7.18}$$

where $[a, b]$ is the product in L. Then $U(L)$ is the quotient algebra $\mathbf{k}\langle L\rangle/J$.

Of course, this definition generalizes the same definition for color Lie algebras and Lie superalgebras. We see that in general $U(L)$ is still a quadratic algebra.

Theorem 7.2 *The algebra $U(L)$ has a natural structure of a braided Hopf algebra. In which case, the braiding of $U(L)$ is involutive, and*

$$\Delta^b(a) = 1\underline{\otimes}a + a\underline{\otimes}1, \quad a \in L.$$

The map $\iota : a \mapsto a + J$, $a \in L$ is a homomorphism of Lie τ-algebras, $\iota : L \to U(L)^{(-)}$.

Proof Consider $\mathbf{k}\langle L\rangle$ as the free braided Hopf algebra described in Theorems 6.1 and 6.2 with $V \leftarrow L$. We note, first, that the extended on $\mathbf{k}\langle L\rangle$ braiding is still involutive. Indeed, if $u \in L^{\otimes r}$, $v \in L^{\otimes(n-r)}$, then according to definition (6.18), we have

$$(u\underline{\otimes}v)\tau^2 = (u \otimes v)\nu_r^{1,n}\theta_{n-r}\tau = (u \otimes v)\nu_r^{1,n}\nu_{n-r}^{1,n}\theta_r = u\underline{\otimes}v$$

because decompositions (6.16) and (6.17) with conditions $\tau_i^2 = \text{id}$, $1 \le i < n$ imply

$$\nu_r^{1,n}\nu_{n-r}^{1,n} = (\tau_r\tau_{r-1}\cdots\tau_1)(\tau_{r+1}\tau_r\cdots\tau_2)\cdots(\tau_{n-1}\tau_{n-2}\cdots\tau_{n-r})$$
$$\times(\tau_{n-r}\tau_{n-r+1}\cdots\tau_{n-1})(\tau_{n-r-1}\tau_{n-r}\cdots\tau_{n-2})\cdots(\tau_1\tau_2\cdots\tau_r) = \text{id}.$$

Further, we shall prove that J is a braided Hopf ideal. Due to Lemma 6.10, it suffices to verify that the generating space $W = \{a\otimes b - (a\otimes b)\tau - [a,b] \mid a, b \in L\}$ satisfies conditions 2–5 of Definition 6.1.

Let $\mathbf{m} : L\underline{\otimes}L \to L$ be the τ-Lie multiplication of L, and let $\mathbf{m}' : L\underline{\otimes}L \to L \otimes L$ be the concatenation product of $\mathbf{k}\langle L\rangle$ restricted to $L\underline{\otimes}L$. In this case, the space W coincides with the image of the operator $\Omega = (\text{id} - \tau)\,\mathbf{m}' - \mathbf{m}$.

If $u \in L^{\otimes m}$, $a, b \in L$ and $w = a \otimes b - (a \otimes b)\tau - [a,b] = (a\underline{\otimes}b)\cdot\Omega$, then

$$(u\underline{\otimes}w)\tau = (u\underline{\otimes}a\underline{\otimes}b)\cdot[(\text{id} - \tau_2)(\text{id}\underline{\otimes}\mathbf{m}')\tau - (\text{id}\underline{\otimes}\,\mathbf{m})\tau].$$

As both $\mathbf{m}$ and $\mathbf{m}'$ are braided products, we have $(\text{id}\underline{\otimes}\mathbf{m})\tau = \tau_1\tau_2(\mathbf{m}\underline{\otimes}\text{id})$, and $(\text{id}\underline{\otimes}\mathbf{m}')\tau = \tau_1\tau_2(\mathbf{m}'\underline{\otimes}\text{id})$. The braid relation implies $(\text{id}-\tau_2)\tau_1\tau_2 = \tau_1\tau_2(\text{id}-\tau_1)$. Hence, we obtain

$$\begin{aligned}(u\underline{\otimes}w)\tau &= (u\underline{\otimes}a\underline{\otimes}b)\cdot\tau_1\tau_2[(\text{id} - \tau_1)(\mathbf{m}'\underline{\otimes}\text{id}) - (\mathbf{m}\underline{\otimes}\,\text{id})]\\ &= (u\underline{\otimes}a\underline{\otimes}b)\cdot\tau_1\tau_2(\Omega\underline{\otimes}\,\text{id}) \in (L\underline{\otimes}L\underline{\otimes}L^{\otimes m})(\Omega\underline{\otimes}\,\text{id})\\ &= \text{im}\,\Omega\underline{\otimes}L^{\otimes m} = W\underline{\otimes}L^{\otimes m};\end{aligned}$$

that is, W is right categorical. In perfect analogy, we have

$$(w\underline{\otimes}u)\tau = (a\underline{\otimes}b\underline{\otimes}u)\cdot\tau_2\tau_1[(\text{id} - \tau_2)(\text{id}\underline{\otimes}\mathbf{m}') - (\text{id}\underline{\otimes}\,\mathbf{m})] \in L^{\otimes m}\underline{\otimes}\text{im}\,\Omega.$$

Thus, W is a categorical space, and the second condition fulfills.

Consider an element $u = a\otimes b-(a\otimes b)\tau = (a\otimes b)(\,\mathrm{id}-\tau)$. Lemma 6.3 implies $\Delta^b(u) = \sum_{r=0}^{2} u\cdot\Phi_r^{(1,2)}\,\theta_r$, where according to (6.23) we have $\Phi_0^{(1,2)} = \Phi_2^{(1,2)} = \mathrm{id}$, and $\Phi_1^{(1,2)} = [1;1]+[1;2] = \mathrm{id}+\tau$. In particular $(\,\mathrm{id}-\tau)\Phi_1^{(1,2)} = 0$, and therefore u is primitive.

The element $[a,b]$ is primitive too, because it belongs to L. As the set $\mathrm{Prim}\,\mathbf{k}\langle L\rangle$ of all primitive elements is a linear space, we obtain $W\subseteq \mathrm{Prim}\,\mathbf{k}\langle L\rangle$. At the same time, every primitive element w satisfies $\varepsilon(w) = 0, \sigma^b(w) = -w$. Hence, W satisfies third fourth and fifth conditions as well.

Finally, the product in $U(L)^{(-)}$ of elements a, b equals $a\otimes b-(a\otimes b)\tau$. In $\mathbf{k}\,\langle L\rangle$, the latter element equals $[a,b]$ modulo W. Hence $a\otimes b-(a\otimes b)\tau = [a,b]$ in $U(L)$, and the map ι is a homomorphism of τ-Lie algebras. □

7.3.2 Embedding into the Universal Enveloping Algebra

Our next goal is to understand when the map ι is injective. Because the braiding τ is involutive, $\tau^2 = \mathrm{id}$, Theorem 1.4 implies that the local action of the braid monoid B_n on $L^{\otimes n}$ is reduced to an action of the symmetric group S_n, so that $u\cdot s_i = u\cdot t_i$, where $t_i = (i, i+1)$ is the transposition of indices $i\leftrightarrow i+1$. Given r, $1\le r\le n$, we fix the notations

$$e_r = \frac{1}{r!}\sum_{\pi\in S_r}\pi,\quad e^{(r)} = \frac{1}{(n-r)!}\sum_{\pi\in S_n^{(r)}}\pi, \tag{7.19}$$

where by definition $S_n^{(r)}$ is a subgroup of all permutations that leave fixed each one of the indices $1,2,\ldots,r$.

Lemma 7.3 *The left annihilator of* $(1-t_1)e^{(1)}$ *in* $\mathbf{k}[S_n]$ *equals*

$$\mathbf{k}[S_n](1+t_1)+\mathbf{k}[S_n](1+t_1t_2+t_2t_1)+\sum_{i=3}^{n-1}\mathbf{k}[S_n](1-t_i). \tag{7.20}$$

Proof We have, first,

$$\sum_{i=3}^{n-1}\mathbf{k}[S_n](1-t_i)(1-t_1)e^{(1)} = \sum_{i=3}^{n-1}\mathbf{k}[S_n](1-t_1)(1-t_i)e^{(1)} = 0;$$

then,

$$(1+t_1t_2+t_2t_1)(1-t_1)e^{(1)} = (1+t_1t_2+t_2t_1)(1-t_1)\frac{1}{2}(1+t_2)e^{(1)} = 0$$

because $(1+t_1t_2+t_2t_1)(1-t_1+t_2-t_1t_2)=0$; and next, $\mathbf{k}[S_n](1+t_1)(1-t_1)e^{(1)}=0$, for $t_1^2=1$.

Conversely, denote by I the left ideal (7.20). In the left module $\mathbf{k}[S_n]/I$ we have the following relations

$$xt_1\equiv -x,\ xt_1t_2\equiv -x-xt_2t_1\equiv -x+xt_2,\ xt_i\equiv x,\ \ i>2, \tag{7.21}$$

where $x\in\mathbf{k}[S_n]$. Using these relations, let us demonstrate that each $\Xi\in\mathbf{k}[S_n]$ has a representation

$$\Xi\equiv\sum_{i=1}^{n-1}\alpha_i[2;i+1]\ (\text{mod}\, I),\ \ \alpha_i\in\mathbf{k}, \tag{7.22}$$

where $[2;i+1]=(2,3,\ldots,i+1)$ is a cyclic permutation $2\to 3\to\ldots\to i+1\to 2$.

Relations (7.21) allow one to reduce the length of some words in t_i, $i\geq 1$. Let w be an irreducible by (7.21) word. If w is not empty then it ends by t_2. The second relation of (7.21) shows that before t_2 may stand only t_3 since otherwise this letter commutes with t_2 and the last relation of (7.21) reduces the length. Let k be the maximal number such that w ends with $t_kt_{k-1}\cdots t_3t_2$; that is, $w=\ldots t_st_kt_{k-1}\cdots t_3t_2$, $k\geq 3$, $s\neq k+1$. If $s>k+1$, then t_s commutes with all followed it factors and the last of (7.21) applies. If $s<k$ then we may move t_s to the right so that we get $w=\ldots t_kt_{k-1}\cdots \underline{t_st_{s+1}t_s}t_{s-1}\cdots t_3t_2$. The braid relation yields $w=\ldots t_kt_{k-1}\cdots t_{s+1}t_s\underline{t_{s+1}}t_{s-1}\cdots t_3t_2$. Now we may move the underlined t_{s+1} to the right and apply the last of (7.21). Thus all irreducible words are $w=t_kt_{k-1}\cdots t_3t_2=[2;k+1]$, $1<k\leq n-1$, which proves (7.22).

Finally, if $\Xi\,(1-t_1)e^{(1)}=0$ with Ξ given in (7.22), then

$$0=\sum_{i=1}^{n-1}\alpha_i[2;i+1](1-t_1)e^{(1)}=(\sum_{i=1}^{n-1}\alpha_i)-\sum_{i=1}^{n-1}\alpha_i[2;i+1]t_1e^{(1)}.$$

All elements $1,[2;i+1]t_1=[1;i+1]$, $1\leq i<n$ belong to the different left co-sets of the decomposition given in Corollary 1.2 with $r=1$, whereas $e^{(1)}\in S_n^{(1)}$. Hence we obtain $\alpha_i=0$, $1\leq i<n$, and $\Xi\in I$. □

Theorem 7.3 *If L is a Lie τ-algebra over a field $\mathbf{k}$ of zero characteristic, then the natural homomorphism $\iota:L\to U(L)^{(-)}$ is injective.*

Proof Denote by $\mathbf{m}_i$, $1\leq i<n$ the linear map

$$\mathbf{m}_i=\text{id}^{\otimes(i-1)}\otimes\mathbf{m}\otimes id^{\otimes(n-i-1)}:L^{\otimes n}\to L^{\otimes(n-1)}. \tag{7.23}$$

Then axioms (7.13) imply the following commutation rules:

$$\begin{aligned}
&(a)\ \mathbf{m}_j t_i = t_i \mathbf{m}_j, && \text{if } j > i+1;\\
&(b)\ \mathbf{m}_{i+1} t_i = t_i t_{i+1} \mathbf{m}_i;\\
&(c)\ \mathbf{m}_i t_i = t_{i+1} t_i \mathbf{m}_{i+1};\\
&(d)\ \mathbf{m}_j t_i = t_{i+1} \mathbf{m}_j, && \text{if } j < i.
\end{aligned} \tag{7.24}$$

The Jacoby identity also provides some sort of commutation rules for $\mathbf{m}_i$'s:

$$\begin{aligned}
&(e)\ \mathbf{m}_j \mathbf{m}_i = \mathbf{m}_i \mathbf{m}_{j-1}, && \text{if } j > i+1;\\
&(f)\ \mathbf{m}_{i+1} \mathbf{m}_i = (1 - t_{i+1}) \mathbf{m}_i \mathbf{m}_i;\\
&(g)\ \mathbf{m}_i \mathbf{m}_i = (1 - t_i) \mathbf{m}_{i+1} \mathbf{m}_i;\\
&(h)\ \mathbf{m}_j \mathbf{m}_i = \mathbf{m}_{i+1} \mathbf{m}_j, && \text{if } j < i,
\end{aligned} \tag{7.25}$$

whereas the skew-symmetry yields

$$t_j \mathbf{m}_j = -\mathbf{m}_j. \tag{7.26}$$

If $l \in L$ and $\iota(l) = 0$, then in $\mathbf{k}\langle L \rangle$ we have a representation

$$l = \sum v_i (a_i - a_i \tau - a_i \mathbf{m}) w_i, \quad a_i \in L^{\otimes 2}, v_i \in L^{\otimes n_i}, w_i \in L^{\otimes k_i}. \tag{7.27}$$

We have

$$v_i (a_i - a_i \tau - a_i \mathbf{m}) w_i = u \cdot (1 - t_{n_i+1}) - u\, \mathbf{m}_{n_i+1},$$

where $u = v_i \otimes a_i \otimes w_i \in L^{\otimes (n_i+k_i+2)}$. Because $u \cdot (1 - t_{n_i+1}) \in L^{\otimes (n_i+k_i+2)}$ and $u\, \mathbf{m}_{n_i+1} \in L^{\otimes (n_i+k_i+1)}$, the equality (7.27) splits into the following system of homogeneous equations.

$$\sum_{i=1}^{n-1} u_n^i \cdot (1 - t_i) = 0; \tag{7.28}$$

$$\sum_{i=1}^{n-1} u_n^i \mathbf{m}_i + \sum_{i=1}^{n-2} u_{n-1}^i \cdot (1 - t_i) = 0; \tag{7.29}$$

$$\sum_{i=1}^{n-2} u_{n-1}^i \mathbf{m}_i + \sum_{i=1}^{n-3} u_{n-2}^i \cdot (1 - t_i) = 0; \tag{7.30}$$

....

$$u_3^1 \mathbf{m}_1 + u_3^2 \mathbf{m}_2 + u_2^1 \cdot (1 - t_1) = 0; \tag{7.31}$$

$$u_2^1 \mathbf{m}_1 = l, \tag{7.32}$$

where $n = \max(n_i + k_i + 2)$, $u_r^s \in L^{\otimes r}$, $1 \leq r \leq n$, $1 \leq s < r$.

In order to show that $l = 0$, we are going to perform induction on n. If $n = 2$ we have two equalities: $u_2^1(1-t_1) = 0$, and $l = u_2^1\mathbf{m}_1$. According to the skew-symmetry axiom $u_2^1(1-t_1)\mathbf{m}_1 = 2u_2^1\mathbf{m}_1$, hence $l = 0$.

We make the inductive step by downward induction on $k = \min\{i \mid u_n^i \neq 0\}$. If $k = n-1$ then Eqs. (7.28) and (7.29) take up the form

$$u_n^{n-1} \cdot (1 - t_{n-1}) = 0;$$

$$u_n^{n-1}\mathbf{m}_{n-1} + \sum_{i=1}^{n-2} u_{n-1}^i \cdot (1 - t_i) = 0. \tag{7.33}$$

The skew-symmetry axiom yields $2u_n^{n-1}\mathbf{m}_{n-1} = u_n^{n-1} \cdot (1 - t_{n-1})\mathbf{m}_{n-1} = 0$. Thus, we may apply the inductive supposition to Eqs. (7.30)–(7.33).

Let $k = \min\{i \mid u_n^i \neq 0\} < n-1$, $n \geq 3$. If a set of elements

$$\{w_s^i \in L^{\otimes s} \mid 1 < s \leq n,\ 1 \leq i < s\}$$

satisfies the system of Eqs. (7.28)–(7.31) and $w_2^1\mathbf{m}_1 = 0$, then the set

$$\{u_s^i - w_s^i \mid 1 < s \leq n, 1 \leq i < s\}$$

still satisfies (7.28)–(7.32) because all operators are linear. Therefore, to complete the inductive step of the downward induction, it remains to find a solution $\{w_s^i\}$ of (7.28)–(7.31) with $w_2^1\mathbf{m}_1 = 0$, such that $w_n^k = u_n^k$, $w_n^{k-1} = w_n^{k-2} = \ldots = 0$.

Recall that we have fixed the notation

$$e^{(k)} = \frac{1}{(n-k)!} \sum_{\pi \in S_n^{(k)}} \pi.$$

Since $(1 - t_i)e^{(k)} = 0$, $k < i < n$, Eq. (7.28) implies $u_n^k \cdot (1 - t_k)e^{(k)} = 0$. By Lemma 7.3 applied to $S_n^{(k-1)}$, the left annihilator I of $(1 - t_k)e^{(k)}$ in $\mathbf{k}[S_n^{(k-1)}]$ equals

$$\mathbf{k}[S_n^{(k-1)}](1+t_k) + \mathbf{k}[S_n^{(k-1)}](1+t_kt_{k+1}+t_{k+1}t_k) + \sum_{i=k+2}^{n-1} \mathbf{k}[S_n^{(k-1)}](1-t_i). \tag{7.34}$$

As a left ideal of a semisimple algebra, this annihilator has the form $I = \mathbf{k}[S_n^{(k-1)}]f$, where $f \in I$ is an idempotent. In this case, $(1 - t_k)e^{(k)}\mathbf{k}[S_n^{(k-1)}] = (1-f)\mathbf{k}[S_n^{(k-1)}]$. Therefore, $u_n^k \cdot (1-f)\mathbf{k}[S_n^{(k-1)}] = 0$; that is, $u_n^k = u_n^k \cdot f$. Let

$$f = r_k(1 + t_k) + r_{k+1}(1 + t_kt_{k+1} + t_{k+1}t_k) + \sum_{i=k+2}^{n-1} r_i(1 - t_i), \quad r_\lambda \in \mathbf{k}[S_n^{(k-1)}].$$

We have received the following decomposition

$$u_n^k = u_n^k \cdot r_k(1+t_k) + u_n^k \cdot r_{k+1}(1+t_kt_{k+1}+t_{k+1}t_k) + \sum_{i=k+2}^{n-1} u_n^k \cdot r_i(1-t_i). \quad (7.35)$$

Let us put

$$\begin{aligned} w_n^k &= u_n^k, \\ w_n^{k+1} &= -u_n^k \cdot r_{k+1}(1+t_kt_{k+1}+t_{k+1}t_k), \\ w_n^i &= -u_n^k \cdot r_i(1-t_k), \ \ k+2 \le i < n. \end{aligned} \quad (7.36)$$

Since $1-t_k$ commutes with all $1-t_i$, $k+2 \le i < n$, and

$$(1+t_kt_{k+1}+t_{k+1}t_k)(1-t_k) = (1+t_kt_{k+1}+t_{k+1}t_k)(1-t_{k+1}),$$

Eq. (7.35) implies

$$\sum_{i=k}^{n-1} w_n^i(1-t_i) = 0, \quad (7.37)$$

and

$$\begin{aligned} \sum_{i=k}^{n-1} w_n^i \mathbf{m}_i = {} & u_n^k \cdot r_{k+1}(1+t_kt_{k+1}+t_{k+1}t_k)\mathbf{m}_k + \sum_{i=k+2}^{n-1} u_n^k \cdot r_i(1-t_i)\mathbf{m}_k \\ & - u_n^k \cdot r_{k+1}(1+t_kt_{k+1}+t_{k+1}t_k)\mathbf{m}_{k+1} - \sum_{i=k+2}^{n-1} u_n^k \cdot r_i(1-t_k)\mathbf{m}_i. \end{aligned} \quad (7.38)$$

Let us define

$$\begin{aligned} w_{n-1}^j &= 0, \ \ j<k; \\ w_{n-1}^k &= u_n^k \cdot \Big(\sum_{i=k+2}^{n-1} r_i\mathbf{m}_i - r_{k+1}(1+t_kt_{k+1}+t_{k+1}t_k)\mathbf{m}_k\Big); \\ w_{n-1}^{i-1} &= -u_n^k \cdot r_i\mathbf{m}_k, \ \ k+2 \le i \le n-1. \end{aligned} \quad (7.39)$$

The defining relations in t_i imply

$$(1+t_kt_{k+1}+t_{k+1}t_k)t_{k+1}t_k = (1+t_kt_{k+1}+t_{k+1}t_k),$$

while axiom (7.13) yields $t_{k+1}t_k\mathbf{m}_{k+1} = \mathbf{m}_k t_k$. Hence, we have

$$(1 + t_k t_{k+1} + t_{k+1}t_k)(\mathbf{m}_k - \mathbf{m}_{k+1}) = (1 + t_k t_{k+1} + t_{k+1}t_k)\mathbf{m}_k(1 - t_k).$$

By means of this formula and the commutation rules

$$(1 - t_i)\mathbf{m}_k = \mathbf{m}_k(1 - t_{i-1}), \quad (1 - t_k)\mathbf{m}_i = \mathbf{m}_i(1 - t_k), \quad k + 2 \leq i,$$

equality (7.38) implies

$$\sum_{i=1}^{n-1} w_n^i \mathbf{m}_i + \sum_{i=1}^{n-2} w_{n-1}^i (1 - t_i) = 0. \tag{7.40}$$

The Jacobi identity and $\mathbf{m}_k\mathbf{m}_{i-1} = \mathbf{m}_i\mathbf{m}_k$, $k + 2 \leq i < n$ provide the equality

$$\sum_{i=1}^{n-2} w_{n-1}^i \mathbf{m}_i = 0. \tag{7.41}$$

Therefore, if we put $w_s^i = 0, 1 < s \leq n-2, 1 \leq i < s$, then equalities (7.37), (7.40), and (7.41) demonstrate that the set $\{w_s^i\}$ is a solution of (7.28)–(7.31) that satisfies

$$w_n^k = u_n^k, \; w_n^{k-1} = w_n^{k-2} = \ldots = 0.$$

If $n > 3$ then $w_2^1\mathbf{m}_1 = 0$ because $w_2^1 = 0$. If $n = 3$ then (7.41) says $w_2^1\mathbf{m}_1 = 0$. □

7.3.3 PBW Isomorphism

In order to understand the relation between L and $U(L)$, it is important to consider the simplest Lie τ-algebra L^0 defined on the same braided space L. This is the Lie τ-algebra with the zero multiplication: $[u, v]_0 = 0$. Certainly, the algebra $U(L^0)$ is uniquely defined by the symmetry τ. For example, if τ is the ordinary flip $(u \otimes v)^\tau = v \otimes u$, then $U(L^0)$ is nothing more than the algebra of commutative polynomials in a basis of L or, in invariant terms, this is the symmetric algebra of the space L. If L is a color Lie algebra then $U(L^0)$ is generated by variables x_i, $i \in I$ which are related to a homogeneous basis $l_i \in L_{g_i}$ of L, and commute according to the rule

$$x_i x_s = q_{is} x_s x_i, \quad q_{is} = \alpha(g_i, g_s).$$

This is the so called *algebra of quantum polynomials* (of course, here $q_{is}q_{si} = 1$). By these reasons, in general, the algebra $U(L^0)$ may be regarded as an algebra of "commutative" polynomials in a given symmetric category.

In the general case of Lie τ-algebras, the Poincaré-Birkhoff-Witt Theorem in its constructive form is not valid any more; that is, $U(L)$ not always has a basis $u_1^{n_1}u_2^{n_2}\cdots u_m^{n_m}$ defined by an ordered basis $u_1 < u_2 < \cdots < u_m < \cdots$ of L.

Example 7.1 The simplest example is given by the Lie τ-algebra with zero multiplication and $\tau = \mathrm{id} \otimes \mathrm{id}$. In this particular case, the ideal J generated by

$$\{u \otimes v - (u \otimes v)\tau - [u, v] \mid u, v \in L\}$$

is zero. Hence, $U(L)$ is the free associative algebra $\mathbf{k}\langle L\rangle$ which certainly has no basis of the form $u_1^{n_1}u_2^{n_2}\cdots u_m^{n_m}$, where $\{u_i\}$ is an ordered basis of L.

Even if L is a color Lie algebra, the M. Scheunert theorem demonstrate that some of the basis elements of L are of height 2 in $U(L)$, see Theorem 7.1.

Nevertheless, two invariant (independent of a fixed basis) forms of the Poincaré-Birkhoff-Witt Theorem are known which may be generalized to the Lie τ-algebras over a field of characteristic zero. The PBW-theorem in one of these forms provides an isomorphism of coalgebras $U(L) \cong U(L^0)$ where, as above, L^0 is the Lie τ-algebra with the zero multiplication defined on the same space L.

Theorem 7.4 *If the characteristic of the ground field* $\mathbf{k}$ *is zero, then the linear map* $\eta : U(L^0) \to U(L)$ *defined as*

$$\eta : \varphi^0(u) \mapsto \varphi(u \cdot e_n), \quad u \in L^{\otimes n},$$

is an isomorphism of coalgebras. Here $\varphi^0 : \mathbf{k}\langle L\rangle \to U(L^0)$ *and* $\varphi : \mathbf{k}\langle L\rangle \to U(L)$ *are the natural homomorphisms appearing in the definition of the universal enveloping algebra, and*

$$e_n = \frac{1}{n!}\sum_{\pi \in S_n} \pi \in \mathbf{k}[S_n].$$

Proof Let us demonstrate, first, that in $\mathbf{k}\,[S_n]$ the following inclusion holds:

$$1 - e_n \in \sum_{i=1}^{n-1} \mathbf{k}\,[S_n](1 - t_i). \tag{7.42}$$

For each $\pi \in S_n$, we have $\pi t_i = -\pi(1 - t_i) + \pi \equiv \pi \pmod{A}$, where A is the right hand side of (7.42). Every element of the symmetric group is a product of transpositions t_i, $1 \le i < n$. Hence $\pi \equiv 1 \pmod{A}$, whereas

$$e_n = \frac{1}{n}\sum_{\pi \in S_n} \pi \equiv \frac{n \cdot 1}{n} \pmod{A}.$$

This is equivalent to (7.42).

Next, we have to demonstrate that η is a well-defined linear map. By definition, $\ker(\varphi^0)$ is generated by the space

$$W^0 = \{a \otimes b - (a \otimes b)\tau \mid a, b \in L\} = L^{\otimes 2}(\mathrm{id} - \tau).$$

In particular, $\ker(\varphi^0)$ (but not $\ker(\varphi)$) is a homogeneous ideal, whereas its nth component takes the form

$$\ker(\varphi^0)_n = \ker(\varphi^0) \cap L^{\otimes n} = \sum_{i=1}^{n-1} L^{\otimes n}(\mathrm{id} - \tau_i).$$

At the same time, $(\mathrm{id} - \tau_i)e_n = 0$, $1 \le i < n$. Thus, $\varphi^0(u) = 0$, $u \in L^{\otimes n}$ implies $u \cdot e_n = 0$ and $\varphi(u \cdot e_n) = 0$. Hence, η is a well-defined linear map.

Inclusion (7.42) implies that

$$L^{\otimes n}(1 - e_n) \subseteq \sum_{i=1}^{n-1} L^{\otimes n}\mathbf{k}[S_n](1 - \tau_i) \subseteq \sum_{i=1}^{n-1} L^{\otimes n}(1 - \tau_i) = \ker(\varphi^0)_n.$$

Therefore, for each $u \in L^{\otimes n}$, we have

$$\varphi^0(u) = \varphi^0(u \cdot e_n + u \cdot (1 - e_n)) = \varphi^0(u \cdot e_n).$$

As both φ^0 and φ are coalgebra maps, $\eta : \varphi^0(u \cdot e_n) \mapsto \varphi(u \cdot e_n)$, $u \in L^{\otimes n}$ is so too.

Let us check that η is an epimorphism. To this end, it suffices to show that

$$\varphi(L^{\otimes n}) \subseteq \sum_{i=1}^{n} \varphi(L^{\otimes i} e_i).$$

If $n = 1$, then $e_1 = 1$ and the inclusion is evident. In general case, for each $u \in L^{\otimes n}$ we have $u - u\tau_i \equiv u \cdot \mathbf{m}_i (\mathrm{mod}\, J)$, whereas $u \cdot \mathbf{m}_i \in L^{\otimes (n-1)}$. Here as above $\mathbf{m}_i = \mathrm{id}^{i-1} \otimes \mathbf{m} \otimes \mathrm{id}^{n-i-1}$ and $\mathbf{m}$ is the τ-Lie multiplication of L. Now, inclusion (7.42) yields

$$u - u \cdot e_n = u(1 - e_n) \in \sum_{i=1}^{n-1} L^{\otimes n} \cdot \mathbf{k}[S_n](1 - t_i) \subseteq L^{\otimes (n-1)} \pmod{J}.$$

Therefore $\varphi(u) - \varphi(u \cdot e_n) \in \varphi(L^{\otimes (n-1)})$, and evident induction applies.

It remains to demonstrate that the kernel of η is zero. Because $\varphi^0(u) = 0$, $u \in L^{\otimes i}$ implies $u \cdot e_i = 0$, it suffices to check that the intersection of $\ker(\varphi)$ with $\sum_{i=1}^{\infty} L^{\otimes i} e_i$

is zero. By induction on n, we shall prove that

$$\ker(\varphi) \cap \sum_{i=1}^{n} L^{\otimes i} e_i = 0.$$

When $n = 1$, this is precisely the statement of Theorem 7.3.

Let $n > 1$, and let $w = \sum_{i=1}^{n} u_i \cdot e_i$, $u_i \in L^{\otimes i}$, $u_n \cdot e_n \neq 0$. By Lemma 6.3, taking into account equalities $e_i \cdot \pi = e_i$ with $\pi \in S_i$, and counting the number of terms of the operator $\Phi_r^{(1,i)}$ defined in (6.23), we have

$$\Delta^b(u_i \cdot e_i) = \sum_{r=0}^{i} u \cdot e_i \cdot \Phi_r^{(1,i)}\, \theta_r = \sum_{r=0}^{i} \binom{i}{r} u \cdot e_i\, \theta_r, \quad u \in L^{\otimes i}.$$

Clearly, the idempotents $e_r, e^{(r)}$, $1 \leq r \leq i$ defined in (7.19) satisfy $e_i = e_i e_r = e_i e^{(r)}$. Therefore if $\varphi(w) = 0$, then

$$\begin{aligned} 0 &= \Delta^b(\varphi(w)) - 1\underline{\otimes}\varphi(w) - \varphi(w)\underline{\otimes}1 \\ &= \sum_{i=1}^{n} \sum_{r=1}^{i-1} \binom{i}{r} u_i \cdot e_i\, \theta_r (e_r \underline{\otimes} e^{(r)})(\varphi \underline{\otimes} \varphi). \end{aligned} \tag{7.43}$$

By the inductive supposition the restriction of $\varphi\underline{\otimes}\varphi$ on $\sum_{r=1}^{n-1} L^{\otimes i} e_r \underline{\otimes} \sum_{r=1}^{n-1} L^{\otimes i} e_r$ is injective. Hence

$$0 = \sum_{i=1}^{n} \sum_{r=1}^{i-1} \binom{i}{r} u_i \cdot e_i\, \theta_r (e_r \underline{\otimes} e^{(r)}) = \sum_{i=1}^{n} \sum_{r=1}^{i-1} \binom{i}{r} u_i \cdot e_i\, \theta_r.$$

Applying the concatenation product, we obtain $\sum_{i=1}^{n} (2^i - 2) u_i e_i = 0$, which implies $(2^n - 2) u_n e_n = 0$. A contradiction. □

Corollary 7.1 *If the characteristic of the ground field is zero, then* Prim $(U(L)) = L$.

Proof Let $w = \sum_{i=1}^{n} u_i \cdot e_i$, $u_i \in L^{\otimes i}$. If $\varphi(w)$ is a primitive element, then equality (7.43) is valid, which implies $\sum_{i=1}^{n} (2^i - 2) u_i e_i = 0$. Hence $(2^i - 2) u_i e_i = 0$, $1 < i \leq n$, and $w = w_1 \in L$. □

The proven theorem does not provide a basis for $U(L)$ in an explicit form, but it shows that in order to construct such a basis it is sufficient to find a basis of the algebra $U(L^0)$, the algebra of τ-commutative polynomials. In particular, we see that the basis of $U(L)$ is independent of the Lie operation on L and, instead, it is completely defined by the symmetry τ.

Corollary 7.2 *If the characteristic of the ground field is zero, then $U(L^0)$ is isomorphic to the Nichols algebra $\mathcal{B}(L)$ as a braided Hopf algebra.*

Proof The set of defining relations (7.18) for $U(L^0)$ consists of homogeneous quadratic polynomials $uv - \sum v_i u_i$, where $(u \otimes v)\tau = \sum_i v_i \otimes u_i$. By Theorem 7.2 there exists a natural homomorphism of braided Hopf algebras $\varphi : \mathbf{k}\langle L\rangle \to U(L^0)$. The kernel of φ is contained in the ideal Λ^2, and whence, due to Lemma 6.15, $\ker \varphi \subseteq \ker \Omega$. Therefore there exists a braided Hopf algebra homomorphism $\xi : U(L^0) \to \mathcal{B}(L)$. In this case $\ker \xi \cap L = 0$. By the above corollary, we have $\mathrm{Prim}\,(U(L^0)) = L$. In particular, $\ker \xi$ has no nonzero primitive elements. Thus, Theorem 6.3 implies $\ker \xi = 0$. □

Another invariant form of the PBW theorem claims that the graded algebra associated with $U(L)$ is isomorphic to the algebra of τ-commutative polynomials.

Theorem 7.5 *If the characteristic of the ground field is zero, then the graded algebra associated with $U(L)$ filtered by*

$$\mathbf{k}\cdot 1 \subseteq \mathbf{k}\cdot 1 + L \subseteq (\mathbf{k}\cdot 1 + L)^2 \subseteq \cdots \subseteq (\mathbf{k}\cdot 1 + L)^n \subseteq \cdots \subseteq U(L) \tag{7.44}$$

is isomorphic to $U(L^0)$ as a braided Hopf algebra.

Proof Let $\{a_i \,|\, i \in I\}$ be a basis of L. Consider the natural homomorphism $\varphi : \mathbf{k}\langle X\rangle \to U(L)$, where $X = \{x_i \,|\, i \in I\}$ and $x_i \mapsto a_i$. Filtration (7.44) is precisely the filtration defined by the formal degree $d(x_i) = 1$, see Sect. 1.6. Due to Definition 7.1 the kernel J of φ is generated by polynomials

$$x_i x_s - \sum_{k,m} \alpha_{i,s}^{k,m} x_k x_m - [x_i, x_s],$$

where $(x_i \otimes x_s)\tau = \sum_{k,m} \alpha_{i,s}^{k,m} x_k \otimes x_m$. The leading components of these polynomials are the defining relations of $U(L^0)$. Hence epimorphism (1.104) takes the form

$$\overline{\varphi} : U(L^0) \longrightarrow \mathrm{gr}\, U(L). \tag{7.45}$$

If $\ker \overline{\varphi} \neq 0$, then by Theorem 6.3 it contains a nonzero primitive element. Corollary 7.1 applied to L^0 states that $\mathrm{Prim}\, U(L^0) = L$. At the same time, the embedding theorem (Theorem 7.3) implies that $\ker \varphi \cap L = 0$, whereas Proposition 1.12 shows that $\ker \overline{\varphi} \cap L = \ker \varphi \cap L = 0$. A contradiction. □

7.4 Free Lie τ-Algebra

Along this section we shall assume that the characteristic of the ground filed $\mathbf{k}$ is zero. We shall consider more thoroughly the free Lie τ-algebra $L\langle V\rangle$ freely generated by a braided space V with an involutive braiding. In particular, we prove that its universal enveloping algebra is the free braided Hopf algebra $\mathbf{k}\langle V\rangle$ and $L\langle V\rangle$ as a subalgebra of $\mathbf{k}\langle V\rangle^{(-)}$ coincides with the space of all primitive elements, $L\langle V\rangle = \operatorname{Prim}\mathbf{k}\langle V\rangle^{(-)}$. Recall that the Lie τ-algebra $\mathbf{k}\langle V\rangle^{(-)}$ by definition is the space $\mathbf{k}\langle V\rangle$ with multiplication $[u, v] = uv - \sum v_i u_i$, where $(u \otimes v)\tau = \sum_i v_i \otimes u_i$, see Lemma 7.2. The following lemma demonstrates that the space $\operatorname{Prim}\mathbf{k}\langle V\rangle$ is closed with respect to the bracketing.

Lemma 7.4 *If u, v are primitive polynomials, then $[u, v]$ is so as well.*

Proof By definition of braided coproduct we have

$$\Delta^b(uv) = (u\underline{\otimes}1 + 1\underline{\otimes}u)(v\underline{\otimes}1 + 1\underline{\otimes}v) = uv\underline{\otimes}1 + 1\underline{\otimes}uv + u\underline{\otimes}v + (u\underline{\otimes}v)\tau.$$

This implies

$$\begin{aligned}\Delta^b([u, v]) &= \Delta^b(uv - \sum v_i u_i) \\ &= [u, v]\underline{\otimes}1 + 1\underline{\otimes}[u, v] + u\underline{\otimes}v + (u\underline{\otimes}v)\tau - \sum_i v_i\underline{\otimes}u_i - (\sum_i v_i\underline{\otimes}u_i)\tau \\ &= [u, v]\underline{\otimes}1 + 1\underline{\otimes}[u, v]\end{aligned}$$

because $(\sum_i v_i\underline{\otimes}u_i)\tau = (u\underline{\otimes}v)\tau^2 = u\underline{\otimes}v.$ □

We are reminded the definition of the free object.

Definition 7.2 A Lie τ-algebra $L\langle V\rangle$ generated by a braided subspace V is said to be *free Lie τ-algebra* if every homomorphism of braided spaces $\varphi : V \to L'$ into a Lie τ-algebra L' extends to a homomorphism of Lie τ-algebras $\varphi : L\langle V\rangle \to L'$.

The main idea of the proofs below is that the local action of the braid monoid on the n-fold tensor product $V^{\otimes n}$ by Theorem 1.4 reduces to the action of the symmetric group. This allows us to formulate the following general *principle*:

If a theorem is valid for ordinary Lie algebras and its statement may be interpreted as a property of the group algebra $\mathbf{k}[S_n]$ under the local action, then this theorem is valid for an arbitrary generalized Lie algebra.

Therefore once we have an interpretation, we need to check the validity of a theorem only for multilinear (noncommutative) polynomials. Somehow this provides the linearization process applied to an arbitrary involutive braiding.

Of course the above principle does not allow us to generalize all the theorems since there exist some important properties of Lie algebras that are not valid for generalized ones. For example, the PBW-theorem in constructive form, see Example 7.1.

Theorem 7.6 (**τ-Friedrichs Criteria**) *The algebra* $\mathrm{Prim}\,\mathbf{k}\langle V\rangle^{(-)}$ *is generated by* V *as a generalized Lie algebra. More precisely, an element* $v \in V^{\otimes n}$ *is primitive if and only if it has a representation*

$$v = \sum_{\mathbf{i}=(i_1,i_2,\ldots,i_n)} \alpha_{\mathbf{i}} [\ldots[[x_{i_1}, x_{i_2}], x_{i_3}], \ldots x_{i_n}], \quad \alpha_{\mathbf{i}} \in \mathbf{k}, \tag{7.46}$$

where $X = \{x_i \mid i \in I\}$ *is a fixed basis of* V.

Proof By Lemma 6.3 the braided coproduct has the form

$$\Delta^b(u) = \sum_{r=0}^{n} \left[u \cdot \Phi_r^{(1,n)}\right] \theta_r, \quad u \in V^{\otimes n}, \tag{7.47}$$

where $(u \otimes v)\theta_r = u\underline{\otimes} v$, $u \in V^{\otimes r}$ and the operators $\Phi_r^{(1,n)}$ are defined in (6.23):

$$\Phi_r^{(1,n)} = \sum_{1 \le k_1 < k_2 < \ldots < k_r \le n} [1; k_1][2; k_2] \cdots [r; k_r],$$

with

$$[k; k] = \mathrm{id}; \quad [m; k] = \tau_{k-1}\tau_{k-2}\tau_{k-3} \cdots \tau_{m+1}\tau_m, \quad m < k.$$

Therefore an element $v \in V^{\otimes n}$ is primitive if and only if $u \cdot \Phi_r^{(1,n)} = 0$, $1 \le r < n$.

We claim that the left annihilator in $\mathbf{k}[S_n]$ of all $\Phi_r^{(1,n)}$, $1 \le r < n$ equals the left ideal generated by the following element:

$$\Psi_n = (\mathrm{id} - [1; 2])(\mathrm{id} - [1; 3]) \cdots (\mathrm{id} - [1; n]). \tag{7.48}$$

Indeed, let $\Xi \cdot \Phi_r^{(1,n)} = 0$, $1 \le r < n$. Consider a linear space Z of dimension n. In the classical case, when τ is the ordinary flip $u \otimes v \mapsto v \otimes u$, the local action of $\mathbf{k}[S_n]$ is a faithful action on the subspace of all multilinear (noncommutative) polynomials of $Z^{\otimes n}$,

$$z_1 z_2 \ldots z_n \cdot \Omega = \sum_{\pi \in S_n} \alpha_\pi z_{\pi^{-1}(1)} z_{\pi^{-1}(2)} \ldots z_{\pi^{-1}(n)}, \quad \Omega = \sum \alpha_\pi \pi \in \mathbf{k}[S_n]. \tag{7.49}$$

Now we have $(z_1 z_2 \cdots z_n \cdot \Xi) \cdot \Phi_r^{(1,n)} = 0$, $1 \le r < n$. Hence $z_1 z_2 \cdots z_n \cdot \Xi$ is a primitive multilinear polynomial of $Z^{\otimes n}$. By the classical Friedrichs criteria $z_1 z_2 \cdots z_n \cdot \Xi$ is a linear combination of the form (7.46) with ordinary commutator. In terms of the local action this means

$$z_1 z_2 \cdots z_n \cdot \Xi = \sum_{\pi \in S_n} \alpha_\pi (z_1 z_2 \cdots z_n \cdot \pi) \Psi_n.$$

Thus $\Xi = (\sum_{\pi \in S_n} \alpha_\pi \pi) \cdot \Psi_n$, which proves the claim.

By Maschke theorem, $\mathbf{k}[S_n]$ is a semisimple algebra, whence by Lemma 1.39 there exists an idempotent $e \in \mathbf{k}[S_n]\,\Psi_n$ such that $\mathbf{k}[S_n]\,\Psi_n = \mathbf{k}[S_n]e$, whereas $\sum_r \Phi_r^{(1,n)}\mathbf{k}[S_n] = (1-e)\mathbf{k}[S_n]$. In particular, if $u \in V^{\otimes n}$ is a primitive element, then $u \cdot \Phi_r^{(1,n)} = 0$, $1 \leq r < n$, and therefore $u \cdot (1-e) = 0$; that is, $ue = u$. Hence $u \in u \cdot \mathbf{k}[S_n]\,\Psi_n$, which implies representation (7.46). □

Theorem 7.7 *The space of primitive elements* $\operatorname{Prim}\mathbf{k}\langle V\rangle$ *with the brackets* $[u, v] = u \otimes v - (u \otimes v)\tau$ *is the free Lie* τ*-algebra freely generated by* V.

Proof By Theorem 7.6 the Lie τ-algebra $\operatorname{Prim}\mathbf{k}\langle V\rangle^{(-)}$ is generated by V as a Lie τ-algebra. Let L be an arbitrary Lie τ-algebra and $\varphi : V \to L$ a linear map that respects the braiding. The map φ has an extension up to a homomorphism of braided Hopf algebras $\tilde{\varphi} : \mathbf{k}\langle V\rangle \to U(L)$. The restriction of $\tilde{\varphi}$ on $\operatorname{Prim}\mathbf{k}\langle V\rangle^{(-)}$ is the required homomorphism of Lie τ-algebras $\overline{\varphi} : \operatorname{Prim}\mathbf{k}\langle V\rangle^{(-)} \to \iota(L)$ because $\iota(L) \cong L$ due to Theorem 7.3. □

Now we are going to prove a number of auxiliary statements in order to show that every braided Hopf subalgebra of $\mathbf{k}\langle V\rangle$ is generated by primitive elements.

Recall that the free braided Hopf algebra $\mathbf{k}\langle X\rangle$ has a right and a left coordinate differential calculi, see Sect. 6.3. Let as above C and C^* denote the subalgebras of constants for the right and left calculi, respectively.

Lemma 7.5 *If* $\tau^2 = id$, *then* $C = C^*$.

Proof Comparing coproduct formula (7.47) and decomposition (6.41), we see that $u \in V^{\otimes n} \cap C$ if and only if $u \cdot \Phi_1^{(1,n)} = 0$, where

$$\Phi_1^{(1,n)} = \sum_{i=1}^{n}[i; n] = \tau_{n-1}\tau_{n-2}\cdots\tau_2\tau_1 + \ldots + \tau_2\tau_1 + \tau_1 + \text{id}. \tag{7.50}$$

Similarly, (7.47) and decomposition (6.44) imply that $u \in V^{\otimes n} \cap C^*$ if and only if $u \cdot \Phi_{n-1}^{(1,n)}$, where by definition

$$\Phi_{n-1}^{(1,n)} = \sum_{1\leq k_1<k_2<\ldots<k_{n-1}\leq n} [1; k_1][2; k_2]\cdots[n-1; k_{n-1}].$$

Lemma 1.14 allows one to replace the area $\{1 \leq k_1 < k_2 < \ldots < k_{n-1} \leq n\}$ by its complement $\{1 \leq i_1 \leq n\}$, so that $\Phi_{n-1}^{(1,n)} = \sum_{i=1}^{n}[n; i]$. Since $\tau_i^2 = \text{id}$, it follows that $[n; i][1; n] = [1; i]$, which implies $\Phi_{n-1}^{(1,n)} \cdot [1; n] = \Phi_1^{(1,n)}$. In particular the kernels of $\Phi_{n-1}^{(1,n)}$ and $\Phi_1^{(1,n)}$ coincide. □

Lemma 7.6 *The operator* $\Phi_1 = \Phi_1^{(1,n)}$ *satisfies an equation of the following form*

$$\Phi_1 = \alpha_2\,\Phi_1^2 + \alpha_3\,\Phi_1^3 + \cdots + \alpha_m\,\Phi_1^m, \quad \alpha_i \in \mathbf{k}. \tag{7.51}$$

Proof We prove the statement in two steps.

Step 1. We show that $\Phi_1 \in \Phi_1^2 \mathbf{k}[S_n]$. Suppose in contrary that $\Phi_1 \notin \Phi_1^2 \mathbf{k}[S_n]$. Then by Corollary 1.3 there exists $\Xi \in \mathbf{k}[S_n]$, such that $\Xi\,\Phi_1 \neq 0$, $\Xi\,\Phi_1^2 = 0$.

Consider a linear space Z of dimension n with ordinary flip $u \otimes v \mapsto v \otimes u$. As we have mentioned before, the local action (7.49) of $\mathbf{k}[S_n]$ is a faithful action on the subspace of all multilinear polynomials of $Z^{\otimes n}$. We have $z_1 z_2 \dots z_n \cdot \Xi\,\Phi_1 \neq 0$. Hence $u = z_1 z_2 \dots z_n \cdot \Xi$ is not a constant, see (7.50).

By (6.41) and (7.47) applied to $\mathbf{k}\langle Z\rangle$, we have $(u \cdot \Phi_1)\theta_1 = \sum_i z_i \underline{\otimes} (\partial u/\partial z_i)$. Therefore the element

$$c = \sum_{i=1}^{n} z_i \frac{\partial u}{\partial z_i} \tag{7.52}$$

satisfies $c = u \cdot \Phi_1$, whereas $c \cdot \Phi_1 = u \cdot \Xi\,\Phi_1^2 = 0$. Thus c is a constant. Differentiating the equality (7.52) by z_k, we obtain

$$0 = \frac{\partial u}{\partial z_k} + \sum_{i=1}^{n} z_i \frac{\partial^2 u}{\partial z_i \partial z_k}.$$

Starting with this equality we may prove by induction

$$\frac{\partial^m u}{\partial z_k \partial z_s \cdots \partial z_r} = -\frac{1}{m} \sum_{i=1}^{n} z_i \frac{\partial^{m+1} u}{\partial z_i \partial z_k \partial z_s \cdots \partial z_r}. \tag{7.53}$$

Indeed, in order to move from m to $m+1$ it suffices to differentiate (7.53) by z_t and remember that in the classical case the partial derivations commute.

As $u \in Z^{\otimes n}$, all its partial derivatives of order $n+1$ are zero. Hence recurrence formula (7.53) by downward induction shows that all partial derivatives, including the first ones, are zero; that is u is a constant. A contradiction.

Step 2. The element Φ_1 as an element of a finite dimensional algebra is a root of some polynomial

$$\alpha_0 + \alpha_1 \Phi_1 + \alpha_2 \Phi_1^2 + \alpha_3 \Phi_1^3 + \cdots + \alpha_m \Phi_1^m = 0.$$

If $\alpha_0 \neq 0$, then we may multiply this equation by $\alpha_0^{-1}\Phi_1$ in order to get (7.51). Let k be the minimal number with $\alpha_k \neq 0$. We have

$$(\alpha_k + \alpha_{k+1}\Phi_1 + \cdots + \alpha_m \Phi_1^{m-k})\Phi_1^k = 0.$$

In the first step we have seen that $\Phi_1 = \Phi_1^2\,\Xi$ for a suitable $\Xi \in \mathbf{k}[S_n]$. Hence the multiplication by Ξ^{k-1} from the right yields

$$(\alpha_k + \alpha_{k+1}\Phi_1 + \cdots + \alpha_m \Phi_1^{m-k})\Phi_1 = 0,$$

which required. □

Theorem 7.8 *The set C of all constants of* $\mathbf{k}\langle X\rangle$ *is a free algebra freely generated by a categorical subspace of homogeneous primitive elements of degree* ≥ 2. *In particular, C itself is a free braided Hopf algebra* $\mathbf{k}\langle Y\rangle$.

Proof By Corollary 6.3 the space C and all of its homogeneous components are categorical subspaces. Moreover $C = C^*$ is a right and left coideal due to Corollary 6.2. Thus C is a homogeneous categorical Hopf subalgebra. By Theorem 6.9 it satisfies the weak algorithm.

By Proposition 1.9, it remains to check that C has a categorical weak algebra basis of primitive elements. In line with the construction of the weak algebra basis (see Lemma 1.38), it suffices to find a decomposition

$$C_n = Y_n \oplus \sum_{i=1}^{n-1} C_{n-i}C_i, \tag{7.54}$$

where Y_n is a suitable categorical subspace of primitive elements.

First, we prove that C coincides with the subalgebra A generated by all primitive elements of degree ≥ 2. Every primitive element of degree ≥ 2 is a constant due to Corollary 6.1, whence $A \subseteq C$.

By Theorem 7.6 the algebra A is generated by all long skew commutators $[\dots[x_{i_1}, x_{i_2}], \dots x_{i_k}]$, $k \geq 2$. A product of the long skew commutators,

$$[\dots[x_{i_1}, x_{i_2}], \dots x_{i_k}] \cdot [\dots[x_{i_{k+1}}, x_{i_{k+2}}], \dots x_{i_s}] \cdots [\dots[x_{i_{t+1}}, x_{i_{t+2}}], \dots x_{i_n}],$$

may be written in terms of the local action as follows:

$$x_{i_1} x_{i_2} \cdots x_{i_n} \cdot \Psi_{1,k} \Psi_{k+1,s} \cdots \Psi_{t+1,n},$$

where

$$\Psi_{a,b} = (\mathrm{id} - [a; a+1])(\mathrm{id} - [a; a+2]) \cdots (\mathrm{id} - [a; b]).$$

Thus in terms of the group algebra the required statement says that the left annihilator of Φ_1 in $\mathbf{k}[S_n]$ equals the left ideal of $\mathbf{k}[S_n]$ generated by all products $\Psi_{1,k}\Psi_{k+1,s} \cdots \Psi_{t+1,n}$. According to the *principle*, it suffices to check that $C \subseteq A$ for the particular case when $V = Z$ has the trivial braiding $u \otimes v \mapsto v \otimes u$.

In the classical case, $\mathbf{k}\langle Z\rangle$ is the universal enveloping algebra of the Lie algebra $\mathrm{Prim}\,\mathbf{k}\langle Z\rangle^{(-)}$. By the PBW-theorem every element has a unique representation

$$u = \sum_{\mathbf{i}=(i_1, i_2, \dots, i_m)} \alpha_{\mathbf{i}}\, z_1^{i_1} z_2^{i_2} \cdots z_m^{i_m} \cdot c_{\mathbf{i}},$$

where $c_{\mathbf{i}}$ are words in primitive elements of degree ≥ 2; that is $c_{\mathbf{i}} \in A$. In particular $c_{\mathbf{i}}$ are constants, hence we have

$$\frac{\partial u}{\partial z_1} = \sum_{\mathbf{i}=(i_1,i_2,\ldots,i_m)} \alpha_{\mathbf{i}}\, i_1 z_1^{i_1-1} z_2^{i_2} \cdots z_m^{i_m} c_{\mathbf{i}} \neq 0,$$

provided that $i_1 \neq 0$. Thus if u is a constant, then $i_1 = i_2 = \ldots = i_m = 0$, and $u \in A$, which proves that $C = A$.

Next, consider the left ideal E generated in $\mathbf{k}[S_n]$ by all proper products

$$\Psi_{1,k}\Psi_{k+1,s} \cdots \Psi_{t+1,n}, \quad 1 < k < \ldots < t < n, \;\; k \neq n,$$

and let E_1 be the left ideal generated by Ψ_n given in (7.48). Since $\mathbf{k}[S_n]$ is semisimple, it follows that $E + E_1 = E \oplus E_2$ with $E_2 \subseteq E_1$; that is, there exist orthogonal idempotents e, f such that $E = \mathbf{k}[S_n]e$, $E_1 + E = \mathbf{k}[S_n](f + e)$, $f \in E_1$. Thus we arrive to a decomposition

$$A_n = V^{\otimes n} \cdot (E_1 + E) = V^{\otimes n} \cdot f \oplus V^{\otimes n} \cdot e.$$

This provides the required decomposition (7.54) since $C = A$, and $V^{\otimes n} \cdot f$ is a categorical (Lemma 6.11) subspace of primitive elements ($f \in E_1$). □

Theorem 7.9 *Every braided subbialgebra of* $\mathbf{k}\langle V\rangle$ *is generated by the primitive elements, and it is a conservative coalgebra in* $\mathbf{k}\langle V\rangle$.

Proof By Lemmas 6.19 and 6.20 it suffices to check that each braided subbialgebra A is conservative in $\mathbf{k}\langle V\rangle$ as a coalgebra; that is $\Delta^o(u) \in A \underline{\otimes} A$ implies that there exists $a \in A$ such that $u - a$ is a primitive element.

We will prove this statement by induction on the formal degree $d_V(u)$, with $d_V(x_i) = 1$. More precisely, the induction supposition is the following:
For every involutive braided space W, and for every braided subbialgebra $A \subseteq \mathbf{k}\langle W\rangle$, if $\Delta^o(u) \in A\underline{\otimes}A$ and $d_W(u) < n$, then $u - u' \in A$ for some primitive $u' \in \mathbf{k}\langle W\rangle$.

By Lemma 6.19 we may suppose that both u and A are homogeneous, $u \in V^{\otimes n}$. Formula (7.47) demonstrates that $u \cdot \Phi_1 \in A$ because $u \cdot \Phi_1$ appears from $[u \cdot \Phi_1]\theta_1$ by replacing $\underline{\otimes}$ with the multiplication in A. Lemma 7.6 implies

$$[u \cdot \Phi_1]\theta_1 = \left[[u \cdot (\sum_{i=2}^{m} \alpha_i \Phi_1^{i-1})]\Phi_1 \right] \theta_1.$$

Therefore $u - w$ is a constant, where $w = \sum_{i=2}^{m} \alpha_i u \cdot \Phi_1^{i-1} \in A$. By Theorem 7.8 the element $u - w$ belongs to C freely generated by a categorical subspace of primitive elements of degree ≥ 2. The formal degree of $u - w$ with respect to free generators of C is smaller than that with respect to x_i. Although $u - w$ may be inhomogeneous in the new free generators, we may apply the induction supposition to $u - w$ and subbialgebra $C \cap A \subseteq C$. Thus there exists $u_1 \in C \cap A$ such that $u - w - u_1$ is primitive. Since $w + u_1 \in A$, the theorem is proved. □

Theorem 7.10 *Every biideal of* $\mathbf{k}\langle V\rangle$ *is conservative in* $\mathbf{k}\langle V\rangle$ *as a coideal.*

Proof The proof is quite similar to that of Theorem 7.9. We start with the induction supposition:
For every involutive braided space W, and for every biideal $J \subseteq \mathbf{k}\langle W\rangle$, *if* $\Delta^o(u) \in J\underline{\otimes}\mathbf{k}\langle W\rangle + \mathbf{k}\langle W\rangle\underline{\otimes}J$ *and* $d_W(u) < n$, *then* $u - u' \in J$ *for some primitive* $u' \in \mathbf{k}\langle W\rangle$.

By Lemma 6.21 we may suppose that both u and J are homogeneous, $u \in V^{\otimes n}$. Then we shall note that formula (7.47) implies $J \cdot \Phi_1 \subseteq J$, and next almost literally follow the above proof of Theorem 7.9. □

Now we are ready to describe subalgebras of free Lie τ-algebras. Recall that according to Theorem 7.7 every free Lie τ-algebra has a form $\mathrm{Prim}\,\mathbf{k}\langle V\rangle^{(-)}$.

Theorem 7.11 *Every right categorical Lie* τ*-subalgebra L of a free Lie* τ*-algebra* $\mathrm{Prim}\,\mathbf{k}\langle V\rangle^{(-)}$ *has a subspace* $W \subseteq L$ *such that* $L = \mathrm{Prim}\,\mathbf{k}\langle W\rangle^{(-)}$, *where* $\mathbf{k}\langle W\rangle$ *is a right categorical associative subalgebra freely generated by W in* $\mathbf{k}\langle V\rangle$. *If the subspace W may be chosen to be braided, then L itself is a free Lie* τ*-algebra.*

Proof Let U denotes an associative subalgebra generated by L in $\mathbf{k}\langle V\rangle$. Since L is a right categorical and braided subspace of primitive elements, it follows that U is a right categorical subbialgebra of $\mathbf{k}\langle V\rangle$. By Theorem 6.9, the algebra U satisfies the weak algorithm, whereas by Theorem 7.9 and Lemma 6.20 it has a weak algebra basis of primitive elements. Proposition 1.9 implies that U is freely generated by a subspace W of primitive elements. It remains to show that $L = \mathrm{Prim}\,U$. The inclusion $L \subseteq \mathrm{Prim}\,U$ is evident.

Consider a free associative τ-algebra $\mathbf{k}\langle L\rangle$ freely generated by the braided space L. The identical map $L \to L$ has an extension up to an epimorphism of associative τ-algebras $\varphi : \mathbf{k}\langle L\rangle \to U$. Since elements from L are primitive both in $\mathbf{k}\langle L\rangle$ and in U, it follows that φ is a homomorphism of braided bi-algebras. By Theorem 7.10, the kernel of φ is a conservative coideal in $\mathbf{k}\langle L\rangle$. Therefore every primitive element $u \in U$ has a primitive pre-image $w \in \mathrm{Prim}\,\mathbf{k}\langle L\rangle$, $\varphi(w) = u$. According to the Theorem 7.6, the element w has a representation

$$w = \sum_{\mathbf{i}=(i_1,i_2,\ldots,i_n)} [\ldots[g_{i_1}, g_{i_2}], \ldots g_{i_n}], \quad g_j \in L^{\otimes 1} \subseteq \mathbf{k}\langle L\rangle.$$

If we apply φ to both sides of this equality, then we obtain

$$u = \varphi(w) = \sum_{\mathbf{i}} [\ldots[\varphi(g_{i_1}), \varphi(g_{i_2})], \ldots \varphi(g_{i_n})] \in L,$$

whence $\mathrm{Prim}\,U \subseteq L$.

If W appears to be braided, then by Theorem 7.7, L is a free Lie τ-algebra. □

Corollary 7.3 (A.A. Mikhalev, A.S. Shtern) *Every subalgebra of the free color Lie superalgebra is free.*

Proof We need to check that the space W in Theorem 7.11 is braided. Every homogeneous subspace with respect to the grading defined by the coloring group G is braided. Since every graded subspace has a graded complement in any graded overspace, it follows that we may suppose that all W_n in the construction of the weak algebra basis are G-homogeneous, hence W is braided. □

7.5 Chapter Notes

The first generalization of Lie algebras appeared in a famous paper by Milnor and Moore [175] who demonstrated that each connected graded Hopf algebra over a field of characteristic zero is isomorphic with the universal enveloping algebra of a graded Lie algebra. The Lie algebras graded by the additive two-element group $G = \{1, 0\}$, $1 + 1 = 0$ were later renamed "Lie superalgebras" due to the development of "supermathematics" that arose from certain demands by quantum mechanics and nuclear physics [25–27]. The monograph *The Theory of Lie Superalgebras* by Scheunert [204] provides an algebraic introduction to the subject.

One of the most important achievements in this respect is the Kac classification of simple finite-dimensional Lie superalgebras [113–115, 117]. More recent developments of infinite-dimensional Lie superalgebras can be found in the works by Bahturin et al. [14, 16]. The most recent book *Lie Superalgebras and Enveloping Algebras* by Ian M. Musson [179] was published in 2012.

Scheunert introduced color Lie algebras in [202]. Notably, each color Lie algebra may be obtained from a superalgebra using a cocycle deformation of the bracket. The process involves changing the bracket of a color Lie algebra by replacing $[x, y]$ with $[x, y]_\sigma = \sigma(g, h)[x, y]$, where σ is a nonzero scalar that depends on the degrees of x and y. If $\sigma(g, h)$ is a 2-cocycle of the group G, then the new bracket also satisfies the anticommutativity property and Jacobi identity, although with a different commutation factor. By selecting a suitable σ, one can always ensure that the new bracket satisfies the identities of a Lie superalgebra. This process is known as a "discoloration" technique as proposed by Scheunert.

The concept of a Lie τ-algebra was introduced by Gurevich [88]. It later appeared in a geometrical context in a paper by Manin [166]. Lie algebras in symmetric monoidal categories are the Lie τ-algebras as defined by D. Gurevich. A standard method of obtaining a symmetric monoidal category is to consider all modules over a triangular Hopf algebra or all comodules over a cotriangular Hopf algebra. Hopf algebras and Lie algebras in distinct symmetric categories were studied in [15, 17, 18, 50, 76, 131, 132, 144, 215]. In [143], Kochetov extended Scheunerts "discoloration" technique to Lie algebras for the categories of (co)modules over (co)triangular Hopf algebras. Certain works by Etingof and Gelaki [68–70, 85] are dedicated to the classification of finite dimensional triangular and cotriangular Hopf algebras.

The category of Lie τ-algebras over a field of characteristic zero is equivalent to the category of the connected τ-cocommutative braided Hopf algebras through the enveloping construction [132]. This statement generalizes the classical Kostant-Cartier-Milnor-Moore Theorem [220, Theorem 3.10]. In [167], Masuoka generalized two classical category equivalences: formal groups with finite-dimensional Lie algebras, and unipotent algebraic affine groups with finite-dimensional nilpotent Lie algebras. He proved that over a field of characteristic zero, the category of Lie τ-coalgebras is equivalent to the category of complete τ-commutative Hopf algebras and that the category of locally nilpotent Lie τ-coalgebras is equivalent to the category of connected τ-commutative Hopf algebras. In [13], Ardizzoni et al. considered braided bialgebras of Hecke type in a similar manner.

Ion investigated PBW isomorphisms for symmetrically braided Hopf algebras (not necessarily τ-commutative or τ-cocommutative). He demonstrated that in characteristic zero, for any connected symmetrically braided Hopf algebra, the associated graded algebra with respect to the coradical filtration is τ-commutative, and therefore it is a Nichols algebra [103].

The proof of the embedding theorem in the book is credited to the author [132]. The Gurevich theorem [89] states that $U(L^0)$ is a Koszul algebra. Based on this theorem, the embedding theorem may be derived from a PBW theorem for quadratic algebras of the Koszul type [132]. This PBW-type theorem was obtained by Braverman and Gaitsgory [42] using algebraic deformation theory. It also appears in a new book by Polishchuk and Positselsky [189, Chap. 5], which focuses on the finite-dimensional case.

Gomez and Majid [86] proposed axioms of a *left quantum Lie algebra* $\mathfrak{g}$ with binary brackets [,] and braiding τ that appear logically within the context of Woronowicz's bicovariant differential calculi over a Hopf algebra [230]. The Jacobi identity remains essentially unchanged:

$$(\mathrm{id} \otimes \mathbf{m})\mathbf{m} = (\mathbf{m} \otimes \mathrm{id})\mathbf{m} + \tau_1(\mathrm{id} \otimes \mathbf{m})\mathbf{m}.$$

The antisymmetry condition becomes a conditional identity (quasi-identity):

$$U\tau = U \Longrightarrow U\mathbf{m} = 0, \qquad U \in \mathfrak{g}^{\otimes 2}.$$

The second axiom of braided algebra (7.13) remains unchanged, whereas the first axiom is transformed as follows:

$$(\mathbf{m} \otimes \mathrm{id})\tau - \tau_2\tau_1(\mathrm{id} \otimes \mathbf{m}) = \tau_2(\mathbf{m} \otimes \mathrm{id}) - \tau_1(\mathrm{id} \otimes \mathbf{m})\tau.$$

If $\tau^2 = \mathrm{id}$, then these axioms are equivalent to the axioms of Lie τ-algebra (7.13), (7.14); however, in general, the braiding is not supposed to be involutive. In this case, the embedding problem remains unsolved. See the detailed discussion in [131, Sect. 5].

Ardizzoni [7, 9, 10] proposed a further modification of axioms postulating the embedding of $\mathfrak{g}$ into $U(\mathfrak{g})$. The latter axiom plays the role of an implicit Jacobi identity. The general theory of algebraic systems states that the conditions for the embedding have generally assumed the form of quasi-identities, i.e., the implications of the form

$$f_1 = 0 \,\&f_2 = 0 \,\& \ldots \&f_n = 0 \Longrightarrow g = 0$$

(as Gomez-Majid antisymmetry is), but not the form of identities. One should remember that the quasi-varieties, the classes of algebras defined by quasi-identities, are not closed with respect to the homomorphic images.

Ardizzoni and Stumbo [12] applied this approach to investigate the structure of primitively generated connected braided bi-algebras whose braided vector space of primitive elements defines a quadratic Nichols algebra, but τ is not necessarily involutive or of Hecke type.

The investigations of the subobjects of free objects were inspired by the famous Nielsen-Schreier theorem [182, 205]: every subgroup of a free group is free. Shirshov [209] and independently of him Witt [229] proved that every subalgebra of a free Lie algebra is free. This result was later generalized to Lie superalgebras by Shtern [214] and to color Lie algebras by Mikhalev [170, 171]. To some extent, Theorem 7.11 is a τ-version of the Shirshov-Witt theorem. As previously mentioned, Lie algebras in the braided category of left (co)modules over a (co)triangular Hopf algebra are important examples of Lie τ-algebras. Each Lie subalgebra in a category is automatically categorical. Hence, Theorem 7.11 applies to free Lie algebras in those braided categories.

The Shirshov-Witt Theorem, as well as its generalization to color Lie superalgebras, remains valid for the field $\mathbf{k}$ of positive characteristics. It is also valid in the restricted version [172, 229]. Therefore, it would be insightful to understand the extent to which Theorem 7.11 and other results (for example, the τ-Friedrichs criterion) remain valid for positive characteristics.

The free braided algebra with an involutive braiding has the structure of a *twisted algebra* as introduced by Barratt [20], or the structure of a $\mathbf{k}\Sigma_*$-algebra [217]. The free algebra with the braided coproduct is not a $\mathbf{k}\Sigma_*$-coalgebra. The theory of *twisted Lie algebras* in the category of *tensor species* ($\mathbf{k}\Sigma_*$-Lie algebras) has been subject to a similar conceptual development. Barratt's main theorem in [20] is that a free $Z\Sigma_*$-Lie algebra (in this case, $\mathbf{k} = Z$ is the ring of integer numbers) is embedded in its enveloping $Z\Sigma_*$-algebra. Joyal [108] established the Poincaré-Birkhoff-Witt theorem for enveloping algebras. Stover proved that a Kostant-Cartier-Milnor-Moore Theorem also holds [217]. The problem regarding whether any $\mathbf{k}\Sigma_*$-subalgebra of a free Lie $\mathbf{k}\Sigma_*$-algebra is free has not been considered yet.

Chapter 8
Algebra of Primitive Nonassociative Polynomials

Abstract In this chapter, we consider nonassociative primitive polynomials as operations for nonassociative Lie theory in a similar manner as how we considered the skew-primitive polynomials as operations for quantum Lie theory in Chaps. 4 and 5. Many of the well-known generalizations of Lie algebras involve only one or two operations. For instance, Malcev algebras have one binary bracket; Lie triple systems have one ternary bracket; Bol and Lie-Yamaguti algebras have one binary bracket and one ternary bracket; and Akivis algebras have two operations, an antisymmetric binary bracket and a ternary bracket (related to commutator and associator), with only one identity that relates the two operations and generalizes the Jacobi identity. The notion of Akivis algebra initially appears to be a proper analog to Lie algebras for the theory of nonassociative products. However, the question raised by K.H. Hofmann and K. Strambach of whether the commutator and associator are the only primitive operations in a nonassociative bialgebra was answered negatively. If true, it would have corroborated the fundamental role of Akivis algebras for nonassociative Lie theory. In 2002, I.P. Shestakov and U.U. Umirbaev discovered infinitely many independent operations, thus proving the theorems considered in this chapter. These results demonstrate that Shestakov-Umirbaev operations together with the commutator form a complete set of nonassociative Lie operations, whereas Theorem 8.3 is a PBW basis theorem for the Lie theory of nonassociative products.

In this chapter, we consider nonassociative primitive polynomials as operations for nonassociative Lie theory in a similar manner as how we considered the skew-primitive polynomials as operations for quantum Lie theory in Chaps. 4 and 5. Many of the well-known generalizations of Lie algebras involve only one or two operations. For instance, Malcev algebras have one binary bracket; Lie triple systems have one ternary bracket; Bol and Lie-Yamaguti algebras have one binary bracket and one ternary bracket; and Akivis algebras have two operations, an antisymmetric binary bracket and a ternary bracket (related to commutator and associator), with only one identity that relates the two operations and generalizes the Jacobi identity. In 2002, I.P. Shestakov and U.U. Umirbaev discovered infinitely many independent operations. Theorems considered in this chapter demonstrate that Shestakov-Umirbaev operations together with the commutator form a complete set

V. Kharchenko, *Quantum Lie Theory*, Lecture Notes in Mathematics 2150,
DOI 10.1007/978-3-319-22704-7_8

of nonassociative Lie operations, whereas Theorem 8.3 is a PBW basis theorem for the Lie theory of nonassociative products.

8.1 Nonassociative Polynomials

Recall that a nonassociative word is a word where the parenthesis are arranged to show how the multiplication applies. Sometimes it is more convenient to variate a designation of the parenthesis, for example instead of $(xy)z$ one may write $xy \cdot z$, whereas $((z(xy))t)v$ takes the form $\{(z \cdot xy)t\}v$. Besides this, a right-normed nonassociative word,

$$u = ((\dots((x_1x_2)x_3)\dots)x_n),$$

has a simplified notation without parenthesis,

$$u = x_1x_2x_3\dots x_n.$$

In the theory of nonassociative algebras, the commutator $[x, y] \stackrel{df}{=} xy - yx$ and the associator $(x, y, z) \stackrel{df}{=} xy \cdot z - x \cdot yz$ play distinguished role.

Lemma 8.1 *In each (nonassociative) algebra the following identities hold:*

$$[xy, z] - x[y, z] - [x, z]y = (x, y, z) - (x, z, y) + (z, x, y), \tag{8.1}$$

$$(x, y, zt) = (x, yz, t) - (xy, z, t) + x(y, z, t) + (x, y, z)t. \tag{8.2}$$

Proof We have

$$\begin{aligned}[xy, z] - x[y, z] - [x, z]y &= xy \cdot z - z \cdot xy - x \cdot yz + x \cdot zy - xz \cdot y + zx \cdot y \\ &= (xy \cdot z - x \cdot yz) - (z \cdot xy - zx \cdot y) + (x \cdot zy - xz \cdot y) \\ &= (x, y, z) - (x, z, y) + (z, x, y),\end{aligned}$$

and

$$\begin{aligned}&(x, yz, t) - (xy, z, t) + x(y, z, t) + (x, y, z)t \\ &= (x \cdot yz)t - x(yz \cdot t) - (xy \cdot z)t + (xy)(zt) \\ &\quad + x(yz \cdot t) - x(y \cdot zt) + (xy \cdot z)t - (x \cdot yz)t \\ &= (xy)(zt) - x(y \cdot zt) = (x, y, zt).\end{aligned}$$

□

A *free nonassociative algebra* $\mathbf{k}\{X\}$ in variables $X = \{x_i \mid i \in I\}$ is the algebra of nonassociative polynomials in X with the concatenation product. By definition each nonassociative polynomial is a linear combination of nonassociative words. On $\mathbf{k}\{X\}$ we fix a coproduct $\Delta : \mathbf{k}\{X\} \to \mathbf{k}\{X\} \otimes \mathbf{k}\{X\}$, which is a homomorphism of algebras, such that the variables are primitive:

$$\Delta(x_i) = x_i \otimes 1 + 1 \otimes x_i.$$

In this chapter, our aim is to understand the algebraic structure of the space $\mathfrak{A}$ of all primitive nonassociative polynomials,

$$\mathfrak{A} = \{f \in \mathbf{k}\{X\} \mid \Delta(f) = f \otimes 1 + 1 \otimes f\}.$$

Lemma 8.2 *The space $\mathfrak{A}$ is closed with respect to commutators and associators.*

Proof We have to check that commutator and associator of primitive elements are primitive. Let u, v are primitive nonassociative polynomials. We have

$$\begin{aligned}\Delta([u,v]) =& \Delta(u)\Delta(v) - \Delta(v)\Delta(u)\\ =&(1\otimes u + u\otimes 1)(1\otimes v + v\otimes 1) - (1\otimes v + v\otimes 1)(1\otimes u + u\otimes 1)\\ =&1\otimes uv + v\otimes u + u\otimes v + uv\otimes 1\\ &- 1\otimes vu + u\otimes v + v\otimes u + vu\otimes 1\\ =&(uv - vu)\otimes 1 + 1\otimes (uv - vu) = [u,v]\otimes 1 + 1\otimes [u,v].\end{aligned}$$

Similarly, if w is another primitive polynomial, then

$$\begin{aligned}\Delta(uv\cdot w - u\cdot vw) =& \Delta(u)\Delta(v)\cdot\Delta(w) - \Delta(u)\cdot\Delta(v)\Delta(w)\\ =&(1\otimes uv + v\otimes u + u\otimes v + uv\otimes 1)(1\otimes w + w\otimes 1)\\ &- (1\otimes u + u\otimes 1)(1\otimes vw + w\otimes v + v\otimes w + vw\otimes 1)\\ =&1\otimes uv\cdot w + uv\cdot w\otimes 1 - 1\otimes u\cdot vw - u\cdot vw\otimes 1\\ =&1\otimes (u,v,w) + (u,v,w)\otimes 1.\end{aligned}$$

□

Definition 8.1 A vector space is called an *Akivis algebra* if it is endowed with an anticommutative bilinear operation $[x, y]$ and a trilinear operation (x, y, z) that satisfy the following *nonassociative Jacobi identity*:

$$\begin{aligned}&[[x,y],z] + [[y,z],x] + [[z,x],y]\\ &= (x,y,z) + (y,z,x) + (z,x,y) - (y,x,z) - (x,z,y) - (z,y,x).\end{aligned}$$

Lemma 8.3 *The commutator and associator of an arbitrary nonassociative algebra satisfy the nonassociative Jacobi identity.*

Proof We have $[[x,y],z] = [xy - yx, z] = xy \cdot z - z \cdot xy - yx \cdot z + z \cdot yx$. Therefore

$$\begin{aligned}[[x,y],z] + [[y,z],x] + [[z,x],y] =& (\underline{xy \cdot z} - \boxed{z \cdot xy} - \underbrace{yx \cdot z} + \overline{z \cdot yx}) \\ &+ (\overbrace{yz \cdot x} - \underline{x \cdot yz} - \overline{zy \cdot x} + x \cdot zy) \\ &+ (\boxed{zx \cdot y} - \overbrace{y \cdot zx} - xz \cdot y + \underbrace{y \cdot xz}).\end{aligned}$$

Combining similarly marked terms, we obtain the required equality. □

Lemma 8.2 demonstrates that $\mathfrak{A}$ has the commutator as an anticommutative bilinear operation and the associator as a trilinear operation. By Lemma 8.3 the space of all primitive nonassociative polynomials is an Akivis algebra with respect to associator and commutator. Nevertheless, these two operations do not exhaust the algebraic structure of $\mathfrak{A}$. The following lemma provides a simplest example of an unary operation that can not be expressed in terms of commutator and associator.

Lemma 8.4 *The polynomial*

$$f = x^4 - x^2 \cdot x^2 - 2x(x,x,x), \quad x \in X$$

is primitive, but it does not belong to the Akivis subalgebra of $\mathfrak{A}$ *generated by* X.

Proof Here, in line with our conventions, x^4 stands for a right-normed nonassociative word $(((xx)x)x)$, whereas $x^2 \cdot x^2 = (xx)(xx)$. It is easy to prove by induction on length that a Newton formula for right-normed nonassociative words in one variable holds,

$$\Delta(x^n) = \sum_{k=0}^{n} \binom{n}{k} x^k \otimes x^{n-k}.$$

In particular, we have

$$\Delta(x^4) = x^4 \otimes 1 + 4x^3 \otimes x + 6x^2 \otimes x^2 + 4x \otimes x^3 + 1 \otimes x^4.$$

The equality $\Delta(x^2) = x^2 \otimes 1 + 2x \otimes x + 1 \otimes x^2$ implies

$$\begin{aligned}\Delta(x^2 \cdot x^2) =& \Delta(x^2)^2 = x^2 \cdot x^2 \otimes 1 + 6x^2 \otimes x^2 + 1 \otimes x^2 \cdot x^2 \\ &+ 2x \otimes x^3 + 2x \otimes x \cdot x^2 + 2x^3 \otimes x + x \cdot x^2 \otimes x.\end{aligned}$$

Considering that (x,x,x) is a primitive polynomial, we obtain

$$\Delta(x(x,x,x)) = x(x,x,x) \otimes 1 + x \otimes (x,x,x) + (x,x,x) \otimes x + 1 \otimes x(x,x,x).$$

Taking into account the equality $(x,x,x) = x^3 - x\cdot x^2$, we see that almost all terms in the decomposition of $\Delta(f)$ cancel, so that $\Delta(f) = f\otimes 1 + 1\otimes f$.

There exists only one superposition of degree four, $p(x) = [(x,x,x),x]$, of the Akivis operations in one variable. Certainly, f is not proportional to p. □

8.2 Shestakov-Umirbaev Operations

The latter example is a particular case of *Shestakov-Umirbaev operations* that we are going to define on the primitive nonassociative polynomials.

Given $m,n \geq 1$, let $\mathbf{U} = (u_1, u_2, \dots, u_m)$ and $\mathbf{V} = (v_1, v_2, \dots, v_n)$ be sequences of nonassociative polynomials, and let $U = u_1u_2\cdots u_m$, $V = v_1v_2\cdots v_n$ be the corresponding right-normed products. The Shestakov-Umirbaev operations are defined inductively as follows:

$$p(\mathbf{U};\mathbf{V};w) = (U,V,w) - \sum U_{(1)}V_{(1)}\cdot p(\mathbf{U}_{(2)};\mathbf{V}_{(2)};w), \tag{8.3}$$

where (U,V,w) is the associator. Here Sweedler's notation is extended so as to mean that the sum is taken over all partitions of the sequences $\mathbf{U}$ and $\mathbf{V}$ into pairs of subsequences, $\mathbf{U} = \mathbf{U}_{(1)} \cup \mathbf{U}_{(2)}$ and $\mathbf{V} = \mathbf{V}_{(1)} \cup \mathbf{V}_{(2)}$ such that $|\mathbf{U}_{(1)}| + |\mathbf{V}_{(1)}| \geq 1$, $\mathbf{U}_{(2)} \neq \emptyset$, $\mathbf{V}_{(2)} \neq \emptyset$; the expressions $U_{(1)}$ and $V_{(1)}$ are the right-normed products of the elements of $\mathbf{U}_{(1)}$ and $\mathbf{V}_{(1)}$ respectively.

For instance, the operation which corresponds to $m = 2$, $n = 1$ is the associator. The operations corresponding to $m = 2$, $n = 1$ and $m = 1$, $n = 2$ are

$$p(u_1,u_2;v;w) = (u_1u_2,v,w) - u_1(u_2,v,w) - u_2(u_1,v,w)$$

and, respectively,

$$p(u;v_1,v_2;w) = (u,v_1v_2,w) - v_1(u,v_2,w) - v_2(u,v_1,w).$$

If we put by definition $p(\emptyset;\mathbf{V};w) = p(\mathbf{U};\emptyset;w) = 0$, then definition (8.3) reduces to a decomposition of the associator:

$$(U,V,w) = \sum U_{(1)}V_{(1)}\cdot p(\mathbf{U}_{(2)};\mathbf{V}_{(2)};w), \tag{8.4}$$

where the sum is taken over all partitions of the sequences $\mathbf{U}$ and $\mathbf{V}$ into pairs of subsequences (including empty ones), in which case as usual $\mathbf{U}_{(1)} = \emptyset$, $\mathbf{V}_{(1)} = \emptyset$ imply $U_{(1)} = 1$, $V_{(1)} = 1$ as products of empty sets of factors. This decomposition is convenient due to the following statement, where $f\circ g$ stands for $f\otimes g + g\otimes f$, so that a polynomial f is primitive if and only if $\Delta(f) = 1\circ f$.

Lemma 8.5 *If the polynomial w and all polynomials in the sequences* **U**, **V** *are primitive, then the following equalities hold:*

$$1)\ \Delta(U) = \sum U_{(1)} \otimes U_{(2)};$$

$$2)\ \Delta(UV) = \sum U_{(1)}V_{(1)} \otimes U_{(2)}V_{(2)};$$

$$3)\ \Delta(UVw) = \sum U_{(1)}V_{(1)} \circ (U_{(2)}V_{(2)}w);$$

$$4)\ \Delta(U \cdot Vw) = \sum U_{(1)}V_{(1)} \circ (U_{(2)} \cdot V_{(2)}w);$$

$$5)\ \Delta((U, V, w)) = \sum U_{(1)}V_{(1)} \circ (U_{(2)}, V_{(2)}, w),$$

where the sums are taken over all partitions of **U** *and* **V** *into pairs of subsequences.*

Proof We demonstrate equality 1) by induction on the length of the sequence **U**. If the sequence has just one element, u_1, then there are two possible partitions: $\mathbf{U}_{(1)} = (u_1)$, $\mathbf{U}_{(2)} = \emptyset$ and $\mathbf{U}_{(1)} = \emptyset$, $\mathbf{U}_{(2)} = (u_1)$. Respectively equality 1) reduces to a correct equality $\Delta(u_1) = u_1 \otimes 1 + 1 \otimes u_1$.

Assume that equality 1) is valid for all sequences of length m. Consider an arbitrary sequence $\mathbf{W} = (u_1, u_2, \ldots, u_m, u_{m+1})$ of length $m + 1$. Each partition $\mathbf{U} = \mathbf{U}_{(1)} \cup \mathbf{U}_{(2)}$ of the subsequence $\mathbf{U} = (u_1, u_2, \ldots, u_m)$ defines two partitions of $\mathbf{W}$:

$$\mathbf{W}_{(1)} = \mathbf{U}_{(1)} \cup \{u_{m+1}\},\ \mathbf{W}_{(2)} = \mathbf{U}_{(2)},\ \text{and}\ \mathbf{W}_{(1)} = \mathbf{U}_{(1)},\ \mathbf{W}_{(2)} = \mathbf{U}_{(2)} \cup \{u_{m+1}\}.$$

For the former partition $W_{(1)} = U_{(1)}u_{m+1}$, $W_{(2)} = U_{(2)}$, whereas for the latter one $W_{(1)} = U_{(1)}$, $W_{(2)} = U_{(2)}u_{m+1}$. Using induction supposition, we obtain the required equality:

$$\begin{aligned}\Delta(Ux_{m+1}) =& \Delta(U)\Delta(u_{m+1}) = (\sum_{(\mathbf{U})} U_{(1)} \otimes U_{(2)})(u_{m+1} \otimes 1 + 1 \otimes u_{m+1}) \\ &= \sum_{(\mathbf{U})} U_{(1)}u_{m+1} \otimes U_{(2)} + \sum_{(\mathbf{U})} U_{(1)} \otimes U_{(2)}u_{m+1} = \sum_{(\mathbf{W})} W_{(1)} \otimes W_{(2)}.\end{aligned}$$

The second statement of the lemma follows immediately from 1) because Δ is a homomorphism of algebras. To check 3), we note that the coproduct in 2) is cocommutative

$$\Delta(UV) = \sum U_{(1)}V_{(1)} \otimes U_{(2)}V_{(2)} = \sum U_{(2)}V_{(2)} \otimes U_{(1)}V_{(1)}$$

because each partition $\mathbf{U} = \mathbf{A} \cup \mathbf{B}$ defines a partition $\mathbf{U} = \mathbf{B} \cup \mathbf{A}$ and vice versa. Thus

$$\begin{aligned}\Delta(UVw) &= \Delta(UV)\Delta(w) = \Delta(UV)(1 \otimes w) + \Delta(UV)(w \otimes 1)\\ &= \sum U_{(1)}V_{(1)} \otimes U_{(2)}V_{(2)}w + \sum U_{(2)}V_{(2)}w \otimes U_{(1)}V_{(1)}\\ &= \sum U_{(1)}V_{(1)} \circ U_{(2)}V_{(2)}w,\end{aligned}$$

which gives 3). In perfect analogy, one gets 4). Equality 5) follows immediately from 3) and 4) by linearity. □

Theorem 8.1 *If nonassociative polynomials*

$$u_1,\ u_2,\ \dots,\ u_m,\ v_1,\ v_2,\ \dots,\ v_n,\ w$$

are primitive then so is the polynomial

$$p(u_1, u_2, \dots, u_m; v_1, v_2, \dots, v_n; w).$$

Proof We perform induction on $m + n$. If either m or n is zero, then by definition $p(\mathbf{U}; \mathbf{V}; w) = 0$ and we have nothing to prove. Assume that for all sequences $\mathbf{U}'$, $\mathbf{V}'$ of lengths m', n', respectively, with $m' + n' < m + n$, the element $p(\mathbf{U}'; \mathbf{V}'; w)$ is primitive. By definition $p(\mathbf{U}; \mathbf{V}; w) = (U, V, w) - f$, where

$$f = \sum U_{(1)}V_{(1)} \cdot p(\mathbf{U}_{(2)}; \mathbf{V}_{(2)}; w)$$

and the sum is taken over all partitions of the sequences $\mathbf{U}$ and $\mathbf{V}$ into pairs of subsequences such that $|\mathbf{U}_{(1)}| + |\mathbf{V}_{(1)}| \geq 1$. In the above formula the elements $p(\mathbf{U}_{(2)}; \mathbf{V}_{(2)}; w)$ are primitive by the induction assumption. Using equality 3) of Lemma 8.5 with $w \leftarrow p(\mathbf{U}_{(2)}; \mathbf{V}_{(2)}; w)$, we obtain

$$\begin{aligned}\Delta(f) &= \sum U_{(1)(1)}V_{(1)(1)} \circ (U_{(1)(2)}V_{(1)(2)}p(\mathbf{U}_{(2)}; \mathbf{V}_{(2)}; w))\\ &= \sum U_{(1)}V_{(1)} \circ (U_{(2)}V_{(2)}p(\mathbf{U}_{(3)}; \mathbf{V}_{(3)}; w)),\end{aligned} \tag{8.5}$$

where the latter sum is taken over all partitions of the sequences $\mathbf{U}$ and $\mathbf{V}$ into triples of subsequences, $\mathbf{U} = \mathbf{U}_{(1)} \cup \mathbf{U}_{(2)} \cup \mathbf{U}_{(3)}$, $\mathbf{U} = \mathbf{V}_{(1)} \cup \mathbf{V}_{(2)} \cup \mathbf{V}_{(3)}$, such that $|\mathbf{U}_{(1)}| + |\mathbf{V}_{(1)}| + |\mathbf{U}_{(2)}| + |\mathbf{V}_{(2)}| \geq 1$. Let us distinguish the partitions with

$\mathbf{U}_{(1)} = \mathbf{V}_{(1)} = \emptyset$. In this case sum (8.5) splits as follows:

$$= \sum_{|\mathbf{U}_{(1)}|+|\mathbf{V}_{(1)}|\geq 1} 1 \circ U_{(1)} V_{(1)} p(\mathbf{U}_{(2)}; \mathbf{V}_{(2)}; w))$$

$$+ \sum_{|\mathbf{U}_{(1)}|+|\mathbf{V}_{(1)}|\geq 1} U_{(1)} V_{(1)} \circ (\sum_{(\mathbf{U}_{(2)}), (\mathbf{V}_{(2)})} U_{(2)(1)} V_{(2)(1)} p(\mathbf{U}_{(2)(2)}; \mathbf{V}_{(2)(2)}; w)).$$

Now the definition of f and representation of the associator (8.4) with $U \leftarrow U_{(2)}$, $V \leftarrow V_{(2)}$ imply

$$\Delta(f) = 1 \circ f + \sum_{|\mathbf{U}_{(1)}|+|\mathbf{V}_{(1)}|\geq 1} U_{(1)} V_{(1)} \circ (U_{(2)} V_{(2),w}).$$

Applying equality 5) of Lemma 8.5, we obtain

$$\begin{aligned} \Delta((U, V, w)) &= \sum U_{(1)} V_{(1)} \circ (U_{(2)}, V_{(2)}, w) \\ &= 1 \circ (U, V, w) + \sum_{|\mathbf{U}_{(1)}|+|\mathbf{V}_{(1)}|\geq 1} U_{(1)} V_{(1)} \circ (U_{(2)} V_{(2),w}). \end{aligned}$$

Consequently,

$$\Delta(p(U, V, w)) = \Delta((U, V, w)) - \Delta(f) = 1 \circ (U, V, w) - 1 \circ f = 1 \circ p(U, V, w);$$

that is, $p(U, V, w)$ is primitive. □

8.3 Lie Algebra of Nonassociative Products

In this section we prove a fundamental result of Shestakov and Umirbaev that the defined in the above section primitive operations together with the commutator form a complete set of nonassociative Lie operations. The proof includes a some sort of PBW basis construction for free nonassociative algebra over primitive polynomials which is formulated in Theorem 8.3.

Theorem 8.2 *If the characteristic of the ground field is zero, then the space $\mathfrak{A}$ of all primitive nonassociative polynomials is generated by X as an algebra with operations $p_{m,n}$, $m, n \geq 1$ and $[u, v]$.*

We shall derive this theorem from the following statement.

Proposition 8.1 *Let $\mathfrak{P}$ be the minimal subspace of $\mathbf{k}\{X\}$ that contains X and is closed with respect to all operations $p_{m,n}$, $m, n \geq 1$ and $[u, v]$. Consider an arbitrary completely ordered basis $B = \{e_\alpha \mid \alpha \in A\}$ of $\mathfrak{P}$. If the characteristic of the ground* *field is zero, then the set of all right-normed words of the type*

$$e_1 e_2 \cdots e_{i-1} e_i, \quad e_1 \leq e_2 \leq \cdots \leq e_{i-1} \leq e_i, \quad e_k \in B,\ 1 \leq k \leq i \tag{8.6}$$

forms a basis of the algebra $\mathbf{k}\{X\}$.

Proof Let C_i, $i \geq 1$ denotes the space spanned by all words (8.6) of length less than or equal to i.

Lemma 8.6 *The spaces* C_i, $1 \leq i$ *satisfy the following conditions:*

$$(C_l, C_s, C_k) \subseteq C_{i+s+k-2}, \tag{8.7}$$

$$[C_i, C_s] \subseteq C_{i+s-1}, \tag{8.8}$$

$$C_i \cdot C_s \subseteq C_{i+s}. \tag{8.9}$$

Proof We perform induction on $n = i + s + k$. If $n = 3$, then (8.7) follows from the fact that each associator (e_p, e_q, e_r) belongs to $C_1 = \mathfrak{P}$. Inclusion $[C_1, C_1] \subseteq C_1$ is evident due to the fact that $\mathfrak{P} = C_1$ is closed with respect to the operation $[u, v]$. If $e_1 \leq e_2$ then by definition $e_1 \cdot e_2 \in C_2$, whereas $e_2 \cdot e_1 = e_1 \cdot e_2 + [e_2, e_1] \in C_2 + C_1 \subseteq C_2$. Thus, we have the base of induction.

Assume that inclusions (8.7)–(8.9) fulfill for $i + s + k < n$. Let u, v, w be words (8.6) of lengths i, s, k respectively and $i + s + k = n$.

If $k = 1$; that is, $w = e_\alpha$, than formula (8.4) and the induction assumption imply that $(u, v, w) = (u, v, e_\alpha) \in C_{n-2}$.

If $k > 1$, then $w = w_1 e_k$, where w_1 is a word (8.6) of length $k - 1$. By Lemma 8.1 identity (8.2) with $x \leftarrow u, y \leftarrow v, z \leftarrow w_1, t \leftarrow e_k$ is valid:

$$(u, v, w) = (u, vw_1, e_k) - (uv, w_1, e_k) + u(v, w_1, e_k) + (u, v, w_1)e_k. \tag{8.10}$$

By induction supposition (8.9), we have $vw_1 \in C_{s+k-1}$, $uv \in C_{i+s}$. These two inclusions and already considered case $k = 1$ imply

$$(u, vw_1, e_k) \in C_{n-2}, \quad (uv, w_1, e_k) \in C_{n-2}.$$

Induction supposition (8.7) yields $(v, w_1, e_k) \in C_{s+k-2}$, and $(u, v, w_1) \in C_{n-3}$. Hence, again by induction supposition (8.9), we have

$$u(v, w_1, e_k) \in C_i C_{s+k-2} \subseteq C_{n-2}, \quad (u, v, w_1)e_k \in C_{n-3} C_1 \subseteq C_{n-2}.$$

Thus, all terms of (8.10) belong to C_{n-2}, which completes the proof of (8.7).

Consider inclusion (8.8). Because $[u, v] = -[v, u]$, without loss of generality we may suppose that $i > 1$; that is, $u = u_1 e_i$, $u_1 \in C_{i-1}$. Identity (8.1) with $x \leftarrow u_1$,

$y \leftarrow e_i, z \leftarrow v$ yields

$$[u, v] = u_1[e_i, v] + [u_1, v]e_i + (u_1, e_i, v) - (u_1, v, e_i) + (v, u_1, e_i).$$

In this decomposition, all associators belong to C_{i+s-2} due to already proven inclusion (8.7). Induction assumption (8.8) implies $[e_i, v] \in C_s$ and $[u_1, v] \in C_{i+s-2}$, whereas induction assumption (8.9) yields

$$u_1[e_i, v] \in C_{i-1}C_s \subseteq C_{i+s-1} \text{ and } [u_1, v]e_i \in C_{i+s-2}C_1 \subseteq C_{i+s-1}.$$

This completes the proof of (8.8).

Let us turn to (8.9). Consider firstly the case $s = 1$. Let $u = e_1e_2 \dots e_i$, $v = e_q$. If $e_i \leq e_q$ then the word $uv = ue_q$ is a word of type (8.6), and by definition it belongs to C_{i+1}. If t is a minimal number such that $e_q < e_t$, then by the same reason $e_1e_2 \dots e_{t-1}e_qe_t \dots e_i \in C_{i+1}$. Hence, it suffices to demonstrate that for all t, $1 \leq t \leq i$ the following relation is valid:

$$uv \equiv e_1e_2 \dots e_{t-1}e_qe_t \dots e_i \pmod{C_i}. \tag{8.11}$$

We perform downward induction on t. If $t = i$, then $u = u_1e_i$, $u_1 \in C_{i-1}$ and

$$uv = u_1e_i \cdot e_q = u_1 \cdot e_ie_q + (u_1, e_i, e_q).$$

The latter associator belongs to $C_{i-1} \subseteq C_i$ due to (8.7). Further,

$$u_1 \cdot e_ie_q = u_1(e_qe_i + [e_i, e_q]),$$

in which case $u_1[e_i, e_q] \in C_{i-1}C_1 \subseteq C_i$ due to induction assumption (8.9). Hence, taking into account (8.7), we have

$$uv \equiv u_1 \cdot e_qe_i = u_1e_q \cdot e_i - (u_1, e_q, e_i) \equiv u_1e_q \cdot e_i \pmod{C_i},$$

which completes the proof of (8.11) with $t = i$. If $t < i$, then already proven (8.11) with $i \leftarrow t$ reads:

$$e_1e_2 \dots e_{t-1}e_te_q \equiv e_1e_2 \dots e_{t-1}e_qe_t \pmod{C_t}.$$

This implies

$$e_1e_2 \dots e_{t-1}e_te_qe_{t+1} \dots e_i \equiv e_1e_2 \dots e_{t-1}e_qe_te_{t+1} \dots e_i \pmod{C_t \underbrace{C_1C_1 \cdots C_1}_{i-t}}.$$

Induction assumption on t yields

$$uv \equiv e_1e_2 \dots e_{t-1}e_te_q \dots e_i \pmod{C_i}.$$

It remains to note that $C_t \underbrace{C_1 C_1 \cdots C_1}_{i-t} \subseteq C_i$ due to induction assumption (8.9). Relation (8.11) and, hence, (8.9) with $s = 1$ are proven.

If $s > 1$, then $v = v_1 e_s$, $v_1 \in C_{s-1}$, $e_s \in B$. We have

$$uv = u \cdot v_1 e_s = uv_1 \cdot e_s + (u, v_1, e_s).$$

The latter associator belongs to $C_{i+s-2} \subseteq C_{i+s}$ due to (8.7), whereas $uv_1 \in C_{i+s-1}$ by the induction assumption. Finally, $uv_1 \cdot e_s \in C_{i+s-1} C_1 \subseteq C_{i+s}$ due to already considered case "$s = 1$". □

Let us return to Proposition 8.1. Inclusion (8.9) demonstrate that $\bigcup_{i \geq 1} C_i$ is closed with respect to the concatenation product. Because $\bigcup_{i \geq 1} C_i$ contains X, we have $\bigcup_{i \geq 1} C_i = \mathbf{k}\{X\}$; that is, the words of type (8.6) span $\mathbf{k}\{X\}$.

It remains to show that words of type (8.6) are linearly independent. We perform induction on length. The words of length one are linearly independent by definition. The word of length 0 (the empty product) equals 1. If

$$\alpha \cdot 1 + \sum_i \alpha_i e_i = 0, \quad e_i \in B, \ \alpha, \alpha_i \in \mathbf{k},$$

then $\alpha \cdot 1$ is a primitive element. Hence

$$\alpha \cdot 1 \otimes 1 = \Delta(\alpha \cdot 1) = 1 \otimes \alpha \cdot 1 + \alpha \cdot 1 \otimes 1.$$

This implies $\alpha = 0$, and therefore $\alpha_i = 0$ for all i.

Assume that words of type (8.6) with length $< n$ are linearly independent. In this case the tensors $u \otimes v \in \mathfrak{B} \otimes \mathfrak{B}$, where u, v are words of type (8.6) with length $< n$, are linearly independent as well. Consider an arbitrary linear combination of words with length $\leq n$:

$$f = \alpha \cdot 1 + \sum_i \alpha_i e_i + \sum_{k_1, k_2, \ldots, k_s} \alpha_{k_1, k_2, \ldots, k_s} e_1^{k_1} e_2^{k_2} \cdots e_s^{k_s}, \tag{8.12}$$

where the latter sum is taken over all sequences $(k_1, k_2, \ldots, k_s)$ such that

$$1 < k_1 + k_2 + \cdots + k_s \leq n,$$

and $\{e_1, e_2, \ldots, e_\alpha, \ldots\} = B$ is the basis of $\mathfrak{B} = C_1$. In this case,

$$\begin{aligned} \Delta(f) - f \otimes 1 - 1 \otimes f = & -\alpha 1 \otimes 1 \\ & + \sum_{t=1}^{s} e_t \otimes \Big(\sum_{k_1, k_2, \ldots, k_s} \alpha_{k_1, k_2, \ldots, k_s} k_t e_1^{k_1} e_2^{k_2} \cdots e_t^{k_t - 1} \cdots e_s^{k_s} \Big) \\ & + \sum \beta_r v_r \otimes w_r = 0, \end{aligned} \tag{8.13}$$

where w_r are words of type (8.6) with length $< n-1$, and v_r are words of type (8.6) with $1 < |v_r| < n$. Because by the induction supposition all tensors are linearly independent, $f = 0$ implies

$$\alpha = 0, \quad \sum_{k_1, k_2, \ldots, k_s} \alpha_{k_1, k_2, \ldots, k_s} k_t e_1^{k_1} e_2^{k_2} \cdots e_t^{k_t - 1} \cdots e_s^{k_s} = 0. \tag{8.14}$$

Again by the induction supposition, the latter equality yields $\alpha_{k_1, k_2, \ldots, k_s} k_t = 0$. As char $\mathbf{k} = 0$, we obtain $\alpha_{k_1, k_2, \ldots, k_s} = 0$. Thus, the linear dependence reduces to $\sum_i \alpha_i e_i = 0$. A contradiction. □

Now we are ready to demonstrate Theorem 8.2. We have to show that $\mathfrak{B} = \mathfrak{A}$; that is, each primitive polynomial $f \in \mathbf{k}\{X\}$ belongs to $\mathfrak{B} = C_1$. By Proposition 8.1 the element f has a representation (8.12). As f is primitive, we have equality (8.13), which implies (8.14). Consequently, all coefficients $\alpha_{k_1, k_2, \ldots, k_s}$ in (8.12) are zero, and representation (8.12) reduces to $f = \sum_i \alpha_i e_i$, which is required.

The proven equality $\mathfrak{B} = \mathfrak{A}$ allows us to reformulate Proposition 8.1 thus:

Theorem 8.3 *Each basis of the space $\mathfrak{A}$ of all primitive nonassociative polynomials forms a set of PBW generators for the free nonassociative algebra* $\mathbf{k}\{X\}$.

8.4 Chapter Notes

Lie theory for nonassociative products appeared as its own subject in the works of Malcev, who constructed the tangent structures corresponding to Moufang loops. For some time Akivis algebras were considered possible analog of Lie algebras for nonassociative products. Although the definition of an Akivis algebra involves only two operations and is quite elegant, the category of Akivis algebras is not equivalent to that of formal loops. Hence, it is not suitable as a basis for nonassociative Lie theory.

A motivation for the development of the machinery of nonassociative Hopf algebras was the question of whether the commutator and associator are the only primitive operations in a non-associative bialgebra. It appeared as a conjecture in the paper by Hofmann and Strambach [100]; if true, it would imply an important role for the Akivis algebras in nonassociative Lie theory. This conjecture was refuted by Shestakov and Umirbaev in [208], where they demonstrated the theorems included in this chapter.

An important advancement in the Lie theory of nonassociative products was the introduction of a *hiperalgebra* by Mikheev and Sabinin, now called a *Sabinin algebra*, which is the most general form of the tangent structure for loops, see [169, 198, 199]. Lie, Maltcev, Bol, Lie-Yamaguti algebras, and Lie triple systems are specific instances of Sabinin algebras. Sabinin algebras have an infinite set of independent operations. There are three different natural constructions of operations in a Sabinin algebra. Two of those constructions were devised by Sabinin and Mikheev.

The third set of operations is the Shestakov-Umirbaev operations considered in this chapter. However, the complete set of axioms for Sabinin algebras in terms of Shestakov-Umirbaev operations remains unknown.

Malcev algebras have universal enveloping algebras that have highly similar properties as typical cocommutative Hopf algebras [188]. Moreover, a similar construction can be completed for Bol algebras [186] and, more generally, for all Sabinin algebras [187]. The role of nonassociative Hopf algebras in the fundamental questions of Lie theory, such as integration, was clarified in [177]. We refer the reader to a recent survey of developments in the Lie theory for nonassociative products [178] which describes the current understanding of the subject in relation to recent works, many of which use nonassociative Hopf algebras as the main tool.

To our knowledge, the quantum aspects of the nonassociative Lie theory, such as the structure of primitive nonassociative polynomials in symmetric categories or the structure of skew-primitive polynomials in free nonassociative character Hopf algebras, have not been elaborated.

References

1. Abe, E.: Hopf Algebras. Cambridge University Press, Cambridge (1980)
2. Andruskiewitsch, N., Graña, M.: Braided Hopf algebras over non abelian finite group. Bol. Acad. Nac. Cienc. Cordoba **63**, 46–78 (1999)
3. Andruskiewitsch, N., Schneider, H.-J.: Pointed Hopf algebras. In: Montgomery, S., Schneider, H.-J. (eds.) New Directions in Hopf Algebras. MSRI Publications, vol. 43, pp. 1–69. Cambridge University Press, Cambridge (2002)
4. Andruskiewitsch, N., Schneider, H.-J.: On the classification of finite-dimensional pointed Hopf algebras. Ann. Math. **171**(2)(1), 375–417 (2010)
5. Andruskiewitsch, N., Heckenberger, I., Schneider, H.-J.: The Nichols algebra of a semisimple Yetter-Drinfeld module. Am. J. Math. **132**(6), 1493–1547 (2010)
6. Apel J., Klaus, U.: FELIX (1991). http://felix.hgb-leipzig.de
7. Ardizzoni, A.: A Milnor-Moore type theorem for primitively generated braided bialgebras. J. Algebra **327**, 337–365 (2011)
8. Ardizzoni, A.: On the combinatorial rank of a graded braided bialgebra. J. Pure Appl. Algebra **215**(9), 2043–2054 (2011)
9. Ardizzoni, A.: On primitively generated braided bialgebras. Algebr. Represent. Theor. **15**(4), 639–673 (2012)
10. Ardizzoni, A.: Universal enveloping algebras of the PBW type. Glasgow Math. J. **54**, 9–26 (2012)
11. Ardizzoni, A., Menini, C.: Associated graded algebras and coalgebras. Commun. Algebra **40**(3), 862–896 (2012)
12. Ardizzoni, A., Stumbo, F.: Quadratic lie algebras. Commun. Algebra **39**(8), 2723–2751 (2011)
13. Ardizzoni, A., Menini, C., Ştefan, D.: Braided bialgebras of Hecke type. J. Algebra **321**, 847–865 (2009)
14. Bahturin, Y.A., Mikhalev, A.A., Petrogradsky, V.M., Zaicev, M.V.: Infinite-Dimensional Lie Superalgebras. De Gruyter Expositions in Mathematics, vol. 7. Walter de Gruyter, Berlin (1992)
15. Bahturin, Y., Fishman, D., Montgomery, S.: On the generalized Lie structure of associative algebras. Israel J. Math. **96**, 27–48 (1996)
16. Bahturin, Y., Mikhalev, A.A., Zaicev, M.: Infinite-Dimensional Lie Superalgebras. Handbook of Algebra, vol. 2, pp. 579–614. North-Holland, Amsterdam (2000)
17. Bahturin, Y., Fishman, D., Montgomery, S.: Bicharacters, twistings and Scheunert's theorem for Hopf algebras. J. Algebra **236**, 246–276 (2001)

V. Kharchenko, *Quantum Lie Theory*, Lecture Notes in Mathematics 2150,
DOI 10.1007/978-3-319-22704-7

18. Bahturin, Y., Kochetov, M., Montgomery, S.: Polycharacters of cocommutative Hopf algebras. Can. Math. Bull. **45**, 11–24 (2002)
19. Bai, X., Hu, N.: Two-parameter quantum groups of exceptional type E-series and convex PBW-type basis. Algebra Colloq. **15**(4), 619–636 (2008)
20. Barratt, M.G.: Twisted Lie algebras. In: Barratt, M.G., Mahowald, M.E. (eds.) Geometric Applications of Homotopy Theory II. Lecture Notes in Mathematics, vol. 658, pp. 9–15. Springer, Berlin (1978)
21. Bautista, C.: A Poincaré-Birkhoff-Witt theorem for generalized Lie color algebras. J. Math. Phys. **39**(7), 3828–3843 (1998)
22. Benkart, G., Witherspoon, S.: Representations of two-parameter quantum groups and Schur-Weyl duality. Hopf Algebras. Lecture Notes in Pure and Applied Mathematics, vol. 237, pp. 65–92. Dekker, New York (2004)
23. Benkart, G., Witherspoon, S.: Two-parameter quantum groups and Drinfel'd doubles. Algebr. Represent. Theor. **7**(3), 261–286 (2004)
24. Benkart, G., Kang, S.-J., Melville, D.: Quantized enveloping algebras for Borcherds superalgebras. Trans. Am. Math. Soc. **350**(8), 3297–3319 (1998)
25. Berezin, F.A.: Method of the Secondary Quantification. Nauka, Moscow (1965)
26. Berezin, F.A.: Mathematical foundations of the supersymmetric field theories. Nucl. Phys. **29**(6), 1670–1687 (1979)
27. Berezin, F.A.: Introduction to Superanalysis. Mathematical Physics and Applied Mathematics, vol. 9. D. Reidel, Dordrecht/Boston, MA (1987)
28. Berger, R.: The quantum Poincaré-Birkhoff-Witt theorem. Commun. Math. Phys. **143**(2), 215–234 (1992)
29. Bergeron, N., Gao, Y., Hu, N.: Drinfel'd doubles and Lusztig's symmetries of two-parameter quantum groups. J. Algebra **301**, 378–405 (2006)
30. Bergman, G.M.: The diamond lemma for ring theory. Adv. Math. **29**, 178–218 (1978)
31. Boerner, H.: Representation of Groups with Special Consideration for the Needs of Modern Physics. North-Holland, New York, NY (1970)
32. Bokut, L.A.: Imbedding into simple associative algebras. Algebra Log. **15**, 117–142 (1976)
33. Bokut, L.A.: Associative Rings I. Library of the Department of Algebra and Logic, vol. 18. Novosibirsk State University, Novosibirsk (1977)
34. Bokut, L.A.: Gröbner–Shirshov basis for the braid group in the Artin-Garside generators. J. Symb. Comput. **43**(6–7), 397–405 (2008)
35. Bokut, L.A., Chen, Y.: Gröbner–Shirshov bases and their calculation. Bull. Math. Sci. **4**(3), 325–395 (2014)
36. Bokut, L.A., Kolesnikov, P.S.: Gröbner–Shirshov bases from their inception to the present time. Int. J. Math. Sci. NY **116**(1), 2894–2916 (2003). Translation from Zap. Nauchn. Sem. POMI **272** 26–67 (2000)
37. Bokut', L.A., Kukin, G.P.: Algorithmic and Combinatorial Algebra. Mathematics and Its Applications, vol. 255. Kluwer, Dordrecht/Boston/London (1994)
38. Bokut, L.A., Vesnin, A.: Gröbner–Shirshov bases for some braid groups. J. Symb. Comput. **41**(3–4), 357–371 (2006)
39. Borowiec, A., Kharchenko, V.K.: Algebraic approach to calculuses with partial derivatives. Sib. Adv. Math. **5**(2), 10–37 (1995)
40. Borowiec, A., Kharchenko, V.K.: First order optimum calculi. Bulletin de la Société des sciences et des lettres de Łódź **45**, 75–88 (1995). Recher. Deform. XIX
41. Borowiec, A., Kharchenko, V.K., Oziewicz, Z.: On free differentials on associative algebras. In: Gonzáles, A. (ed.) Non Associative Algebras and Its Applications. Mathematics and Applications, pp. 43–56. Kluwer, Dordrecht (1994)
42. Braverman, A., Gaitsgory, D.: Poincaré-Birkhoff-Witt theorem for quadratic algebras of Koszul type. J. Algebra **181**, 315–328 (1996)
43. Buchberger, B.: An algorithm for finding a basis for the residue class ring of a zero-dimensional ideal. Ph.D. Thesis, University of Innsbruck (1965)

44. Buchberger, B.: Ein algorithmisches Kriterium für die Lösbarkeit eines algebraischen Gleichungssystems. Aequtiones Math. **4**, 374–383 (1970)
45. Cameron, P.J.: Permutation Groups, London Mathematical Society. Student Texts, vol. 45. University Press, Cambridge (1999)
46. Chaichian, M., Demichev, A.: Introduction to Quantum Groups. World Scientific, Singapore (1996)
47. Chari, V., Xi, N.: Monomial basis of quantized enveloping algebras. In: Recent Development in Quantum Affine Algebras and Related Topics. Contemporary Mathematics, vol. 248, pp. 69–81. American Mathematical Society, Providence, RI (1999)
48. Chen, Y.Q., Shao, H.S., Shum, K.P.: On Rosso–Yamane theorem on PBW basis of $U_q(A_N)$. CUBO Math. J. **10**, 171–194 (2008)
49. Chuang, C.-L.: Identities with skew derivations. J. Algebra **224**, 292–335 (2000)
50. Cohen, M., Fishman, D., Westreich, S.: Schur's double centralizer theorem for triangular Hopf algebras. Proc. Am. Math. Soc. **122**(1), 19–29 (1994)
51. Cohn, P.M.: Sur le critère de Friedrichs pour les commutateur dans une algèbre associative libre. C. R. Acad. Sci. Paris **239**(13), 743–745 (1954)
52. Cohn, P.M.: The class of rings embeddable in skew fields. Bull. Lond. Math. Soc. **6**, 147–148 (1974)
53. Cohn, P.M.: Free Rings and Their Relations, 2nd edn. Academic Press, London (1985)
54. Connes, A.: Non-commutative Geometry. Academic Press, New York (1994)
55. Constantini, M., Varagnolo, M.: Quantum double and multiparameter quantum groups. Commun. Algebra **22**(15), 6305–6321 (1994)
56. Cotta-Ramusino, P., Rinaldi, M.: Multiparameter quantum groups related to link diagrams. Commun. Math. Phys. **142**, 589–604 (1991)
57. Curtis, C.W., Reiner, I.: Representation Theory of Finite Groups and Associative Algebras. Interscience/Wiley, New York/London (1962)
58. De Concini, C., Kac, V.G., Procesi, C.: Quantum coadjoint action. J. Am. Math. Soc. **5**, 151–189 (1992)
59. Deng, B., Du, J.: Bases of quantized enveloping algebras. Pac. J. Math. **220**(1), 33–48 (2005)
60. Deng, B., Du, J.: Monomial bases for quantum affine $\mathfrak{sl}_n$. Adv. Math. **191**(2), 276–304 (2005)
61. Diaconis, P.: Group Representations in Probability and Statistics. Lecture Notes Monograph Series, vol. 11. Institute of Mathematical Statistics, Hayward, CA (1988)
62. Diaz Sosa, M.L., Kharchenko, V.K.: Combinatorial rank of $u_q(\mathfrak{so}_{2n})$. J. Algebra **448**, 48–73 (2016)
63. Dito, G., Flato, M., Sternheimer, D., Takhtadjian, L.: Deformation quantization and Nambu mechanics. Commun. Math. Phys. **183**, 1–22 (1997)
64. Dixon, D., Mortimer, B.: Permutation Groups. Springer, Berlin (1996)
65. Drinfeld, V.G.: Hopf algebras and the Yang–Baxter equation. Soviet Math. Dokl. **32**, 254–258 (1985)
66. Drinfeld, V.G.: Quantum groups. In: Proceedings of the International Congress of Mathematics, Berkeley, CA, vol. 1, pp. 798–820 (1986)
67. Drinfeld, V.G.: On almost cocommutative Hopf algebras. Algebra i Analiz **1**(2), 30–46 (1989). English translation: Leningr. Math. J. **1**, 321–342 (1990)
68. Etingof, P., Gelaki, S.: The classification of finite-dimensional triangular Hopf algebras over an algebraically closed field. Int. Math. Res. Not. **5**, 223–234 (2000)
69. Etingof, P., Gelaki, S.: On cotriangular Hopf algebras. Am. J. Math. **123**, 699–713 (2001)
70. Etingof, P., Gelaki, S.: The classification of triangular semisimple and cosemisimple Hopf algebras over an algebraically closed field of characteristic zero. Mosc. Math. J. **3**(1), 37–43 (2003)
71. Etingof, P., Schiffman, O.: Lectures on Quantum Groups. International Press Incorporated, Boston (1998)
72. FELIX – an assistant for algebraists. In: Watt, S.M. (ed.) ISSACi91, pp. 382–389. ACM Press, New York (1991)

73. Ferreira, V.O., Murakami, L.S.I., Paques, A.: A Hopf-Galois correspondence for free algebras. J. Algebra **276** 407–416 (2004)
74. Filippov, V.T.: n-Lie algebras. Sib. Math. J. **26**(6), 126–140 (1985)
75. Finkelshtein, D.: On relations between commutators. Commun. Pure. Appl. Math. **8**, 245–250 (1955)
76. Fischman, D., Montgomery, S.: A Schur double centralizer theorem for cotriangular Hopf algebras and generalized Lie algebras. J. Algebra **168**, 594–614 (1994)
77. Flores de Chela, D.: Quantum symmetric algebras as braided Hopf algebras. In: De la Peña, J.A., Vallejo, E., Atakishieyev, N. (eds.) Algebraic Structures and Their Representations. Contemporary Mathematics, vol. 376, pp. 261–271. American Mathematical Society, Providence, RI (2005)
78. Flores de Chela, D., Green, J.A.: Quantum symmetric algebras II. J. Algebra **269**, 610–631 (2003)
79. Fox, R.H.: Free differential calculus. I. Ann. Math. **57**(3), 517–559 (1953); II. Ann. Math. **59**, 196–210 (1954); III. Ann. Math. **64**, 407–419 (1956); IV. Ann. Math. **68**, 81–95 (1958); IV. Ann. Math. **71**, 408–422 (1960)
80. Friedrichs, K.O.: Mathematical aspects of the quantum theory of fields V. Commun. Pure Applied Math. **6**, 1–72 (1953)
81. Frønsdal, C.: On the Classification of q-algebras. Lett. Math. Phys. **53**, 105–120 (2000)
82. Frønsdal, C., Galindo, A.: The ideals of free differential algebras. J. Algebra **222**, 708–746 (1999)
83. Gabber, O., Kac, V.G.: On defining relations of certain infinite-dimensional Lie algebras. Bull. Am. Math. Soc. **5**, 185–189 (1981)
84. Gavarini, F.: A PBW basis for Lusztig's form of untwisted affine quantum groups. Commun. Algebra **27**(2), 903–918 (1999)
85. Gelaki, S.: On the classification of finite-dimensional triangular Hopf Algebras. In: Montgomery, S., Schneider, H.-J. (eds.) New Directions in Hopf Algebras. Cambridge University Press, Cambridge, MSRI Publications, vol. 43, pp. 69–116 (2002)
86. Gomez, X., Majid, S.: Braided Lie algebras and bicovariant differential calculi over co-quasitriangular Hopf algebras. J. Algebra **261**(2), 334–388 (2003)
87. Graña, M., Heckenberger, I.: On a factorization of graded Hopf algebras using Lyndon words. J. Algebra **314**(1), 324–343 (2007)
88. Gurevich, D.: Generalized translation operators on Lie Groups. Sov. J. Contemp. Math. Anal. **18**, 57–70 (1983). Izvestiya Akademii Nauk Armyanskoi SSR. Matematica **18**(4), 305–317 (1983)
89. Gurevich, D.: Algebraic aspects of the quantum Yang-Baxter equation. Leningr. Math. J. **2**(4), 801–828 (1991)
90. Heckenberger, I.: Weyl equivalence for rank 2 Nichols algebras of diagonal type. Ann. Univ. Ferrara Sez. VII (NS) **51**(1), 281–289 (2005)
91. Heckenberger, I.: The Weyl groupoid of a Nichols algebra of diagonal type. Invent. Math. **164**, 175–188 (2006)
92. Heckenberger, I.: Classification of arithmetic root systems. Adv. Math. **220**, 59–124 (2009)
93. Heckenberger, I., Kolb, S.: Right coideal subalgebras of the Borel part of a quantized enveloping algebra. Int. Math. Res. Not. **2**, 419–451 (2011)
94. Heckenberger, I., Kolb, S.: Homogeneous right coideal subalgebras of quantized enveloping algebras. Bull. Lond. Math. Soc. **44**(4), 837–848 (2012)
95. Heckenberger, I., Schneider, H.-J.: Root systems and Weyl groupoids for Nichols algebras. Proc. Lond. Math. Soc. **101**(3), 623–654 (2010)
96. Heckenberger, I., Schneider, H.-J.: Right coideal subalgebras of Nichols algebras and the Duflo order on the Weyl groupoid. Isr. J. Math. **197**(1), 139–187 (2013)
97. Heckenberger, I., Yamane, H.: Drinfel'd doubles and Shapovalov determinants. Rev. Un. Mat. Argentina **51**(2), 107–146 (2010)
98. Herstein, I.: Noncommutative Rings. Carus Mathematical Monographs, vol. 15, The Mathematical Association of America, USA (1968)

99. Hironaka, H.: Resolution of singularities of an algebraic variety over a field of characteristic zero. I, II. Ann. Math. (2) **79**, 109–203, 205–326 (1964)
100. Hofmann, K.H., Strambach, K.: Topological and analytic loops. In: Chein, O., Pflugfelder, H., Smith, J.D.H (eds.) Quasigroups and Loops: Theory and Applications, pp. 205–262. Heldermann, Berlin (1990)
101. Hopf, H.: Über die Topologeie der Gruppenmannigfaltigkeiten und ihre Verallgemeinerungen. Ann. Math. **42**, 22–52 (1941)
102. Humphreys, J.E.: Introduction to Lie Algebras and Representation Theory. Springer, New York-Heidelberg-Berlin (1972)
103. Ion, B.: Relative PBW type theorems for symmetrically braided Hopf algebras. Commun. Algebra **39**(7), 2508–2518 (2011)
104. Jacobson, N.: Lie Algebras. Interscience, New York (1962)
105. Jantzen, J.C.: Lectures on Quantum Groups. Graduate Studies in Mathematics, vol. 6. American Mathematical Society, Providence, RI (1996)
106. Jimbo, M.: A q-difference analogue of $U(g)$ and the Yang-Baxter equation. Lett. Math. Phys. **10**, 63–69 (1985)
107. Joseph, A.: Quantum Groups and Their Primitive Ideals. Springer, Berlin, Heidelberg (1995)
108. Joyal, A.: Foncteurs analytiques et espèces de structures. Combinatoire énumérative. Lecture Notes in Mathematics, vol. 1234, pp. 126–159. Springer, Berlin (1986)
109. Joyal, A., Street, R.: Braided tensor categories. Adv. Math. **102**, 20–87 (1993)
110. Jumphreys, J.E.: Linear Algebraic Groups. Graduate Texts in Mathematics **21**. Springer-Verlag, New York (1975)
111. Kac, G.I.: Ring groups and duality principle. Proc. Mosc. Math. Soc. **12**, 259–301 (1963)
112. Kac, V.G.: Simple irreducible graded Lie algebras of finite growth. Izvestija AN USSR (Ser. Mat.) **32**, 1923–1967 (1968). English translation: Math. USSR-Izvestija, **2**, 1271–1311 (1968)
113. Kac, V.G.: Classification of Simple Lie Superalgebras. Funct. Analys i ego prilozh. **9**(3), 91–92 (1975); Letter to the editor, **10**(2), 93 (1976); English translation: Funct. Anal. Appl. **9**, 263–265 (1975)
114. Kac, V.G.: A sketch of Lie superalgebra theory. Commun. Math. Phys. **53**, 31–64 (1977)
115. Kac, V.G.: Lie superalgebras. Adv. Math. **26**(1), 8–96 (1977)
116. Kac, V.G.: Infinite-Dimensional Lie Algebras, 3rd edn. Cambridge University Press, Cambridge (1995)
117. Kac, V.G.: Classification of infinite-dimensional simple linearly compact Lie superalgebras. Adv. Math. **139**, 1–55 (1998)
118. Kang, S.-J.: Quantum deformations of generalized Kac-Moody algebras and their modules. J. Algebra **175**, 1041–1066 (1995)
119. Kashiwara, M.: On crystal bases of the q-analog of universal enveloping algebras. Duke Math. J. **63**(2), 465–516 (1991)
120. Kassel, C.: Quantum Groups. Springer, New York (1995)
121. Kharchenko, V.K.: Automorphisms and Derivations of Associative Rings. Mathematics and Its Applications (Soviet Series), vol. 69. Kluwer, Dordrecht/Boston/London (1991)
122. Kharchenko, V.K.: Noncommutative Galois Theory. Nauchnaja Kniga, Novosibirsk (1996)
123. Kharchenko, V.K.: An algebra of skew primitive elements. Algebra i Logika **37**(2), 181–224 (1998). English translation: Algebra and Logic **37**(2), 101–126 (1998)
124. Kharchenko, V.K.: A quantum analog of the Poincaré-Birkhoff-Witt theorem. Algebra i Logika **38**(4), 476–507 (1999). English translation: Algebra and Logic **38**(4), 259–276 (1999)
125. Kharchenko, V.K.: An existence condition for multilinear quantum operations. J. Algebra **217**, 188–228 (1999)
126. Kharchenko, V.K.: Multilinear quantum Lie operations. Zap. Nauchn. Sem. S.-Petersburg. Otdel. Mat. Inst. Steklov. **272**. Vopr. Teor. Predst. Algebr. i Grupp. **7**, 321–340 (2000)
127. Kharchenko, V.K.: Skew primitive elements in Hopf algebras and related identities. J. Algebra **238**(2), 534–559 (2001)
128. Kharchenko, V.K.: A combinatorial approach to the quantification of Lie algebras. Pac. J. Math. **203**(1), 191–233 (2002)

129. Kharchenko, V.K.: Constants of coordinate differential calculi defined by Yang–Baxter operators. J. Algebra **267**, 96–129 (2003)
130. Kharchenko, V.K.: Multilinear quantum Lie operations. J. Math. Sci. (NY) **116**(1), 3063–3073 (2003)
131. Kharchenko, V.K.: Braided version of Shirshov-Witt theorem. J. Algebra **294**, 196–225 (2005)
132. Kharchenko, V.K.: Connected braided Hopf algebras. J. Algebra **307**, 24–48 (2007)
133. Kharchenko, V.K.: PBW-bases of coideal subalgebras and a freeness theorem. Trans. Am. Math. Soc. **360**(10), 5121–5143 (2008)
134. Kharchenko, V.K.: Triangular decomposition of right coideal subalgebras, J. Algebra **324**(11), 3048–3089 (2010)
135. Kharchenko, V.K.: Right coideal subalgebras of $U_q^+(so_{2n+1})$. J. Eur. Math. Soc. **13**(6), 1677–1735 (2011)
136. Kharchenko, V.K.: Quantizations of Kac–Moody algebras. J. Pure Appl. Algebra **218**(4), 666–683 (2014)
137. Kharchenko, V.K., Andrede Álvarez, A.: On the combinatorial rank of Hopf algebras. Contemp. Math. **376**, 299–308 (2005)
138. Kharchenko, V.K., Díaz Sosa, M.L.: Computing of the combinatorial rank of $u_q(\mathfrak{so}_{2n+1})$. Commun. Algebra **39**(12), 4705–4718 (2011)
139. Kharchenko, V.K., Lara Sagahón, A.V.: Right coideal subalgebras in $U_q(sl_{n+1})$. J. Algebra **319**(6), 2571–2625 (2008)
140. Kharchenko, V.K., Lara Sagahón, A.V., Garza Rivera, J.L.: Computing of the number of right coideal subalgebras of $U_q(so_{2n+1})$. J. Algebra **341**(1), 279–296 (2011); Corrigendum **35**(1), 224–225 (2012)
141. Killing, W.: Die Zusammensetzung der steigen endlichen Transformationsgruppen, I. Math. Ann. **31**, 252–290 (1886)
142. Klimik, A., Schmüdgen, K.: Quantum Groups and Their Representations. Springer, Berlin, Heidelberg (1997)
143. Kochetov, M.: Generalized Lie algebras and cocycle twists. Commun. Algebra **36**, 4032–4051 (2008)
144. Kochetov, M., Radu, O.: Engel's theorem for generalized Lie algebras. Algebr. Represent. Theor. **13**(1), 69–77 (2010)
145. Larson, R.G., Towber, J.: Two dual classes of bialgebras related to the concept of "quantum groups" and "quantum Lie algebras". Commun. Algebra **19**(12), 3295–3345 (1991)
146. Leclerc, B.: Dual canonical bases, quantum shuffles and q-characters. Math. Z. **246**(4), 691–732 (2004)
147. Letzter, G.: Coideal subalgebras and quantum symmetric pairs. In: Montgomery, S., Schneider, H.-J. (eds.) New Directions in Hopf Algebras. Cambridge University Press, Cambridge, MSRI Publications, vol. 43, pp. 117–165 (2002)
148. Lothaire, M.: Combinatorics on Words. Encyclopedia of Mathematics and Its Applications, vol. 17. Addison-Wesley, Reading, MA (1983)
149. Lothaire, M.: Algebraic Combinatorics on Words. Cambridge University Press, Cambridge (2002)
150. Lothaire, M.: Applied Combinatorics on Words. Cambridge University Press, Cambridge (2005)
151. Lusztig, G.: Canonical bases arising from quantized enveloping algebras. J. Am. Math. Soc. **3**(2), 447–498 (1990)
152. Lusztig, G.: Finite-dimensional Hopf algebras arising from quantized enveloping algebras. J. Am. Math. Soc. **3**(1), 257–296 (1990)
153. Lusztig, G.: Introduction to Quantum Groups. Progress in Mathematics, vol. 110. Birkhauser, Boston (1993)
154. Lyndon, R.C.: On Burnside's problem. Trans. Am. Math. Soc. **7**, 202–215 (1954)
155. Lyndon, R.C.: A theorem of Friedrichs. Mich. Math. J. **3**(1), 27–29 (1955–1956)

156. Lyubashenko, V.V.: Hopf algebras and vector symmetries. Sov. Math. Surveys **41**, 153–154 (1986)
157. Magnus, W.: On the exponential solution of differential equations for a linear operator. Commun. Pure Appl. Math. **7**, 649–673 (1954)
158. Majid, S.: Algebras and Hopf Algebras in Braided Categories. Advances in Hopf Algebras. Lecture Notes in Pure Applied Mathematics, vol. 158, pp. 55–105. Dekker, New York (1994)
159. Majid, S.: Crossed products by braided groups and bosonization. J. Algebra **163**, 165–190 (1994)
160. Majid, S.: Foundations of Quantum Group Theory. Cambridge University Press, Cambridge (1995)
161. Mal'tcev, A.I.: On the immersion of an algebraic ring into a field. Math. Ann. **113**, 686–691 (1937)
162. Mal'tcev, A.I.: On the inclusion of associative systems into groups, I. Mat. Sbornik **6**(2), 331–336 (1939)
163. Mal'tcev, A.I.: On the inclusion of associative systems in groups, II. Mat. Sbornik **8**(2), 251–263 (1940)
164. Mal'tcev, A.I.: Algebraic Systems. Nauka, Moscow (1965)
165. Mal'tcev, A.I.: Algorithms and Recursive Functions. Nauka, Moscow (1970)
166. Manin, Y.: Quantum Groups and Non-Commutative Geometry. CRM Publishing, Université de Montréal (1988)
167. Masuoka, A.: Formal groups and unipotent affine groups in non-categorical symmetry. J. Algebra **317**, 226–249 (2007)
168. Moody, R.V.: Lie algebras associated with generalized Cartan matrices. Bull. Am. Math. Soc. **73**, 217–221 (1967)
169. Miheev, P.O., Sabinin, L.V.: Quasigroups and differential geometry. In: Chein, O., Pflugfelder, H., Smith, J.D.H. (eds.) Quasigroups and Loops: Theory and Applications, pp. 357–430. Heldermann, Berlin (1990)
170. Mikhalev, A.A.: Subalgebras of free color Lie superalgebras. Mat. Zametki **37**(5), 653–661 (1985). English translation: Math. Notes **37**, 356–360 (1985)
171. Mikhalev, A.A.: Free color Lie superalgebras. Dokl. Akad. Nauk SSSR **286**(3), 551–554 (1986). English translation: Soviet Math. Dokl. **33**, 136–139 (1986)
172. Mikhalev, A.A.: Subalgebras of free Lie *p*-Superalgebras. Mat. Zametki **43**(2), 178–191 (1988). English translation: Math. Notes **43**, 99–106 (1988)
173. Milinski, A.: Actions of pointed Hopf algebras on prime algebras. Commun. Algebra **23**, 313–333 (1995)
174. Milinski, A.: Operationen punktierter Hopfalgebren auf primen Algebren. Ph.D. thesis, München (1995)
175. Milnor, J.W., Moore, J.C.: On the structure of Hopf algebras. Ann. Math. **81**, 211–264 (1965)
176. Montgomery, S.: Hopf Algebras and Their Actions on Rings. Conference Board of Mathematical Sciences, vol. 82. American Mathematical Society, Providence, RI (1993)
177. Mostovoy, J., Pérez Izquierdo, J.M.: Formal multiplications, bialgebras of distributions and non-associative Lie theory. Transform. Groups **15**, 625–653 (2010)
178. Mostovoy, J., Pérez-Izquierdo, J.M., Shestakov, I.P.: Hopf algebras in non-associative Lie theory. Bull. Math. Sci. **4**(1), 129–173 (2014)
179. Musson, I.M.: Lie Superalgebras and Enveloping Algebras. Graduate Studies in Mathematics, vol. 131. American Mathematical Society, Providence, RI (2012)
180. Nambu, Y.: Generalized Hamilton dynamics. Phys. Rev. D **7**(8), 2405–2412 (1973)
181. Nichols, W.: Bialgebras of type one. Commun. Algebra **6**(15), 1521–1552 (1978)
182. Nielsen, J.: A basis for subgroups of free groups. Math. Scand. **3**, 31–43 (1955)
183. Pareigis, B.: Skew-primitive elements of quantum groups and braided Lie algebras. In: Caenepeel, S., Verschoren, A. (eds.) Rings, Hopf Algebras, and Brauer Groups (Antwerp/Brussels). Lecture Notes in Pure and Applied Mathematics, vol. 197, pp. 219–238. Marcel Dekker, Inc, New York (1998)

184. Pareigis, B.: On Lie algebras in braided categories. In: Budzyński, R., Pusz, W., Zakrzewski, S. (eds.) Quantum Groups and Quantum Spaces, vol. **40**, pp. 139–158. Banach Center Publications, Warszawa (1997)
185. Pareigis, B.: On Lie algebras in the category of Yetter–Drinfeld modules. Appl. Categ. Struct. **6**(2), 152–175 (1998)
186. Pérez Izquierdo, J.M.: An envelope for Bol algebras. J. Algebra **284**, 480–493 (2005)
187. Pérez Izquierdo, J.M.: Algebras, hyperalgebras, nonassociative bialgebras and loops. Adv. Math. **208**, 834–876 (2007)
188. Pérez Izquierdo, J.M., Shestakov, I.P.: An envelope for Malcev algebras. J. Algebra **272**, 379–393 (2004)
189. Polishchuk, A.E., Positselsky, L.E.: Quadratic Algebras. University Lecture Series, vol. 37. American Mathematical Society, Providence, RI (2005)
190. Radford, D.E.: The structure of Hopf algebras with projection. J. Algebra **92**, 322–347 (1985)
191. Radford, D.E.: Hopf Algebras. Series on Knots and Everything, vol. 49. World Scientific, Singapore (2012)
192. Reineke, M.: Feigin's map and monomial bases for quantized enveloping algebras. Math. Z. **237**(3), 639–667 (2001)
193. Reshetikhin, N.: Multiparameter quantum groups and twisted quasitriangular Hopf algebras. Lett. Math. Phys. **20**, 331–335 (1990)
194. Ringel, C.M.: PBW-bases of quantum groups. J. Reine Angew. Math. **470**, 51–88 (1996)
195. Rosso, M.: An analogue of the Poincare–Birkhoff–Witt theorem and the universal R-matrix of $U_q(sl(N+1))$. Commun. Math. Phys. **124**, 307–318 (1989)
196. Rosso, M.: Groupes quantiques et algèbres de battagesquantiques. C.R. Acad. Sci. Paris **320**, 145–148 (1995)
197. Rosso, M.: Quantum groups and quantum Shuffles. Invent. Math. **113**(2) 399–416 (1998)
198. Sabinin, L.V.: Smooth Quasigroups and Loops. Mathematics and Its Applications, vol. 429. Kluwer, Dordrecht/Boston/London (1999)
199. Sabinin, L.V.: Smooth quasigroups and loops: forty-five years of incredible growth. Comment. Math. Univ. Carol. **41**, 377–400 (2000)
200. Sagan, B.E.: The Symmetric Group. Graduate Texts in Mathematics, vol. 203. Springer, New York (2001)
201. Schauenberg, P.: On Coquasitriangular Hopf Algebras and the Quantum Yang–Baxter Equation. Algebra Berichte, vol. 67, Fischer, Munich (1992)
202. Schauenberg, P.: A characterization of the Borel-like subalgebras of quantum enveloping algebras. Commun. Algebra **24**, 2811–2823 (1996)
203. Scheunert, M.: Generalized Lie algebras. J. Math. Phys. **20**, 712–720 (1979)
204. Scheunert, M.: The Theory of Lie Superalgebras. An Introduction. Lecture Notes in Mathematics, vol. 716. Springer, Berlin-Heidelberg-New York (1979)
205. Schreier, O.: Die Untergruppen der freien Gruppen. Abhandlungen Hamburg **5**, 161–183 (1927)
206. Schützenberger, M.P., Sherman, S.: On a formal product over the conjugate classes in a free group. J. Math. Anal. Appl. **7**, 482–488 (1963)
207. Serre, J.-P.: Algébres de Lie Semi-simples Complexes. Benjamin, New York, Amsterdam (1966)
208. Shestakov, I.P., Umirbaev, U.U.: Free Akivis algebras, primitive elements, and hyperalgebras. J. Algebra **250**, 533–548 (2002)
209. Shirshov, A.I.: Subalgebras of free Lie algebras. Matem. Sb. **33**(75), 441–452 (1953)
210. Shirshov, A.I.: On some nonassociative nil-rings and algebraic algebras. Matem. Sbornik **41**(83)(3), 381–394 (1957)
211. Shirshov, A.I.: On free Lie rings. Mat. Sb. **45**(87)(2), 113–122 (1958)
212. Shirshov, A.I.: Some algorithmic problems for Lie algebras. Sib. Math. J. **3**(2), 292–296 (1962)
213. Shneider, S., Sternberg, S.: Quantum Groups. International Press Incorporated, Boston (1993)
214. Shtern, A.S.: Free Lie superalgebras. Sib. Math. J. **27**, 551–554 (1986)

215. Skryabin, S.: Hopf Galois extensions, triangular structures, and Frobenius Lie algebras in prime characteristic. J. Algebra **277**, 96–128 (2004)
216. Stanley, R.P.: Some aspects of group acting on finite posets. J. Combin. Theory Ser. A **32**, 132–161 (1982)
217. Stover, R.: The equivalence of certain categories of twisted Lie algebras over a commutative ring. J. Pure Appl. Algebra **86**, 289–326 (1993)
218. Sturmfels, B.: Gröbner Bases and Convex Polytopes. University Lecture Series, vol. 8. American Mathematical Society, Providence, RI (1996)
219. Sturmfels, B.: Solving Systems of Polynomial Equations. Conference Board of Mathematical Sciences, vol. 97. American Mathematical Society, Providence, RI (2002)
220. Sweedler, M.: Hopf Algebras. Benjamin, New York (1969)
221. Takeuchi, M:. Survey of braided Hopf algebras. In: New Trends in Hopf Algebra Theory. Contemporary Mathematics, vol. 267, pp. 301–324. American Mathematical Society, Providence, RI (2000)
222. Takeuchi, M.: A survey on Nichols algebras. In: De la Peña, J.A., Vallejo, E., Atakishieyev, N. (eds.) Algebraic Structures and Their Representations. Contemporary Mathematics, vol. 376, pp. 105–117. American Mathematical Society, Providence, RI (2005)
223. Takhtadjian, L.: On foundations of the generalized Nambu mechanics. Commun. Math. Phys. **160**, 295–315 (1994)
224. Towber, J.: Multiparameter quantum forms of the enveloping algebra U_{gl_n} related to the Faddeev-Reshetikhin-Takhtajan $U(R)$ constructions. J. Knot Theory Ramif. **4**(5), 263–317 (1995)
225. Ufer, S.: PBW bases for a class of braided Hopf algebras. J. Algebra **280**(1), 84–119 (2004)
226. Wess, J., Zumino, B.: Covariant differential calculus on the quantum hyperplane. Nucl. Phys. B Proc. Suppl. **18**, 302–312 (1991)
227. White, A.T.: Groups, Graphs, and Surfaces. North-Holland Mathematical Series, vol. 8. North-Holland, New York (1988)
228. Wielandt, H.: Finite Permutation Groups. Academic Press, New York (1964)
229. Witt, E.: Die Unterringe der freien Liescher Ringe. Math. Zeitschr. Bd. **64**, 195–216 (1956)
230. Woronowicz, S.L.: Differential calculus on compact matrix pseudogroups (quantum groups). Commun. Math. Phys. **122**, 125–170 (1989)
231. Yamaleev, R.M.: Elements of cubic quantum mechanics. Commun. JINR Dubna E2-88-147, 1–11 (1988)
232. Yamaleev, R.M.: Model of multilinear oscillator in a space of non-integer quantum numbers. Commun. JINR Dubna E2-88-871, 1–10 (1988)
233. Yamaleev, R.M.: Model of multilinear Bose- and Fermi-like oscillator. Commun. JINR Dubna E2-92-66, 1–14 (1992)
234. Yamane, H.: A Poincaré-Birkhoff-Witt theorem for quantized universal enveloping algebras of type A_N. Publ. RIMS. Kyoto Univ. **25**, 503–520 (1989)
235. Yanai, T.: Galois correspondence theorem for Hopf algebra actions. In: Algebraic Structures and Their Representations. Contemporary Mathematics, vol. 376, pp. 393–411. American Mathematical Society, Providence, RI (2005)
236. Zhevlakov, K.A., Slinko, A.M., Shestakov, I.P., Shirshov, A.I.: Rings that are Nearly Associative. Academic, New York (1982)

Index

V. Kharchenko, *Quantum Lie Theory*, Lecture Notes in Mathematics 2150,
DOI 10.1007/978-3-319-22704-7

LECTURE NOTES IN MATHEMATICS **Springer**

Editors in Chief: J.-M. Morel, B. Teissier;

Editorial Policy

1. Lecture Notes aim to report new developments in all areas of mathematics and their applications – quickly, informally and at a high level. Mathematical texts analysing new developments in modelling and numerical simulation are welcome.

   Manuscripts should be reasonably self contained and rounded off. Thus they may, and often will, present not only results of the author but also related work by other people. They may be based on specialised lecture courses. Furthermore, the manuscripts should provide sufficient motivation, examples and applications. This clearly distinguishes Lecture Notes from journal articles or technical reports which normally are very concise. Articles intended for a journal but too long to be accepted by most journals, usually do not have this "lecture notes" character. For similar reasons it is unusual for doctoral theses to be accepted for the Lecture Notes series, though habilitation theses may be appropriate.

2. Besides monographs, multi-author manuscripts resulting from SUMMER SCHOOLS or similar INTENSIVE COURSES are welcome, provided their objective was held to present an active mathematical topic to an audience at the beginning or intermediate graduate level (a list of participants should be provided).

   The resulting manuscript should not be just a collection of course notes, but should require advance planning and coordination among the main lecturers. The subject matter should dictate the structure of the book. This structure should be motivated and explained in a scientific introduction, and the notation, references, index and formulation of results should be, if possible, unified by the editors. Each contribution should have an abstract and an introduction referring to the other contributions. In other words, more preparatory work must go into a multi-authored volume than simply assembling a disparate collection of papers, communicated at the event.

3. Manuscripts should be submitted either online at www.editorialmanager.com/lnm to Springer's mathematics editorial in Heidelberg, or electronically to one of the series editors. Authors should be aware that incomplete or insufficiently close-to-final manuscripts almost always result in longer refereeing times and nevertheless unclear referees' recommendations, making further refereeing of a final draft necessary. The strict minimum amount of material that will be considered should include a detailed outline describing the planned contents of each chapter, a bibliography and several sample chapters. Parallel submission of a manuscript to another publisher while under consideration for LNM is not acceptable and can lead to rejection.

4. In general, **monographs** will be sent out to at least 2 external referees for evaluation.

   A final decision to publish can be made only on the basis of the complete manuscript, however a refereeing process leading to a preliminary decision can be based on a pre-final or incomplete manuscript.

   Volume Editors of **multi-author works** are expected to arrange for the refereeing, to the usual scientific standards, of the individual contributions. If the resulting reports can be forwarded to the LNM Editorial Board, this is very helpful. If no reports are forwarded or if other questions remain unclear in respect of homogeneity etc, the series editors may wish to consult external referees for an overall evaluation of the volume.

5. Manuscripts should in general be submitted in English. Final manuscripts should contain at least 100 pages of mathematical text and should always include

   – a table of contents;
   – an informative introduction, with adequate motivation and perhaps some historical remarks: it should be accessible to a reader not intimately familiar with the topic treated;
   – a subject index: as a rule this is genuinely helpful for the reader.
   – For evaluation purposes, manuscripts should be submitted as pdf files.

6. Careful preparation of the manuscripts will help keep production time short besides ensuring satisfactory appearance of the finished book in print and online. After acceptance of the manuscript authors will be asked to prepare the final LaTeX source files (see LaTeX templates online: https://www.springer.com/gb/authors-editors/book-authors-editors/manuscriptpreparation/5636) plus the corresponding pdf- or zipped ps-file. The LaTeX source files are essential for producing the full-text online version of the book, see http://link.springer.com/bookseries/304 for the existing online volumes of LNM). The technical production of a Lecture Notes volume takes approximately 12 weeks. Additional instructions, if necessary, are available on request from lnm@springer.com.

7. Authors receive a total of 30 free copies of their volume and free access to their book on SpringerLink, but no royalties. They are entitled to a discount of 33.3 % on the price of Springer books purchased for their personal use, if ordering directly from Springer.

8. Commitment to publish is made by a *Publishing Agreement*; contributing authors of multiauthor books are requested to sign a *Consent to Publish form*. Springer-Verlag registers the copyright for each volume. Authors are free to reuse material contained in their LNM volumes in later publications: a brief written (or e-mail) request for formal permission is sufficient.

Addresses:

Professor Jean-Michel Morel, CMLA, École Normale Supérieure de Cachan, France
E-mail: moreljeanmichel@gmail.com

Professor Bernard Teissier, Equipe Géométrie et Dynamique,
Institut de Mathématiques de Jussieu – Paris Rive Gauche, Paris, France
E-mail: bernard.teissier@imj-prg.fr

Springer: Ute McCrory, Mathematics, Heidelberg, Germany,
E-mail: lnm@springer.com

Printed in the United States
By Bookmasters